THE ADAPTIVE CHALLENGE OF CLIMATE CHANGE

This book presents a new perspective on adaptation to climate change. It considers climate change as more than a problem that can be addressed solely through technical expertise. Instead, it approaches climate change as an adaptive challenge that is fundamentally linked to beliefs, values, and worldviews as well as to power, politics, identities, and interests. Drawing on case studies from high-income countries, the book argues that it is time to consider adaptation to climate change as a challenge of social, personal, and political transformations. The authors represent a variety of fields and perspectives, illustrating the importance of interdisciplinary approaches to the problem. The book will be of interest to researchers, policy makers, and advanced students in the environmental sciences, social sciences, and humanities as well as to decision makers and practitioners interested in new ideas about adapting to climate change.

KAREN O'BRIEN is a professor in the Department of Sociology and Human Geography at the University of Oslo, Norway. She has been working on various aspects of climate change for more than twenty-five years, including on impacts, vulnerability, and adaptation and the implications for human security. She also works on the links between global environmental change and globalization. Her current research explores adaptation as a social, cultural, and human process; the relationships between belief system flexibility and adaptive capacity; and the values and visions of youth toward the future in a changing climate. She is on the Future Earth Science Committee and has participated in the Intergovernmental Panel on Climate Change (IPCC) Fourth and Fifth Assessment Reports as well as the IPCC Special Report on extreme events. She has coauthored and coedited numerous books, including *Environmental Change and Globalization: Double Exposures* (2008); *Adapting to Climate Change: Thresholds, Values, Governance* (Cambridge University Press, 2009); *Climate Change, Ethics and Human Security* (Cambridge University Press, 2010); and *Climate Change Adaptation and Development: Transforming Paradigms and Practices* (2015).

ELIN SELBOE holds a PhD in human geography from the University of Oslo, Norway. She has researched local mobilization and politics in Senegal, with her dissertation focusing on the dynamics of social networks and participation in associational life in the context of economic crisis, changes in political Islam, and processes of democratization. She has been researching the politics and social organization of adaptation to climate change in Norway and currently holds a postdoctoral position in the project "Voices of the Future: Values and Visions of Norwegian Youth on Responses to Climate Change."

THE ADAPTIVE CHALLENGE OF CLIMATE CHANGE

Edited by

KAREN O'BRIEN
University of Oslo

ELIN SELBOE
University of Oslo

CAMBRIDGE
UNIVERSITY PRESS

32 Avenue of the Americas, New York, NY 10013-2473, USA

Cambridge University Press is part of the University of Cambridge.

It furthers the University's mission by disseminating knowledge in the pursuit of
education, learning, and research at the highest international levels of excellence.

www.cambridge.org
Information on this title: www.cambridge.org/9781107022980

© Karen O'Brien and Elin Selboe 2015

This publication is in copyright. Subject to statutory exception
and to the provisions of relevant collective licensing agreements,
no reproduction of any part may take place without the written
permission of Cambridge University Press.

First published 2015

A catalog record for this publication is available from the British Library.

Library of Congress Cataloging in Publication Data
The adaptive challenge of climate change / edited by Karen O'Brien, Elin Selboe.
pages cm
Includes bibliographical references and index.
ISBN 978-1-107-02298-0 (hardback)
1. Climatic changes – Social aspects.   2. Climatic changes – Political aspects.
3. Climate change mitigation – Social aspects.   4. Climate change mitigation –
Political aspects.   I. O'Brien, Karen L., editor of compilation.
II. Selboe, Elin, editor of compilation.
QC903.A247   2015
363.738'74–dc23   2015010554

ISBN 978-1-107-02298-0 Hardback

Cambridge University Press has no responsibility for the persistence or accuracy of
URLs for external or third-party Internet Web sites referred to in this publication
and does not guarantee that any content on such Web sites is, or will remain,
accurate or appropriate.

Contents

List of Tables	page vii
List of Figures	ix
Contributors	xi
Preface	xv

1 Climate Change as an Adaptive Challenge   1
   KAREN O'BRIEN AND ELIN SELBOE

2 The Intangibles of Climate Change Adaptation: Philosophy, Ethics, and Values   24
   BERTRAND GUILLAUME AND STIJN NEUTELEERS

3 Urban Climate Change Policy Transitions: Views from New York City and London   41
   WILLIAM SOLECKI, LESLEY PATRICK, AND ZOE SPRIGINGS

4 Planning for Climate Change Adaptation in Urban Areas   63
   JAN ERLING KLAUSEN, INGER-LISE SAGLIE, KNUT-BJØRN STOKKE, AND MARTE WINSVOLD

5 The Challenge of Governing Adaptation in Australia   81
   STEVEN WALLER AND JON BARNETT

6 Emerging Equity and Justice Concerns for Climate Change Adaptation: A Case Study of New York State   98
   PETER VANCURA AND ROBIN LEICHENKO

7 Transforming toward or away from Sustainability? How Conflicting Interests and Aspirations Influence Local Adaptation   118
   SIRI ERIKSEN AND ELIN SELBOE

8 Opportunistic Adaptation: New Discourses on Oil, Equity, and
   Environmental Security   140
   BERIT KRISTOFFERSEN

9 Place Attachment, Identity, and Adaptation   160
   TARA QUINN, IRENE LORENZONI, AND W. NEIL ADGER

10 Values and Traditional Practices in Adaptation to Climate Change:
   Evidence from a Q Method Study in Two Communities in Labrador,
   Canada   171
   JOHANNA WOLF, ILANA ALLICE, AND TREVOR BELL

11 Exploring Vulnerability and Adaptation Narratives among Fishers,
   Farmers, and Municipal Planners in Northern Norway   194
   GRETE K. HOVELSRUD, JENNIFER J. WEST, AND
   HALVOR DANNEVIG

12 Changes in Organizational Culture, Changes in Adaptive Capacity?
   Examples from the Norwegian and Swedish Electricity Sectors   213
   TOR HÅKON INDERBERG

13 From Informant to Actor to Leader: Social-Ecological Inventories as a
   Catalyst for Leadership Development in Participatory Community
   Climate Change Adaptation   230
   BRADLEY MAY

14 Participation and Learning for Climate Change Adaptation: A Case
   Study of the Swedish Forestry Sector   252
   ÅSA GERGER SWARTLING, OSKAR WALLGREN, RICHARD J. T.
   KLEIN, JOHANNA ULMANEN, AND MAJA DAHLIN

15 Integral GIS: Widening the Frame of Reference for Adaptation Planning   271
   LYNN D. ROSENTRATER

16 There Must Be More: Communication to Close the Cultural Divide   287
   SUSANNE C. MOSER AND CAROL L. BERZONSKY

17 Social Transformation: The Real Adaptive Challenge   311
   KAREN O'BRIEN AND ELIN SELBOE

   Index   325

Tables

4.1	Key features of hierarchies, markets, and networks	*page* 66
6.1	Characteristics and production costs and revenues of dairy regions	104
6.2	Small versus large dairy farms in 2007 and 2008	105
6.3	Characteristics of population in Census block groups in and out of one-hundred-year floodplain for the New York Coastal Zone	110
6.4	Profile of the population residing in the one-hundred-year floodplain (ClimAID Region 4)	111
7.1	Restructuring in the agricultural sector	123
10.1	Schwartz's cultural values	174
10.2	Schwartz's individual values	174
10.3	List of Q statements and associated Schwartz values in increasing order of disagreement between discourses	180
10.4	Four discourses based on pooled data from Rigolet and St. Lewis	183
11.1	Summary of research methods	199
12.1	Overview of organizational changes in Norway and Sweden	222
13.1	Aspects of climate change adaptation leadership discussed	238
14.1	Focus group themes, content, and participatory exercises	257
14.2	Summary of stakeholders participating in the study	258
14.3	Selection of stakeholder views on long-term effects of participatory process and learning from follow-up interviews	263
15.1	Comparisons between conventional GIS and qualitative GIS	277
16.1	Typology of engagement with climate change	294

Figures

3.1	A diagram of key drivers and contextual elements of traditional and urban climate change policy transitions.	*page* 56
4.1	A conceptual model for adaptation, combining collective learning and governance mechanisms.	64
4.2	Geographical location of the five case study towns.	69
7.1	The location of Øystre Slidre.	121
7.2	Annual mean temperatures, 1919–2013, at Løken Research Station, Øystre Slidre.	122
8.1	Norway's historical production of oil and gas, 1971–2012.	143
10.1	Flow chart of steps in conducting and analyzing a Q sort.	177
10.2	Q sorting grid.	178
11.1	Map displaying the case study area in northern Norway.	196
13.1	Adaptive collaborative risk management (ACRM).	232
13.2	Niagara region, Ontario, Canada, administrative boundaries.	239
14.1	The case study regions of Västerbotten County and Kronoberg County.	256
15.1	(a) The perspectives, methodological families, and information associated with the integral quadrants and (b) representative knowledge used to understand vulnerability to climate change from an integral perspective.	273
15.2	GIS methods and representative information for assessing vulnerability in an integral framework.	279
15.3	Idealized workflow for assessing vulnerability with integral GIS to identify adaptation needs.	280
16.1	Proportion of the U.S. adult population in the Six Americas, 2008–12.	290
16.2	Shifts in the level of engagement for the Six Americas based on a hypothetical cumulative engagement index.	295
16.3	Discourse continuum from Orwellian manipulation to true dialogue.	300

Contributors

W. Neil Adger is Professor of Geography at the College of Life and Environmental Sciences, and the Theme Leader for Environment and Sustainability in the Humanities and Social Science Strategy, both at the University of Exeter, United Kingdom.

Ilana Allice researched and submitted this chapter when she was a Research Assistant at the Labrador Institute of Memorial University of Newfoundland, Canada.

Jon Barnett is Professor at the School of Geography, University of Melbourne, Australia.

Trevor Bell is University Research Professor of Geography at Memorial University of Newfoundland, St. John's, Canada.

Carol L. Berzonsky is a PhD candidate in the Depth Psychology Department of Pacifica Graduate Institute in Carpinteria, California, United States.

Maja Dahlin is Editor at Webbverkstan and a former Research Associate at the Stockholm Environment Institute, Sweden.

Halvor Dannevig is a Research Fellow at the Western Norway Research Institute in Sogndal, Norway, and a PhD candidate in the Department of Planning, Aalborg University, Denmark.

Siri Eriksen is an Associate Professor at Noragric, Department of International Environment and Development Studies, at the Norwegian University of Life Science in Ås, Norway.

Bertrand Guillaume is an Associate Professor at TU Troyes, France, where he serves as the Chair of the Department of Humanities, Environment, and Technology. He is also an Affiliate at the Max Planck Institute for the History of Science, Germany.

Grete K. Hovelsrud is Professor and Research Director at the Nordland Research Institute in Bodø and a Professor at the University of Nordland, Norway.

Tor Håkon Inderberg is Director of the European Programme and a Senior Research Fellow at the Fridtjof Nansen Institute in Oslo, Norway.

Jan Erling Klausen is a Senior Researcher at the Norwegian Institute for Urban and Regional Research and an Associate Professor of political science at the University of Oslo, Norway.

Richard J. T. Klein is a Senior Research Fellow at the Stockholm Environment Institute and Professor at the Centre for Climate Science and Policy Research at Linköping University, Sweden.

Berit Kristoffersen is a Research Fellow in the Department of Sociology, Political Science, and Social Planning at the University of Tromsø, Norway.

Robin Leichenko is a Professor in the Department of Geography, Rutgers University, New Jersey, United States.

Irene Lorenzoni is a Senior Lecturer in the School of Environmental Sciences at the University of East Anglia, United Kingdom, and a member of the Tyndall Centre for Climate Change Research and of the Science, Society and Sustainability Research Group.

Bradley May is a PhD candidate at the University of Waterloo, Environmental Change and Governance Group, and a Research Associate with Brock University's Environmental Sustainability Research Centre. He was formerly a climate change researcher with Environment Canada.

Susanne C. Moser is an independent researcher and Director of Susanne Moser Research and Consulting in Santa Cruz, California, and a Social Science Research Fellow at Stanford University's Woods Institute for the Environment, Stanford, California, United States.

Stijn Neuteleers is a Researcher at the Centre for Economics and Ethics at the University of Leuven, Belgium.

Karen O'Brien is a Professor in the Department of Sociology and Human Geography at the University of Oslo, Norway.

Lesley Patrick is a PhD candidate in the Earth and Environmental Sciences at the Graduate Center, City University of New York, New York, United States.

Tara Quinn is an Associate Research Fellow at the Environment and Sustainability Institute and a member of the Geography department in the College of Life and Environmental Sciences, both at the University of Exeter, United Kingdom.

Lynn D. Rosentrater is a PhD candidate in the Department of Sociology and Human Geography, University of Oslo, Norway.

Inger-Lise Saglie is a Professor and Head of Department at the Department of Landscape Architecture and Spatial Planning at the Norwegian University of Life Sciences in Ås, Norway.

Elin Selboe is a Postdoctoral Research Fellow in the Department of Sociology and Human Geography at the University of Oslo, Norway.

William Solecki is Professor in the Department of Geography at Hunter College, City University of New York, New York, United States.

Zoe Sprigings researched and submitted this chapter while studying for her master's degree in Disasters, Adaptation, and Development at King's College London. She is now Network Manager for Energy Efficiency at the C40 Cities Climate Leadership Group, London office, United Kingdom.

Knut-Bjørn Stokke is an Associate Professor in the Department of Landscape Architecture and Spatial Planning at the Norwegian University of Life Sciences in Ås, Norway.

Åsa Gerger Swartling is a Senior Research Fellow at the Stockholm Environment Institute, Sweden, and Affiliated Researcher at the Stockholm Resilience Centre, Sweden.

Johanna Ulmanen is a Researcher in the Department for Energy Technology, Section for Systems Analysis, at the SP Technical Research Institute of Sweden, and a former Research Fellow at the Stockholm Environment Institute, Sweden.

Peter Vancura is a PhD candidate in the Department of Geography, Rutgers University, New Jersey, United States.

Steven Waller is a Graduate Researcher in the School of Geography, University of Melbourne, Australia.

Oskar Wallgren is Head of the Environmental Assessment Unit at the WSP Group in Stockholm, Sweden, and former Senior Project Manager at the Stockholm Environment Institute, Sweden.

Jennifer J. West is a Research Fellow at the Center for International Climate and Environmental Research–Oslo and a PhD candidate at Noragric, the Department of International Environment and Development Studies, at the Norwegian University of Life Sciences in Ås, Norway.

Marte Winsvold is a Senior Researcher at the Norwegian Institute for Urban and Regional Research (NIBR) in Oslo, Norway.

Johanna Wolf is a member of the associate faculty in the School of Environment and Sustainability at Royal Roads University and an adjunct Assistant Professor in the School of Environmental Studies at the University of Victoria, Canada.

Preface

Research on climate change adaptation has increased dramatically in recent years, providing new insights into and information on the possibilities and challenges of living with environmental change. However, this growing area of research has also raised important questions about what it really means to adapt to climate change. How do social processes influence the capacity to adapt? What are the limits to adaptation as a response to changing climate conditions? And finally, what are the implications of these limits for human security? These were some of the questions addressed in a large research project on "The Potentials of and Limits to Adaptation in Norway" (PLAN), funded by the Research Council of Norway's NORKLIMA program (2007–12). This project explored how societies adapt to changing climate conditions as well as why they do not. Looking at relationships between behaviors, systems, cultures, values, and worldviews, the research showed that adaptation involves much more than technical responses to changing temperatures and rainfall patterns, exemplified through improved water management, better infrastructure, new regulations, or different types of crops. Recognizing adaptation as a social, cultural, political, and human process, it became clear that climate change is an adaptive challenge – a problem or situation that challenges mind-sets and approaches to change in general. The chapters in this book, written by project participants and an international network of collaborators, present analytical frames and case studies related to climate change responses. The research focuses largely on adaptation yet also suggests a need to redefine this concept, taking into account both the personal and political dimensions of climate change.

Writing this book presented more than a technical problem for us as editors; it was an adaptive challenge that pushed us to question our own assumptions about what adaptation really means. It took some time for us to identify the threads that run through the chapters and to pull out the larger messages. We hope that we have adequately defined and illustrated the nature of the climate change challenge in a way that can contribute constructively to effective responses, whether adaptation, mitigation, or social transformation.

We are grateful to the many people involved in the PLAN project who made this book possible, particularly Siri Mittet, the PLAN project manager. Including all of the authors of the chapters in this book, we would like to thank the researchers, research assistants, students, artists, and international collaborators who provided valuable contributions and inspiration throughout the PLAN project: W. Neil Adger, Virginia Antonijevic, Eva Bakkeslett, Hans Ivar Blystad, Elin Lerum Boasson, Marianne Bruusgaard, Tim Carter, Halvor Dannevig, Jon Vegard Dokken, Per-Ove Eikeland, Torill Engen Skaugen, Siri Eriksen, David Furlow, Inger Hanssen-Bauer, Gail Hochachka, Ida Holth, Gunhild Hoogensen, Grete Hovelsrud, Tor Håkon Inderberg, Jan Erling Klausen, Richard J. T. Klein, Janina Kringstad, Berit Kristoffersen, Irene Lorenzoni, Anne Lucas, Susanne Moser, Mauricio Múnera, Lynn Rosentrater, Cathrine Ruud, Inger-Lise Saglie, Mina Saunte, Jimmy Schmincke, Andreas Segrov, Monica Sharma, Louise Simonsson, Michael Simpson, Ida Skivenes, Knut-Bjørn Stokke, Roger Street, Linda Sygna, Jennifer West, Olga Wilhelmi, Marte Winsvold, and Johanna Wolf. We are also grateful to the Research Council of Norway and the NORKLIMA program, particularly Stine Madland Kaasa and Eivind Hoff-Elimari, for their support. Finally, we thank the chapter reviewers for their valuable feedback and Matt Lloyd and Holly Turner at Cambridge University Press for their continuous support.

1

Climate Change as an Adaptive Challenge

KAREN O'BRIEN AND ELIN SELBOE

How do we adapt to changes for which we ourselves are responsible? This question is philosophical and personal – as well as deeply political – and it is critical to current discussions and debates about human responses to climate change. As climate change adaptation moves to the forefront of policy debates, there is increasing interest in how societies can respond to rapid and unprecedented environmental changes. This includes adapting to what some scientists refer to as nonanalogue ecosystems in a nonanalogue climate (Fox 2007; Williams and Jackson 2007). Although a growing literature on adaptation explores the factors, capacities, and processes that can contribute to successful adaptation, there remains a significant mismatch between the strategies and actions that are being discussed (and to a lesser extent implemented) and the full scope of the problem.

The full scope of the problem is not limited to changes in climate parameters. It is about multiple, interacting processes that can amplify or dampen the social and biophysical impacts of climate change at different spatial and temporal scales and influence the capacity to perceive and respond to change (Leichenko and O'Brien 2008; Hovelsrud and Smit 2010). Biodiversity loss, increasing demands for water, changes in the terms of international trade, urbanization, and other trends create both the foreground and background in which climate change adaptation will occur, both now and in the future. The potential consequences of climate change depend not only on the rate and magnitude of changes in the climate system but also on concurrent transformations in environmental, economic, social, technological, institutional, and political systems (Denton et al. 2014; Pelling 2011).

It can be argued that climate change also calls for deeper transformations, including adapting to the *idea* that humans are responsible for the conditions that will be experienced in the future. This is not trivial, for acknowledging the reality that the dynamics of the climate system are not random or influenced by external sources alone but rather the outcome of human activities and decisions is challenging to many belief systems.

Climate variability and change have thus transcended their meteorological and ecological boundaries to become political issues opened to debate, dispute, contestation, and transformation (Manuel-Navarrete 2010; Manuel-Navarrete and Buzinde 2010). The recognition of a collective capacity to shape global environmental and social conditions implies a fundamental redefinition of the meaning of "climate change adaptation," to include not only responses to the observed and anticipated impacts but also broader and deeper transformations to an equitable and sustainable world (O'Brien 2012).

Such a redefinition is, however, largely absent in the climate change adaptation literature. Adaptation research, policy, and practice to date have been mostly about identifying what to do about climate change impacts, who will do it, how much it will cost, and so on, often with direct reference to the specific parameters that will be influenced by climate change (e.g., temperature, precipitation, sea levels, crop yields) based on different greenhouse gas emissions scenarios (Ford and Berrang-Ford 2011; Biagini et al. 2014). This to-do-list approach is practical; it considers adaptation as a technical problem that can be addressed through expertise, innovation, know-how, skills, and resources (Wise et al. 2014). This may involve changes in management and planning, development of new technologies, institutional reforms, and behavioral changes. Examples include raising bridges, changing building specifications, improving irrigation systems, developing new insurance products, and modifying agricultural practices. This type of response to the direct impacts of climate change can be effective. The to-do-list approach, however, seldom addresses conflicting values, interests, understandings, and approaches to change. Moreover, it turns adaptation into a double-edged sword, for although these measures may be important, they rarely address the wider and deeper systems and structures that are contributing to risk and vulnerability in the first place (Bassett and Fogelman 2013; Eriksen et al. 2015).

This book offers a different perspective on climate change adaptation. Our starting point is that climate change introduces a new type of adaptation challenge for humanity – one that touches on deeper issues related to individual and collective beliefs, values, worldviews, and paradigms, as well as to questions of interests, identities, and power. The particular combinations and constellations of these deeper "adaptive" elements influence and are influenced by sociopolitical processes, and they shape how different people or groups relate not only to the climate system but to each other. These adaptive elements affect current social organization and practices and underlie both collaboration and conflict.

From this perspective, climate change is much more than a technical problem; it represents what Heifetz et al. (2009) refer to as an "adaptive challenge" – a challenge that draws attention to mind-sets, including the assumptions and beliefs that underpin individual and shared attitudes and understandings of change itself. Adaptive challenges are not only personal; they are political, for policies and actions often involve hegemonic and entrenched ways of thinking and acting that may serve certain interests, actors, and priorities over others (Manuel-Navarrete et al. 2011; Pelling 2011; Swyngedouw 2010a, 2010b). Indeed, climate change is not a neutral process. Approaching climate change as an adaptive challenge raises critical questions: Who decides what is "acceptable" versus "dangerous" climate change, and for whom? Who determines which adaptations are good ones and should be pursued? Why is the reduction of greenhouse gas emissions considered political and controversial, whereas adaptation to climate change is viewed as acceptable and desirable? What types of adaptation are equitable, ethical, and sustainable?

The chapters in this book present important insights into the adaptive challenge of climate change and its implications for climate change adaptation. Drawing on diverse methodologies, the contributors approach adaptation as a social, political, cultural, and human process. Many of the chapters point to the significance of differing values, which can make it challenging to identify and agree on the best way forward. Some of the contributions emphasize the political nature of adaptation, including issues of equity and justice that need to be recognized and addressed. Others show examples of the many openings that exist for alternative approaches to adaptation, including participatory approaches and dialogues that acknowledge the adaptive elements. Taken together, the chapters show that climate change adaptation involves much more than adapting to the impacts of climate change. The book concludes by considering how the adaptive challenge can be used as an entry point for much broader and deeper social transformations.

In this introductory chapter, we consider what it means to approach climate change as an adaptive challenge. We first discuss adaptation from an evolutionary perspective, recognizing that not all strategies and measures are likely to have long-term adaptive effects. In looking at adaptation from a broader and deeper perspective, we make an important distinction between adapting to climate change impacts and adapting to the *idea* that humans are changing the global climate system and hence are capable of transforming global systems. We then relate this to the distinction between technical problems and adaptive challenges and discuss why adaptive challenges are inherently both personal and political. Finally, we extract some of the key messages coming out of the contributions to this book, emphasizing the implications for climate change adaptation.

What Is Adaptation?

Adaptation is a response to changing conditions. Whether it is autonomous or planned, reactive or anticipatory, or unconscious or deliberate, adaptation recognizes the reality of change. Although changes are always occurring, human responses to change vary dramatically. In most dictionaries, to adapt is to change to suit different conditions, which can include anything from environmental to political, social, economic, cultural, technological, psychological, or even emotional conditions. Within the climate change literature, adaptation is defined more specifically as "the process of adjustment to actual or expected climate and its effects. In human systems, adaptation seeks to moderate harm or exploit beneficial opportunities. In natural systems, human intervention may facilitate adjustment to expected climate and its effects" (IPCC WGII Glossary 2014, 1).

It is often pointed out that humans have adapted to changes in the past, including environmental changes. Indeed, as a species, *Homo sapiens* has been remarkably adaptable over time. Archaeological records show that we have lived in diverse environmental conditions, surviving and adapting to climate variations and fluctuations (Orlove 2009). Examining the relationship between climate change and modern human evolution, Hetherington and Reid (2010) trace the migrations and global expansion of humans to the last glacial cycle that began 135,000 years ago. This period was characterized by significant global climate change, and it created the conditions for humans to emerge as the dominant mammalian species during the last 10,000 years.

What, then, is adaptation to climate change? Hetherington and Reed (2010, 7) argue that adaptation represents something novel, new, or radical: "for humans, the disruption of a stable climate puts a premium on physiological and behavioural adaptability, leading to the application of novel behaviors and new ideas that might have been obstructed as 'too revolutionary' or too much of a change from traditional practices in previously stable social groups." Ellen (1982) provides examples of four types of novel adaptations in biological organisms: phylogenetic (e.g., sweat gland distribution), physiological (e.g., oxygen uptake in high altitudes), learning (e.g., escaping from a predator), and cultural (e.g., wearing warmer clothes). Both learning and cultural modifications represent behavioral changes that require participation of the central nervous system; with these types of adaptation, the responses of individuals can be transmitted to others independently of genes (Ellen 1982). Adaptation is thus not only a biological process but also a social, cultural, and human one.

Although adaptation is possible from an evolutionary perspective, Hetherington and Reid (2010) recognize constraints to adaptation, particularly if change is not recognized or understood, or if groups or societies are not open and willing to accept difference and change. They point out that "if the change required is too great for the individual, organism or society to manage, or alternatively if they are not willing to adjust, then decline and even extinction prevails" (Hetherington and Reid 2010, 300). Drawing on a comparative history of past adaptations, Orlove (2005) shows that human societies can be fragile and resistant to changing established patterns of action. Ellen (1982) emphasizes that adaptation is more than a passive effect in response to the presence of certain conditions but instead requires that the problem be properly recognized, diagnosed, and responded to effectively. Yet "because of the inadequacy of human sensory perception, cerebral coordination, cultural information and ability to respond, adaptations frequently fall short of a goal or have no adaptive effects at all" (Ellen 1982, 239).

The risk of pursuing adaptations that have no long-term adaptive effects at all is salient to current approaches to climate change, where risks are directly linked to human activities, social and economic policies, and development approaches. A high adaptive capacity, often associated with access to technology, high education levels, economic equity, and strong institutions (see Yohe and Tol 2002), may not always lead to successful adaptation and, in some cases, may contribute to complacency (O'Brien et al. 2007). Bassett and Fogelman (2013) point out that rather than adjusting to changes in the climate, addressing the social structural causes of vulnerability is essential. Eriksen et al. (2015) challenge "development as usual," arguing that adaptation is as much a problem *of* development as *for* development. Indeed, the limits to adaptation are considered by Adger et al. (2009) to be endogenous to the system rather than a result of outside, uncontrollable forces. Although authors such as Diamond (2005) show that failure to adapt can lead to the collapse of societies, a common will to overcome adversity by revising or developing collective strategies for survival is the hallmark of societal resilience (Butzer and Enfield 2012).

In the past, adaptations to environmental changes were largely local, with some groups faring better than others (Ellen 1982; Orlove 2005). However, the scope of the issue is now global and long term, and the impacts of anthropogenic climate change are expected to continue for centuries, even after the rate and amount of greenhouse gas emissions decrease. The IPCC (2014, 17) warns that "without additional mitigation efforts beyond those in place today, and even with adaptation, warming by the end of the 21st century will lead to high to very high risk of severe, widespread, and irreversible impacts globally." This places a premium on adaptation, but what kind of adaptation? What is common to most contemporary discussions and approaches to climate change adaptation is a focus on the challenges of adapting *to* a changing climate. Such adaptations usually start with observed or projected changes in climate parameters, then consider what can be done to minimize impacts (O'Brien et al. 2007). In reviewing the adaptation literature, Bassett and Fogelman (2013) found that 70 percent of the articles focused on adaptation as adjustment to climate stimuli. Importantly, Berrang-Ford et al. (2011) point out that adaptation in the real world is most often in response to multiple processes, including changing economic conditions, health threats, and social change. And this is where adaptation becomes more complicated.

Whether it is considered an adjustment to climate stimuli or multiple stressors, the normalization of climate change adaptation as a good and necessary response to both experienced and projected climate change impacts comes at a cost, for it can draw attention away from critically questioning the changes. In fact, the naturalized ways of thinking about adaptation often serve dominant or hegemonic powers or hide the influence, interests, and agendas of those who are invested in current systems and trajectories of change – and this includes those who are adapting. Following along current global greenhouse gas emissions trajectories, there is a risk that such adaptations will have few long-term adaptive effects, particularly if the limits to adaptation are exceeded, that is, if adaptive actions to avoid intolerable risks are not possible or not currently available (IPCC 2014). For adaptation to be successful, a proper diagnosis of the problem is essential.

Technical Problems versus Adaptive Challenges

To understand the difference between adapting *to* climate change and adapting to the *idea* that humans can influence global systems and future trajectories, it is necessary to step back and consider how people and organizations approach the process of change. According to Heifetz et al. (2009), there are two distinct approaches to any problem involving change. The first is to treat change as a technical problem, that is, one that can be diagnosed and solved by applying established know-how and expertise. Technical problems may call for improved skills, better procedures or management, increased allocation of resources to a problem, more innovation, or new types of governance. Working with a known set of tools and approaches, they often involve doing things better and more effectively. Improved early warning systems, more efficient energy systems, more effective governance across scales, and new institutional arrangements are just some of the important technical responses to climate change (Biagini et al. 2014; Wise et al. 2014). Although technical problems are often complicated and difficult to address, the required skills can be identified, developed, and applied (Kegan and Lahey 2009). It is usually not a lack of options that hinders solutions to technical problems but rather issues such as costs and political priorities. Humans have proven to be remarkably good at solving technical problems, and this would serve as reason for optimism regarding adaptation – if climate change were, in fact, only a technical problem.

The second approach is to see the problem as an "adaptive challenge." Adaptive challenges are uncomfortable and often conflictual conditions or situations that have no predefined solutions and often require changes in mind-sets, priorities, habits, and loyalties (Heifetz et al. 2009). They call for new ways of perceiving systems, relationships, and interactions, often at new levels of mental complexity (Kegan and Lahey 2009). As such, the solutions to adaptive challenges do not follow clear, linear pathways, nor are they amenable to expertise-based management approaches. According to Heifetz et al. (2009, 70), "adaptive challenges are typically grounded in the complexity of values, beliefs, and loyalties rather than technical complexity and stir up intense emotions rather than dispassionate analysis." Adaptive challenges may be experienced by individuals, households, organizations, communities, governing bodies, or institutions dealing with change at any scale.

Addressing an adaptive challenge may often be less about doing something and more about shedding entrenched ways of doing things, which Heifetz et al. (2009) consider to include tolerating losses while gaining new capacities. Adaptive challenges thus require clarity over the values and conditions that are considered worth maintaining. In fact, they are challenges precisely because they draw attention to the individual and collective values, beliefs, and worldviews that often underlie disagreements and conflicts. As Wise et al. (2014) note, technical approaches to climate change adaptation seldom consider the dynamic interactions between values, knowledge, cultures, and institutions. These subjective aspects influence how both systems and change itself are perceived and approached and the types of responses that are prioritized or normalized.

A few specific examples can illustrate the distinction between technical problems and adaptive challenges. Developing panels and storage capacity for solar energy is a technical problem that can be solved through increased investments in research and development. However, producing and implementing solar energy technologies at a rate and scale that can replace fossil fuel energy sources is an adaptive challenge that confronts interests and assumptions related to investments in coal, gas, and oil; opens debates about centralized versus decentralized energy supplies; and introduces cultural issues related to the amount, timing, and accessibility of energy. Moving a population off an island that is vulnerable to sea level rise is a technical problem related to housing, livelihoods, immigration, and citizenship, but it is an adaptive challenge that affects cultural values, identities, and ideas about belonging and social justice (Adger et al. 2013; Barnett and Campbell 2010; Kuruppu and Liverman 2011). Although adaptive challenges often do have important technical aspects, focusing on the technical dimensions alone is considered a recipe for failure (Heifetz et al. 2009).

How can adaptive challenges be identified? Heifetz et al. (2009) describe five general characteristics that can be useful for distinguishing adaptive challenges from technical problems. First among these is a *persistent gap between aspirations and reality*. In relation to climate change, aspirations to avoid dangerous climate change, as stipulated in Article 2 of the UNFCCC, have been unmatched by international efforts to reduce greenhouse gas emissions. In fact, the level of CO_2 emissions from the burning of fossil fuel and cement production reached were the highest in human history in 2013 (Global Carbon Project 2014). Second, adaptive challenges are often revealed when *current responses are inadequate*. Globally, responses to both mitigation and adaptation are currently insufficient compared to what is considered necessary to respond effectively to climate change (IPCC 2014). Third, *adaptive challenges usually require difficult learning*. Although many problems are complex, climate change can be considered a hypercomplex problem that includes dynamic complexity (where cause and effect are distant in space and time), social complexity (where there are conflicting interests and priorities), emergent complexity (where nonlinear outcomes are likely to lead to surprises), and human complexity (including different perceptions and approaches to meaning making) (see Scharmer 2009). This challenges traditional approaches to both education and capacity building (O'Brien et al. 2013).

Fourth, *adaptive challenges require the engagement of new stakeholders*, including people who have not been traditionally associated with the solution. Climate change impacts affect all sectors and groups, and responses call for a plurality of voices and perspectives. Important stakeholders include religious leaders, youths, artists, labor unions, nongovernmental organizations, communities, and many others. Fifth, *adaptive challenges often involve long-term issues* that cannot be addressed by quick fixes. Unlike the ozone hole, climate change is not a problem that can be addressed through scientific expertise leading to a comprehensive international regime (Keohane and Victor 2011). Responding to the causes and consequences of climate change will require adaptive responses for decades to come (IPCC 2014). Finally, adaptive challenges become most clear when *disequilibrium or a sense of crisis is experienced*, marking recognition that technical solutions may be insufficient to address the problem. This last point is clear to many, as climate change is already experienced as a crisis among some groups in many parts of the world (Doherty and Clayton 2011).

Adaptive challenges involve questioning what is natural, common sense, or self-evident, and they often open up for possibilities that were either not articulated, silenced, or dismissed as impossible. As Heifetz et al. (2009) point out, adaptive challenges can be quite tough because they require that people modify the stories they have been telling themselves and others about what they believe in, stand for, and represent. This may involve acknowledging that traditional hierarchies have benefited some much more than others, leaving many people and groups socially vulnerable to current climate variability and change. It may involve recognizing that the successes and accomplishments that define modernity have incurred significant ecological, social, and cultural costs. Moreover, it may involve acknowledging that postmodern values of equity, justice, and fairness are not prioritized by everyone, nor do such values change easily (O'Brien and Wolf 2010). Climate change brings out the limitations of many contemporary approaches to dealing with complex issues, underscoring the dangers of treating it as a technical problem (Kegan and Lahey 2009).

Adaptive Challenges Are Personal

Adaptive challenges are fundamentally about change, and the role of beliefs, values, and worldviews in influencing approaches to climate change cannot be underestimated. Worldviews, which combine beliefs, assumptions, attitudes, values, and ideas into a model of reality, are particularly significant because they filter human perceptions and influence every aspect of how people understand and interact with the world around them (Schlitz et al. 2010). Worldviews influence perceptions of human–environment relationships, including how individuals perceive particular ecological issues and their solutions. They also tend to influence how willing people are to engage with environmental issues through politics (Hedlund-de Witt 2012). In short, adaptive challenges are personal because they call for people to confront the ways that they individually and collectively view the world.

It is well understood in the social sciences and humanities that people view and perceive systems, structures, and knowledge in general through filters influenced by interests, identities, habits, loyalties, and motivations, and not the least by emotions (Bourdieu 1977, 1990, 1998; Norgaard 2011). The positioned and embodied vision of an individual is also conditioned by the historical context and social setting (see Haraway 1996; Rose 1997). Such partial and situated knowledge contributes to a subjective view of the world rather than of the world as an absolute, objective truth. It provides conscious and unconscious justifications for some strategies or outcomes over others, and it often underlies attitudes toward equity and fairness. Adaptive elements affect human relationships at every scale, and they are critical to understanding the adaptive challenge of climate change.

Mind-sets are influenced by individual and shared beliefs, which can protect against risks but also perpetuate them, especially when they validate the systems and structures that contribute to risk and vulnerability. Kegan and Lahey (2009) point out that there is a mismatch between the challenges now facing individuals and groups and their capacities to deal with them, at least with current mind-sets. Interestingly, the suggestion that a problem such as climate change demands increased mental complexity often leads to the conclusion that everybody else must change *their* mind-sets, including their values, beliefs, and worldviews. This turns climate change into a technical problem, with the objective of nudging, manipulating, or changing people's beliefs and values and consequently their behaviors (Shove 2010). Although behavioral changes are an important part of many technical problems, imposing changes on individual and collective beliefs is not always considered a legitimate way of producing lasting change (Freire 1970; Schlitz et al. 2010; Rowson 2011). Beliefs, values, and worldviews have nonetheless been manipulated throughout history, whether through oppression, indoctrination, or "brainwashing" via constant messages received through various media. As Freire (1970) notes, values and worldviews are often imposed on others through cultural invasion rather than through cultural synthesis. This invasive approach often breeds resistance and resentment rather than cooperation and collaboration (Scott 1985).

Although research shows that it is difficult to change beliefs, values, and worldviews, there is substantial evidence that the mind-sets of individuals can and often do change over a lifetime, thus influencing meaning making and shared meanings (Kegan and Lahey 2009; Wilber 2001). Research on adult learning reveals that mental complexity can increase throughout adulthood in stages and over periods of time, giving rise to different ways of knowing the world (Kegan 1994; Kegan and Lahey 2009). Kegan and Lahey (2009) describe the expansion of mental complexity as a change in the relationship between subject and object, that is, between what one *looks through* and what one can *look at*. When an individual or group starts to question assumptions and relationships that have been taken for granted, meaning making shifts and problems and solutions are viewed from a different perspective. Stepping back to reflect on one's own thought processes develops the capacity for self-reflexivity and "metacognitive awareness," which can stimulate worldview transformations (Schlitz et al. 2010).

The problem with change is not always the change itself but the anxiety that it can create, especially the feelings of fear, danger, or not being able to cope with the complexity of a challenge (Kegan and Lahey 2009). Addressing an adaptive challenge often involves engaging in difficult discussions, dialogues, or self-reflection that can potentially surface beliefs that are limiting, "facts" that are taken for granted, assumptions about others, and ideas about what is possible or impossible. Heifetz et al. (2009) recognize that adaptive work can be challenging and that most individuals and organizations avoid dealing with a problem until the sense of disequilibrium becomes so high that they have no other choice. This is indeed a common characteristic of climate change. Norgaard (2011), for example, found that discussions of climate change were associated with a significant degree of helplessness and powerlessness. She argues that silence and inaction on climate change are in most cases not a rejection of information "but the failure to integrate this knowledge into everyday life or to transform it into social action" (Norgaard 2011, 11). Transforming knowledge into action thus calls for political engagement. As we discuss, although adaptive challenges are personal, they are also political.

Adaptive Challenges Are Political

The plurality and differences in individual and collective beliefs, values, and worldviews discussed earlier underpin interests and influence what is considered desirable. They guide actions and shape interpretations and perceptions of both problems and solutions. Although seldom explicitly mentioned, these subjective interpretations of the world inform current responses to climate change, including adaptation. Decisions and actions that are presented as self-evident and inevitable (as though based on rational, objective, and neutral evaluations) often reflect the views and interests of dominant groups and help naturalize and legitimize their power and way of constructing the world (Bourdieu 1977, 1998). Adaptive challenges are political in that they surface differences, contradictions, and antagonisms, whether in relation to climate change or change in general, and they raise important questions about how to decide the types and timing of responses that will be taken, and who will decide them.

Current dominant and consensual or hegemonic approaches to climate change tend to maintain rather than challenge the status quo (Manuel-Navarrete 2010; Pelling 2011; Pelling and Manuel-Navarrete 2011; Pelling et al. 2012; Swyngedouw 2010a, 2010b, 2013). Swyngedouw (2010a) argues that the presentation of climate change as a global humanitarian cause produces a depoliticized imaginary that conceals the interests behind the choice of one trajectory over another. Not only is depoliticization framed around the perceived inevitability of climate change but, he contends, it also is framed around "the perceived inevitability of capitalism and a market economy as the basic organizational structure of the social and economic order, for which there is no alternative" (Swyngedouw 2010a, 215). There is a focus on behavioral changes and technological and managerial solutions to climate change, achieved through compromise and the production of consensus. This often results only in reforms of institutions and practices that support and maintain existing systems rather than more radical transformations (Manuel-Navarrete 2010; Manuel-Navarrete et al. 2012; Pelling 2011; Pelling et al. 2012; Swyngedouw 2010a, 2010b, 2013). Giving primacy to economy and growth intensifies rather than reduces climate change (Klein 2014). Such framings, managerial solutions, and reforms serve current elites and systems and prevent the development of more sustainable alternatives.

Swyngedouw (2010a, 223–24) describes this as a passive revolution, "whereby the elites have not only acknowledged the climate conundrum and, thereby, answered the call of the 'people' to take the climate seriously," but have also convinced the world that their current dominant economic and political systems of capitalism and liberal democracy hold the right and only solution. Furthermore, rather than presenting people as heterogeneous political subjects, Swyngedouw (2010a, 221) notes that they are now constituted as "universal victims, suffering from processes beyond their control." This is reminiscent of Freire's (1970, 73) critique of traditional education, which avoids developing a critical consciousness associated with transformation of the world by instead presenting the world as a given: "The more completely they accept the passive role imposed on them, the more they tend simply to adapt to the world as it is and to the fragmented view of reality deposited in them." Freire (1970) goes on to argue that this serves the interests of the oppressors, who do not care to have the world revealed or transformed, as they can use their humanitarianism to preserve a profitable situation. In short, "the more completely the majority adapt to the purposes which the dominant minority prescribe for them (thereby depriving them of the right to their own purposes), the more easily the minority can continue to prescribe" (76).

The potential for elites and dominant groups to construct a certain view of the world, or prescribe specific social, economic, environmental, and political arrangements, is related to the functioning of symbolic systems (Bourdieu 1991). The exercise of power takes a symbolic form that is connected to the ability to impose meaning in a way that it is taken for granted; consensus (doxa) is established, and "the arbitrariness of power relations . . . is misrecognised as natural, giving legitimacy to domination and reproducing existing power relations" (Stokke and Selboe 2009, 63). Thus symbolic systems involve ways of seeing the world, cultural frames for classification, and ways of communicating. Symbolic systems are highly political because they conceal underlying power relations, when in fact they reflect these and are used as instruments for legitimizing domination; as such, they represent symbolic violence toward dominated groups (Bourdieu 1991).

Likewise, current approaches to vulnerability and adaptation research and policy often take a naturalized view on climate change. Cameron (2012, 104), for example, criticizes these approaches for perpetuating colonial "systems of cognition" and colonial ways of both knowing and understanding northern Indigenous peoples. She argues that the benevolent, well-meaning interventions in the name of climate change prioritize technical interventions that exclude political-economic questions linked to colonialism and the perpetuation of these relations through continued extraction of northern resources. Furthermore, she contends that the shift away from slowing or stopping climate change and instead adapting to it is profoundly political, as it "reroutes political discourse on Inuit relations with climatic change, particularly efforts to hold distant greenhouse gas producers accountable for their impacts on Arctic lands and peoples and to frame climate change in terms of equity, justice, and human rights" (107).

Whereas Cameron (2012) argues for undertaking actions that address the issues and concerns of northerners, recognizing that interventions are never neutral, Swyngedouw (2010a, 228) calls for "foregrounding and naming different socio-environmental futures and recognizing conflict, difference and struggle over the naming and trajectories of these futures." This real or genuine politics is currently pushed to other arenas as conflict and disagreement is silenced or sanctioned. They can be found in various types of collective action like street protests (Swyngedouw 2010b) or the struggles of social movements. Freire (1970, 50), however, recognized that in fact "the concrete situation which begets oppression must be transformed," and he drew attention to the "thought-language with which men and women refer to reality, the levels at which they perceive that reality, and their view of the world" (97).

Thus, what is needed to respond to the adaptive challenge of climate change are critical discourses and radical political action that disclose, question, and transform the legitimized order and its underlying power relations, which constitute and normalize certain imaginaries and practices (Manuel-Navarrete 2010; Swyngedouw 2010a, 2010b, 2013). These issues must be drawn out of the "universe of the undiscussed" and into the "universe of discourse"; confronted with alternative discourses, the arbitrariness of commonsensical political, social, and economic arrangements will be exposed, and they will lose their character as natural phenomena (Bourdieu 1991, 1998). It is precisely in times of crises that the undiscussed is brought into discussion through critiques of dominant systems and power structures and through articulation of alternatives that used to be considered impossible. While this may be a prerequisite for repoliticization of climate change and social transformation, Manuel-Navarrete (2010) also calls for a stronger focus on human agency and flourishing. He links human agency and "the internal transformation through which each individual liberates him or herself" (784) to the process of changing oppressive power structures "and their transformation into fairer and less exploitative social and environmental relations" (783).

This involves an emancipatory-democratic politics that focus on equality and freedom (Swyngedouw 2011). To realize this, it may ironically be necessary to work *with* rather than against values and worldviews, skillfully engaging with people as they are rather than where we think that they should be (O'Brien and Hochachka 2010).

The Adaptive Challenge of Climate Change

The adaptive challenge of climate change is both personal and political because it fundamentally confronts our relationship to both individual and collective change, as well as relationships to each other, to nature and to the future. Climate change is a challenging idea because many societies have traditionally treated climate as an external condition to which they needed to adapt (Hulme 2008). The climate is now no longer a given background for human activities but instead has become a collectively produced (although variable) and deeply sociopolitical phenomenon (Manuel-Navarrete and Buzinde 2010). The chapters that follow in this book discuss climate change adaptation as a social, political, cultural, and human process. Although the chapters and case studies focus on adaptation in high-income countries that are considered to have a high adaptive capacity, the adaptive challenge of climate change is universally relevant. Whether focusing on adaptation in urban areas or the communication of climate change, each chapter highlights adaptive elements that make climate change into more than a technical problem that can be addressed through resources, expertise, improved practices, and reformed institutions. Taken together, the chapters present important insights on how and why adaptation may have to be defined and approached differently.

We start with a philosophical perspective on climate change adaptation. In Chapter 2, Guillaume and Neuteleers explore various ethical issues related to climate change and its implications for justice and responsibility. At the epistemological level, they argue for adopting a global stance, looking at the Earth as a single global system. At the ontological level, they point out that the relationship between nature and humans has fundamentally changed, as we are now impacting the Earth system on the planetary scale. They highlight the potential of virtue ethics – the positive idea of building a desirable society likely to persist and flourish in a changing climate. They conclude that the most challenging adaptation might be the one of adapting our values to the reality of a new climate, including adapting to the idea of anthropogenic climate change and the responsibilities this entails. As such, climate change is philosophically an adaptive challenge.

The next three chapters describe and analyze what is taking place on the ground in relation to climate change adaptation. Working at different scales, the chapters illustrate the importance of networks and learning. They show that although there is a strong tendency to focus on the technical aspects of climate change adaptation, the personal and political aspects of the adaptive challenge are being observed and, in some cases, addressed.

In Chapter 3, Solecki, Patrick, and Sprigings use an urban policy transition frame to assess the role of system tipping points and policy triggers in public policy formation. They analyze climate change adaptation strategies developed in New York City and London as a set of complex interactions among and between policy entrepreneurs, government officials, and members of the scientific-expert community. In reviewing their case studies they present a broad conceptual model of urban climate policy development, where they include traditional policy drivers (activities that foster the demand and capacity for policy action) and dynamic contextual elements (properties that influence the speed and rate of policy action), as well as several drivers and elements specific to increased climate risk and climate nonstationarity. They also show how narratives of sustainability and resilience have affected urban climate policies and brought new questions about the rate and nature of future urban environmental transitions. The chapter suggests the emergence of an adaptive framing of climate change adaptation in these cities.

In Chapter 4, Klausen, Saglie, Stokke, and Winsvold analyze the extent to which climate change considerations have made their way into Norwegian urban planning and development, which is characterized by extensive interaction between public governments and private investors and developers. Through case studies, the authors investigate how different mechanisms of social coordination (hierarchies, markets, and networks) affect collective learning for urban climate change adaptation. They found that rather than being grounded in established routines, policies, or strategies, adaptation in urban planning has been arbitrary and circumstantial, introduced to a project by one or more individuals with a particular awareness or knowledge about climate change issues. They argue that the scarcity of adaptation measures is not due a lack of institutions or arenas for social coordination but rather is systemic. In the absence of political awareness and will, hierarchies will not act; if there is little hope for profit, the markets will fail to deliver adaptation; and even if the networks appear to somewhat improve the situation, this is mostly through stimulating individual learning. They stress that organizational learning depends on clear signals to act from national authorities rather than relying on dedicated individuals with adaptation knowledge. These case studies clearly reveal the importance of addressing the political dimensions of adaptive challenges.

In Chapter 5, Waller and Barnett show that Australian adaptation policies perpetuate patterns of resource use or development that are biased toward existing economic activities rather than toward social and cultural issues. Viewing adaptation as a process of change, they draw on insights from the organizational change literature to analyze adaptation policy in Australia. The Australian case illustrates that adaptive capacity does not necessarily lead to adaptation action: there has been only a slow and ad hoc adaptation response so far, as adaptation policies have been abstract and focused mainly on acquiring sufficiently detailed climate and impact knowledge to economically justify controlling risks. They argue that developing effective adaptation policy seems to be at odds with the prevailing neoliberal and economically rationalist policy regimes in Australia. Leadership, time, and concerted changes to both social and political cultures are needed. A first step would be to start activities that favor the social acceptance of change, such as creating visions, sustaining urgency, and acknowledging the role of cultural and process-oriented aspects of adaptive change. The limits of the technical approach are revealed through this case, as are the opportunities for questioning the hegemonic discourse and considering new alternatives.

The next three chapters highlight some of the important personal and political issues underlying the adaptive challenge of climate change. In Chapter 6, Vancura and Leichenko explore important equity and justice issues related to both vulnerability and adaptation actions in the coastal zone and in the agricultural sector of New York State. They recognize uneven exposure to impacts of climate changes as well as the role of preexisting socioeconomic or spatial inequalities and imbalanced capacities and possibilities for engaging in adaptation strategies and planning. There is a need, they argue, both to incorporate equity and justice dimensions into adaptation discussions and policies to ensure that unintended impacts do not exacerbate vulnerability and to include broad participation in planning processes. This chapter demonstrates that climate change adaptation is not neutral and that important values and power discrepancies are often excluded from policy discussions.

In Chapter 7, Eriksen and Selboe analyze how variations in values, aspirations, and vulnerability make people understand climate change and other stressors differently and inspire different actions and considerations. In a case study of a small municipality in Norway, they show how social cleavages are linked to differing aspirations and values. They discuss how local power relations, negotiations, and collaborations favor certain development discourses and actors over others in a context of depoliticization where the prioritization of certain values and goals is implicit and presented as self-evident and unavoidable. The authors discuss the plurality of values and aspirations and identify a need for participatory processes that provide possibilities for voicing alternative visions. Such processes must acknowledge and make explicit value conflicts and issues of power and ensure transparent prioritization in decision making.

In Chapter 8, Kristoffersen shows how the hegemonic discourse on petroleum policies in Norway is extremely powerful, contributing to a framing of climate change adaptation as an opportunity to realize short-term goals and provide a political foundation for the status quo. She introduces the concept of "opportunistic adaptation" and describes Norwegian state policy to uphold and expand extraction of oil and gas, based on the idea that the economic benefits of climate change should be prioritized over the efforts to address the causes. This has led to a shift from mitigation to economic adaptation, thereby promoting climate change as a problem to be managed but not solved.

Opportunistic adaptation is supported by a discourse on how increased production of oil and gas in Norway is a fundamental prerequisite for economic development in the Global South. Kristoffersen's research demonstrates how economic arguments and Norwegian state interests are downplayed and hidden behind the framing and maintained ethics of helping developing countries.

The next four chapters emphasize the role of values in adaptation processes, linking personal and political dimensions of adaptive challenges. In Chapter 9, Quinn, Lorenzoni, and Adger explore how climate change impacts and adaptions alter the physical and social characteristics of a place, deeply affecting individuals in that locality and their place attachment and identities. Place attachment and identity play a central role in adaptation, influencing how climate change is framed and managed and how responses by governments and other agencies are deemed to be appropriate or fair. Social networks and group identity, as well as personal identity, partly determine how people understand and deal with different types of stressors and risks. The authors argue that decision-making processes, which influence how places change, must be carefully negotiated to ensure sensitive, sustainable, and legitimate adaptation. Social, psychological, symbolic, and emotional issues must be acknowledged and included in responses to climate change, along with the mainstream focus on physical and financial aspects.

In Chapter 10, Wolf, Allice, and Bell examine how people's values relate to their experience of climate change and ideas about adaptation. Through a study of two communities in Labrador, Canada, they explore the role of values and traditional practices in adaptation to climate change. Using Q-methodology, the authors identify four discourses, underpinned by different and at times competing values that support dissimilar adaptation goals. They show that there are diverse views, both within and between communities, on what a changing climate means and how best to adapt to it. Their findings have implications for adaptation planning: planning should result from concrete local needs identified by communities themselves and should acknowledge different values. Where adaptation is not perceived as necessary, for instance, because of a strong belief in local inherent adaptability, planning must stress respect for local views (particularly if directed from outside the community). They conclude that adaptation planning should consider values and find ways to acknowledge and reconcile competing goals.

In Chapter 11, Hovelsrud, West, and Dannevig explore the different vulnerability and adaptation narratives among fishers, farmers, and municipal planners in northern Norway and how they are underpinned by different ways of life or worldviews. They argue that such narratives shape and reflect adaptation and adaptive capacity, as they influence how people perceive and act on climate change. Local adaptation is shown to be influenced by lasting cultural discourses, such as the one common in their study areas: *vi står han av* (we always handle hardship). Such cultural discourses are grounded in local worldviews, perceptions, and experiences and might constitute both a potential for and a limitation to adaptation. A proper understanding of the historic, social, and economic context, as well as of differing values and worldviews that are reflected in narratives and discourses, is vital to understanding local priorities. These "adaptive elements" are in turn vital for developing appropriate climate information to include in local adaptation processes.

In Chapter 12, Inderberg highlights the role of norms, values, and cultural context as important adaptive elements of organizations, as they affect what types of behavior and decisions are regarded as appropriate. He investigates changes in organizational culture and formal structure through a comparative analysis of how reform and large organizational changes in the Norwegian and Swedish electricity grid sectors influence adaptive capacity to extreme weather events that can be associated with climate change. For instance, he demonstrates how cultural changes, characterized by a shift from an engineer-based to an economic-institutional logic or mind-set, changed the basis on which decisions in both countries' electricity sectors were legitimized. In the case of Norway, adaptive capacity was reduced when economic efficiency was promoted over security of supply. It is argued that the influence of culture and values must be acknowledged along with formal organizational factors to properly understand the organizational capacity to adapt to climate change.

The next four chapters look at some of the tools and approaches for dealing with adaptive challenges – including learning processes such as deliberation and participation. Tools to incorporate diverse values and perspectives into planning processes can help to transform climate change vulnerability and adaptation assessments from a technical exercise that can be analyzed with data into a facilitated multistakeholder approach that supports public participation, discussion, and analysis in spatial decision making.

In Chapter 13, May explores the possibility of creating meaningful local collaborative initiatives that include the range of perspectives within communities as part of community climate change adaptation. He discusses a study of a multiactor collaborative process in the Niagara region of Ontario, Canada, which was developed and implemented to reflect the local needs and capacity of the community. Participatory approaches create an opportunity for sense making that is more responsive to the diversity of views, values, and social complexity that is seldom captured in more managerial approaches. He argues that there is a need for social-ecological inventories that capture the wealth of knowledge and relationships that already exist in an area, as these also can be useful in developing community climate change adaptation leadership and capacity for adaptation, possibly also strengthening existing institutions, networks, and emerging leaders. The chapter also explores how informants in such processes become actors and potentially also leaders in championing climate change adaptation.

In Chapter 14, Swartling, Wallgren, Klein, Ulmanen, and Dahlin explore the role of learning and participation in advancing climate change adaptation. Through the case and observation of a deliberative process with stakeholders from the Swedish forestry sector, they investigate how participatory processes can contribute to learning on climate change and adaptation needs and options. Their findings point to increased individual learning, as they found changes in expressed perceptions, values, and knowledge about climate change effects, vulnerability, and potential adaptation measures, as well as self-perceived learning among participants. The process was appreciated by the stakeholders who called for multilevel, multistakeholder arenas for knowledge sharing and collaborations on climate change and adaptation within the Swedish forestry sector. The authors point to the need to further explore how these approaches can be scaled up to promote mainstreaming of participatory learning for adaptation.

In Chapter 15, Rosentrater describes a process for assessing vulnerability that accounts for the subjective judgments that influence decisions about adaptation needs that inform an integral approach to geographic information systems (GIS). Her point of departure is that current conventional GIS and vulnerability maps do not represent the multidimensional nature of vulnerability, and she argues that rule-based knowledge about vulnerability must be augmented with contextual knowledge that situates the experiences and concerns of all adaptation actors. She describes integral GIS as a facilitated multistakeholder approach that supports participation, discussion, and analysis in spatial decision making. It is an approach that is designed to broaden the frame of reference when gathering information about specific planning decisions, as it combines evidence- and values-based dimensions of climate change. The chapter considers how integral GIS allows for exchange of information, comparison of perspectives, and building a collective model of vulnerability that incorporates the interests and values of diverse social groups.

In Chapter 16, Moser and Berzonsky address the very polarized political views on climate change in the United States and how they are based on underlying value commitments and beliefs. They explore the possibilities for meaningful and constructive communication across cultural and ideological divides and point to the need for deeper engagement, particularly with the unheard majority of Americans. They argue that openness and curiosity about the other are prerequisites for finding common ground and moving from cultural divides to interpersonal opening and understanding through deliberation. Although such divides may be difficult to overcome, and dialogue may not be possible with everyone, the help of courageous leaders can create deep productive conversations through true dialogue on the adaptive challenge of climate change.

In Chapter 17, we conclude by considering the types and quality of responses needed to face the adaptive challenge of climate change, emphasizing the relationship between adaptation and social transformations to sustainability. We consider both the political and personal dimensions of adaptive challenges and argue that they must be addressed together by challenging "the given" through a politics of inclusion, participation, and plurality. We stress the importance of allowing for and acknowledging difference, disagreement, and conflict, which are inherent in true democratic politics and critical to adaptive work. This is a challenging process, and one where issues of power, exclusion, and justice must be taken seriously and repeatedly scrutinized. Although resistance to change is common and fear of loss and lack of control can lead to anxiety or denial, there is great potential to overcome individual and collective limitations through reflexivity, dialogue, collaboration, learning, and leadership.

Conclusion

Taken together, the chapters in this book present a new way of thinking about adaptation to climate change. We show that although there are technical aspects to most adaptive challenges, treating adaptation as only a technical problem is likely to fail because it does not engage with subjective elements such as beliefs, values, and worldviews, which influence interests, identities, loyalties, and power relations. Without recognizing both the personal and political dimensions of climate change, the technical approach is likely to have limited long-term effects, not the least because it promotes a neutral, apolitical approach to adaptation that maintains rather than challenges the status quo (Manuel-Navarrete 2010; Swyngedouw 2010b).

Seldom do discussions of adaptation explore deeper issues, including how individuals and societies are adapting the idea that human decisions and actions are capable of transforming systems at a global scale. By questioning assumptions and beliefs, it becomes easier to see alternatives that lie beyond individual and collective blind spots. This is critical in relation to climate change, where the narratives and stories that have developed often make it seem that there are limited political options for addressing the underlying drivers of risk and vulnerability. With adaptive challenges, business-as-usual is often the underlying source of the problem (Heifetz et al. 2009). To successfully adapt to a changing climate in a world with power inequalities and competing interests, there is a need to engage in new conversations about not only climate change, but change itself. The future in a changing climate is increasingly presented as a choice, and the adaptive challenge of climate change demands an answer to the question, "Whose choice?"

Acknowledgments

We would like to thank the authors of this book as well as all the researchers of the PLAN project (Potentials for and Limitations to Adaptation to Climate Change in Norway). The project was funded by the Research Council of Norway through the NORKLIMA program and led by Karen O'Brien at the Department of Human Geography and Sociology at the University of Oslo. We would also like to thank David Manuel-Navarrete and Linda Sygna for valuable comments on earlier drafts of this chapter.

References

Adger, W. N., Barnett, J., Brown, K., Marshall, N. and K. O'Brien. 2013. "Cultural dimensions of climate change impacts and adaptation." *Nature Climate Change* 3 (2): 112–17.

Adger, W. N., Dessai, S., Goulden, M., Hulme, M., Lorenzoni, I., Nelson, D., Naess, L. O., Wolf, J. and A. Wreford. 2009. "Are there social limits to adaptation to climate change?" *Climatic Change* 93 (3): 335–54.

Barnett, S. and J. Campbell. 2010. *Climate Change and Small Island States: Power, Knowledge and the South Pacific*. London: Earthscan.

Bassett, T. J. and C. Fogelman. 2013. "Déjà vu or something new? The adaptation concept in the climate change literature." *Geoforum* 48: 42–53.

Berrang-Ford, L., Ford, J. D. and J. Paterson. 2011. "Are we adapting to climate change?" *Global Environmental Change* 21 (1): 25–33.

Biagini, B., Bierbaum, R., Stults, M., Dobardzic, S. and S. M. McNeeley. 2014. "A typology of adaptation actions: A global look at climate adaptation actions financed through the Global Environment Facility." *Global Environmental Change* 25: 97–108.

Bourdieu, P. 1977. *Outline of a Theory of Practice*. Cambridge: Cambridge University Press.

Bourdieu, P. 1990. *The Logic of Practice*. Stanford, CA: Stanford University Press.

Bourdieu, P. 1991. *Language and Symbolic Power*. Cambridge: Polity Press.

Bourdieu, P. 1998. *Practical Reason: On the Theory of Action*. Cambridge: Polity Press.

Butzer, K. W. and G. H. Enfield. 2012. "Critical perspectives on historical collapse." *PNAS* 109 (10): 3628–31.

Cameron, E. S. 2012. "Securing Indigenous politics: A critique of the vulnerability and adaptation approach to the human dimensions of climate change in the Canadian Arctic." *Global Environmental Change* 22 (1): 103–14.

Denton, F., Wilbanks, T. J., Abeysinghe, A. C., Burton, I., Gao, Q., Lemos, M. C., Masui, T., O'Brien, K. L. and K. Warner. 2014. "Climate-resilient pathways: Adaptation, mitigation, and sustainable development." In *Climate Change 2014: Impacts, Adaptation, and Vulnerability. Part A: Global and Sectoral Aspects. Contribution of Working Group II to the Fifth Assessment Report of the Intergovernmental Panel on Climate Change*, edited by Field, C. B., Barros, V. R., Dokken, D. J. Mach, K. J., Mastrandrea, M. D., Bilir, T. E., Chatterjee, M., et al., 1101–31. Cambridge: Cambridge University Press.

Diamond, J. 2005. *Collapse: How Societies Choose to Fail or Survive*. London: Penguin.

Doherty, T. J. and S. Clayton. 2011. "The psychological impacts of global climate change." *American Psychologist* 66 (4): 265–76.

Ellen, R. 1982. *Environment, Subsistence and System: The Ecology of Small-Scale Social Formations*. Cambridge: Cambridge University Press.

Eriksen, S., Inderberg, T. H., O'Brien, K. and L. Sygna. 2015. "Introduction: Development as usual is not enough." In *Climate Change Adaptation and Development: Transforming Paradigms and Practices*, edited by Inderberg, T. H., Eriksen, S., O'Brien, K. and L. Sygna, 1–18. London: Routledge.

Ford, J. D. and L. Berrang-Ford. 2011. Introduction to *Climate Change Adaptation in Developed Nations: From Theory to Practice*, edited by Ford, J. D. and L. Berrang-Ford, 3–20. Advances in Global Change Research 42. Dordrecht, Netherlands: Springer.

Fox, D. 2007. "Back to the no-analog future?" *Science* 316 (5826): 823–25.

Freire, P. 1970. *The Pedagogy of the Oppressed*. New York: Pantheon.

Global Carbon Project. 2014. "Global Carbon Budget, Highlights, Emissions from fossil fuels and cement." http://www.globalcarbonproject.org/carbonbudget/14/hl-full.htm.

Haraway, D. 1996. "Situated knowledges: The science question in feminism and the privilege of a partial perspective." In *Human Geography: An Essential Anthology*, edited by Agnew, J., Livingstone, D. N. and A. Rodgers, 575–99. Oxford: Blackwell.

Hedlund-de Witt, A. 2012. "Exploring worldviews and their relationships to sustainable lifestyles: Towards a new conceptual and methodological approach." *Ecological Economics* 84: 74–83.

Heifetz, R., Grashow, A. and M. Linsky. 2009. *The Practice of Adaptive Leadership: Tools and Tactics for Changing Your Organization and the World*. Boston, MA: Harvard Business Press.

Hetherington, R. and R. G. B. Reid. 2010. *The Climate Connection: Climate Change and Modern Evolution*. Cambridge: Cambridge University Press.

Hovelsrud, G. and B. Smit, eds. 2010. *Community Adaptation and Vulnerability in Arctic Regions*. Dordrecht, Netherlands: Springer.

Hulme, M. 2008. "The conquering of climate: Discourses of fear and their dissolution." *The Geographical Journal* 174 (1): 5–16.

IPCC. 2014. *Climate Change 2014. Synthesis Report. Summary for Policymakers*, edited by The Core Writing Team, Pachauri, R. K. and L. Meyer. Cambridge: Cambridge University Press. http://www.ipcc.ch/pdf/assessment-report/ar5/syr/AR5_SYR_FINAL_SPM.pdf.

IPCC WG II Glossary. 2014. *Climate Change 2014: Impacts, Adaptation, and Vulnerability, Appendix II: Glossary. Contribution of Working Group II to the Fifth Assessment Report of the Intergovernmental Panel on Climate Change*. Cambridge: Cambridge University Press.

Kegan, R. 1994. *In Over Our Heads: The Mental Demands of Modern Life*. Cambridge, MA: Harvard University Press.

Kegan, R. and L. Lahey. 2009. *Immunity to Change*. Boston, MA: Harvard Business Press.

Keohane, R. O. and D. G. Victor. 2011. "The regime complex for climate change." *Perspectives on Politics* 9 (1): 7–23.

Klein, N. 2014. *This Changes Everything: Capitalism vs the Climate*. London: Allan Lane.

Kuruppu, N. and D. Liverman. 2011. "Mental preparation for climate adaptation: The role of cognition and culture in enhancing adaptive capacity of water management in Kiribati." *Global Environmental Change* 21 (2): 657–69.

Leichenko, R. M. and K. L. O'Brien. 2008. *Environmental Change and Globalization: Double Exposures*. Oxford: Oxford University Press.

Manuel-Navarrete, D. 2010. "Power, realism, and the ideal of human emancipation in a climate of change." *WIREs Climate Change* 1 (6): 781–85.

Manuel-Navarrete, D. and C. N. Buzinde. 2010. "Socio-ecological agency: From 'human exceptionalism' to coping with 'exceptional' global environmental change." In *The International Handbook of Environmental Sociology*, edited by Redclift, M. and G. Woodgate, 136–49. Cheltenham, UK: Edward Elgar.

Manuel-Navarrete, D., Pelling, M. and M. Redclift. 2011. "Critical adaptation to hurricanes in the Mexican Caribbean: Development visions, governance structures, and coping strategies." *Global Environmental Change* 21 (1): 249–58.

Manuel-Navarrete, D., Pelling, M. and M. Redclift. 2012. "Conclusions: Alientation, reclamation and radical vision." In *Climate Change and the Crisis of Capitalism: A Chance to Reclaim Self, Society and Nature*, edited by Pelling, M., Manuel-Navarrete, D. and M. Redclift, 189–98. Abingdon, UK: Routledge.

Norgaard, K. M. 2011. *Living in denial: Climate change, emotions and everyday life*. Cambridge, MA: MIT Press.

O'Brien, K. 2012. "Global environmental change II: From adaptation to deliberate transformation." *Progress in Human Geography* 36 (5): 667–76.

O'Brien, K. and G. Hochachka. 2010. "Integral adaptation to climate change." *Journal of Integral Theory and Practice* 5 (1): 89–102.

O'Brien, K. L. and J. Wolf. 2010. "A values-based approach to vulnerability and adaptation to climate change." *WIREs Climate Change* 1 (2): 232–42.

O'Brien, K., Eriksen, S., Nygaard, L. P. and A. Schjolden. 2007. "Why different interpretations of vulnerability matter in climate change discourses." *Climate Policy* 7 (1): 73–88.

O'Brien, K., Reams, J., Caspari, A., Dugmore, A., Faghihimani, M., Fazey, I., Hackmann, H., et al. 2013. "You say you want a revolution? Transforming education and capacity building in response to global change." *Environmental Science and Policy* 28: 48–59.

Orlove, B. 2005. "Human adaptation to climate change: A review of three historical cases and some general perspectives." *Environmental Science and Policy* 8 (6): 589–600.

Orlove, B. 2009. "The past, the present and some possible futures of adaptation." In *Adapting to Climate Change: Thresholds, Values, Goverance*, edited by Adger, W. N., Lorenzoni, I. and K. L. O'Brien, 131–63. Cambridge: Cambridge University Press.

Pelling, M. 2011. *Adaptation to Climate Change: From Resilience to Transformation*. Abingdon, UK: Routledge.

Pelling, M. and D. Manuel-Navarrete. 2011. "From resilience to transformation: The adaptive cycle in two Mexican urban centers." In *Ecology and Society* 16 (2): 11.

Pelling, M., Manuel-Navarrete, D. and M. Redclift. 2012. "Climate change and the crisis of capitalism." In *Climate Change and the Crisis of Capitalism: A Chance to Reclaim Self, Society and Nature*, edited by Pelling, M., Manuel-Navarrete, D. and M. Redclift, 1–17. Abingdon, UK: Routledge.

Rose, G. 1997. "Situating knowledges: Positionality, reflexivities and other tactics." *Progress in Human Geography* 21 (3): 305–20.

Rowson, J. 2011. "Transforming behavior change: Beyond nudge and neuromania." RSA Report, November. http://www.thersa.org/__data/assets/pdf_file/0006/553542/RSA-Transforming-Behaviour-Change.pdf.

Scharmer, C. O. 2009. *Theory U: Leading from the Future as It Emerges – The Social Technology of Presencing*. San Francisco, CA: Berrett-Koehler.

Schlitz, M. M., Vieten, C. and Miller, E. M. 2010. "Worldview transformation and the development of social consciousness." *Journal of Consciousness Studies* 17 (7–8): 18–36.

Scott, J. 1985. *Weapons of the Weak: Everyday Forms of Peasant Resistance*. New Haven, CT: Yale University Press.

Shove, E. 2010. "Beyond the ABC: Climate change policy and theories of social change." *Environment and Planning A* 42 (6): 1273–85.

Stokke, K. and E. Selboe. 2009. "Symbolic representation as political practice." In *Rethinking Popular Representation*, edited by Törnquist, O., Webster, N. and K. Stokke, 59–78. New York: Palgrave Macmillan.

Swyngedouw, E. 2010a. "Apocalypse forever? Post-political populism and the spectre of climate change." *Theory, Culture, and Society* 27 (2–3): 213–32.

Swyngedouw, E. 2010b. "The impossible sustainability and the post-political condition." In *Making Strategies in Spatial Planning: Knowledge and Values*, edited by Cerreta, M., Concilio, G. and V. Monno, 185–205. Dordrecht, Netherlands: Springer.

Swyngedouw, E. 2011. "Interrogating post-democratization: Reclaiming egalitarian political spaces." *Political Geography* 30 (7): 370–80.

Swyngedouw, E. 2013. "The non-political politics of climate change." *Acme* 12 (1): 1–8.

Wilber, K. 2001. *A Brief History of Everything*. 2nd ed. Dublin: Gateway.

Williams, J. W. and S. T. Jackson. 2007. "Novel climates, no-analog communities, and ecological surprises." *Frontiers in Ecology and the Environment* 5 (9): 475–82.

Wise, R. M., Fazey, I., Stafford Smith, M., Park, S. E., Eakin, H. C., Archer Van Garderen, E. R. M. and B. Campbell. 2014. "Reconceptualising adaptation to climate change as part of pathways of change and response." *Global Environmental Change* 28: 325–36.

Yohe, G. and R. S. J. Tol. 2002. "Indicators for social and economic coping capacity – moving toward a working definition of adaptive capacity." *Global Environmental Change* 12: 25–40.

2

The Intangibles of Climate Change Adaptation

Philosophy, Ethics, and Values

BERTRAND GUILLAUME AND STIJN NEUTELEERS

Introduction

Climate change has a significant negative impact on both current and future generations, and there are good normative arguments to limit these effects (Gardiner et al. 2010). Standard climate change discourse distinguishes between two strategies, namely, mitigation and adaptation (UNFCC 1992). Mitigation is about reducing the anthropogenic or human causes of climate change, and adaptation concerns adjustments to the consequences for ecosystems and human societies. Adaptation, however, is a complex concept, and there are important relationships between adaptation and mitigation (Klein et al. 2007; Simonet 2010). Currently adaptation debates are primarily about the potentials of and limits to physical adaptation and sociopolitical or cultural adaptations (e.g., Adger et al. 2009).

In this chapter, we go beyond this debate and investigate some fundamental issues that have received less attention so far. Looking through a philosophical lens, we present an overview of different perspectives that make adaptation a particularly challenging concept. We proceed in three steps, descending from abstract toward more practically oriented reflections. First, we look at climate change from a broad perspective, investigating epistemological issues (the nature of knowledge) and ontological issues (the nature of the world and of humanity) to understand some of the fundamental dimensions of climate change. Second, we examine important ethical issues and the implications for justice. Here we aim to show that climate change has increased human responsibilities, which then have moral implications. Finally, we discuss the more practical question of moral motivation and value change, as it is one thing to say that values should change and another to actually make these changes.

Epistemology, Ontology, and Meta-Ethics

Epistemology deals with the issue of what we can know and how we can know this. If, following Wittgenstein, one subscribes to the view that what is thinkable is also what is expressible, then adapting to the idea of climate change is a precondition of any other adaptation. In other words, if human beings ultimately "are suspended in language... to communicate experience and ideas to others," as Niels Bohr once put it (cf. Petersen 1985, 301), we will no longer be able to grasp and understand the reality around us (and consequently to act on that understanding) unless we adapt our concepts and language. Without the idea of climate change, there can be neither a debate about it nor any adaptation.

Some social theorists have argued that social, technological, cultural, and political changes in the second half of the twentieth century constitute a radical break with the past and that we are now in need of new theoretical frameworks (Beck et al. 1994). For example, whereas in premodern societies, a particular time was connected to a particular space, both are disconnected in modern societies. This time-space distanciation is a continual process, but its level is significantly higher in the contemporary era than in any previous period and is changing our perspectives about the world (Giddens 1990). Processes such as globalization have spurred debate in social sciences (Beck 1998; Giddens 2002; Held and McGrew 2000; Stiglitz 2002), leading some to explore interactions between globalization and environmental change (Leichenko and O'Brien 2008). Some authors have even argued that issues such as climate change require a shift of paradigm, namely, the change of the entire worldview (assumptions, models, etc.). Schellnhuber (1999), for example, calls for a "Copernican revolution" with regard to the study of the Earth. The first Copernican revolution was facilitated by an increased ability to see details through the invention of microscopes and telescopes. The second one will be somewhat the reverse, that is, characterized by an increased ability to see the whole (again). Changes in the global processes (e.g., biogeochemical cycles, climate change) and changes in science (global networks of satellites and measuring points, experiments such as Biosphere II, simulation modeling, etc.) have facilitated a new understanding of the Earth as a whole, also known as Earth system science (Steffen et al. 2004). From an epistemological point of view, one of the required perspective changes to adapt to climate change would thus involve adopting a more global perspective and systemic view that recognizes "planetary boundaries" (Rockstrom et al. 2009).

An ontological perspective concerns the inquiry into existence and reality. Climate change reflects a fundamental change in the relation between humankind and the surrounding environment. Humankind is now able to change the world in an unprecedented and fundamental sense, emphasized by the concept of the Anthropocene, which suggests that a new geological era is succeeding the previous one, namely, the Holocene (Crutzen 2002; Zalasiewicz et al. 2010). Geologically, the Holocene presents an uncommon period of stability with regard to biogeochemical parameters. Historically, it is more or less the history of human civilization, beginning with the agricultural revolution and the rise of the first cities. In the Anthropocene, human activities are changing the Earth in such a profound way that they rival the great forces of nature. Some of these changes are now seen as permanent, even on geological time scales (Zalasiewicz et al. 2010, 2228).

Although human societies have always influenced their environment, sometimes even to the extent that they have undermined their own ecological preconditions (Diamond 2005; Ponting 1992), the Anthropocene denotes that humans are now able to modify the environment at the planetary scale and to jeopardize the current stability of the Earth system.[1] Some have argued that, with phenomena such as climate change, human influence over nature has reached a stage where the distinction between natural and human histories has begun to collapse (Chakrabarty 2009). McKibben (1990) even claims that environmental change has already led to the "end of nature," for nature is now being affected everywhere by anthropogenic processes.[2] This points to something crucial, namely, a fundamentally new relationship between humans and the biosphere.

Looking at epistemological and ontological dimensions thus suggests two important adaptation challenges: (1) the necessity of looking at the Earth system to understand what is actually happening (epistemological level) and (2) the necessity of conceiving of ourselves as having agency, capable of intervening in essential planetary processes at the global scale (ontological level).

Before discussing ethics itself, it might be useful to think about the meta-ethical level (i.e., the possibility of ethics). The question of whether climate change might influence the sheer possibility of ethical thinking may seem rather extreme. However, the question can be reformulated if we contemplate the most extreme warming scenarios, including scenarios resulting in a runaway greenhouse effect (Hansen 2009; Schellnhuber 1999). Under such scenarios, what is at stake is the potential end of civilization, the human race, or life on Earth. The primary question is not so much what the chances of such scenarios are – probably they are rather low – but rather what their mere possibility says about our ethical responsibilities.

Let us first contemplate the prospect of a "breakdown scenario" (Gallopin et al. 1997), that is, a generalized disintegration of social, cultural, and political institutions. This includes the possibility of human extinction (although still not very likely, it is more plausible than the extinction of all life on Earth). It is clear that most humans would oppose such prospects. According to Sandel (2005, 179–82), this opposition cannot be explained in purely individualistic terms. From an individualistic perspective, the problem of human extinction is similar to the one of murder or genocide, but on a much larger scale. However, there seems to be something beyond this, that is, something that goes much deeper, for the continuity of human civilization in the future seems to be a precondition for meaningful lives today. To be meaningful, current human actions must be remembered tomorrow (Sandel 2005, 180). More

---

[1] The Anthropocene might also be an expression in favor of active planetary management (Crist 2013). The idea of geoengineering (Shepherd 2012) is an instance of such a discourse (cf. Guillaume and Laramée 2012).

[2] This claim is controversial. If one employs another definition of nature, based on the idea of spontaneous processes, for instance ("the growing of a tree," say), the thesis would be harder to maintain (Thompson 2009).

generally, the meaning of our lives depends to a significant extent on our contribution to something that transcends our individual selves, such as art, science, or offspring. It would be strange to dedicate one's whole life, for instance, to science if one did not believe that science will continue and matter after one's death (O'Neill 1993b). A limited future would not only be a loss for future generations but would undermine the possibility for current generations to lead a meaningful life today, or at least one important dimension of it. Therefore, such dire prospects might undermine current thinking about what is valuable, making much of current action appear meaningless.

Finally, let us consider a potential end to life on Earth resulting from climate change, having in mind the so-called rare Earth hypothesis (Ward and Brownlee 2000). This hypothesis states that complex life is uncommon in the universe because its emergence requires a rather improbable combination of conditions. Arguably, if very few planets could indeed offer the long-term stability necessary for the emergence of intelligent life, our destiny would then have "cosmic significance" (Rees 2003, 157). In this case, our responsibility to address climate change would increase exponentially. Such extreme scenarios might lead to defeatism. But the only reason for real defeatism is if we believe that we are basically unable to deal with matters of this extent not only in practice but also *in principle*. For instance, one view suggests that our moral psychology prevents us from processing climate change as an ethical issue (see Jamieson 2010). If the assumption that we cannot see the problem is true, then no moral imperative remains, nor we can be blamed. However, such a vision presents a rather limited view of what humans are and can do (Gardiner 2011b). As long as one believes that influencing the future is possible, climate change does not undermine the meaningfulness of our actions and can, on the contrary, give rise to an enhanced feeling of moral responsibility.

An Ethical Perspective: Expanding the Boundaries of Justice

Ethics deals with the question, How should we live? Ricœur (1990, 202), for example, defines it as "the orientedness towards the good life, with and for others, within just institutions."[3] Many ethical theories and systems have their origins prior to the Industrial Revolution. Since then, the potential impact of human actions has drastically increased, such that over the last decades, many thinkers have called for new ethical theories. Jonas (1984) proposed an ethic of responsibility that can deal better with the new extent of our powers. In a much-cited article, Sylvan (2003) suggested the necessity to develop environmental ethics as a new field of applied ethics. However, one should not be too hasty in demanding radically new concepts, because the reasons for nonaction are not necessarily conceptual ones (Gardiner 2011b).

---

[3] "La visée de la 'vie bonne' avec et pour autrui dans des institutions justes" (Ricoeur 1990, 202).

This sketches the setting for our core consideration: to what extent does climate change call for a change in our ethical thinking? There seems to be two dimensions that are particularly relevant. First, our current actions can have impacts over long distances, both in space (beyond local or national borders) and time (beyond current generation), such that the perception of their consequences is difficult. Second, we participate in a system where harmful consequences are only considered to be side effects and associated responsibilities are consequently diffused. This means, for instance, that we drive cars that burn fossil fuels, contributing to negative impacts on livelihoods in distant locations, but it is not clear who bears responsibility for the consequences. However, uncertainties about consequences and responsibilities do not take away the moral issue. If harming others is morally wrong, then distance in space or time is no longer relevant. In the following sections, we first examine the international and intergenerational challenge to justice, then we discuss the impact of climate change on procedural and cultural justice.

Spatial Relations and International Justice

Traditionally, justice was conceived within the framework of nation-states, and many contemporary theories of justice still use this framework (Dworkin 2000; Miller 1999; Rawls 1999b). For instance, Rawls's conception of justice is primarily based on reciprocity. It examines how justice would require distribution of the productive surplus of a free and equal citizens' society. Because all citizens are part of the same system (they all contribute to some extent to the benefits created in a society), they should all receive a fair share of the surplus of cooperation. An important question, then, will be whether distributive justice should treat compatriots and noncompatriots in a similar way. This issue forms the core of the global justice debate, now one of the most debated issues in political theory (e.g., Brooks 2008). The question is whether domestic principles of distributive justice should be extended to the global level. Distributive justice specifically deals with the fairness of the distribution of benefits (e.g., income, opportunities, and jobs) and burdens (e.g., taxes and health risks) that result from socioeconomic cooperation. Theorists have argued in the direction of global distributive justice, for instance, by extending Rawls's principles of justice to a global scale (Beitz 1979; Pogge 1989). Yet, this "cosmopolitan" stance has been criticized (Miller 2005; Nagel 2005; Rawls 1999a), suggesting that the step from domestic to global distributive justice is less straightforward than appears at first sight.

However, with regard to climate change, one should not get too distracted by the global justice debate itself, which focuses primarily on one type of justice, namely, distributive (or social) justice. The issue of climate change involves types of justice that are more basic, namely, reparative, cooperative, and minimal (global) justice (Van Parijs 2007, 639–41). These three types of justice seem more relevant to climate change than pure distributive justice. *Reparative justice* concerns the reparation or compensation for the damage caused by the transgression of established entitlements (creating environmental pollution in another country, for example). *Cooperative justice* concerns the distribution of burdens and benefits when actors explicitly cooperate to create a public good, such as a stable global climate. Although everyone will clearly benefit from the implementation of the public good (e.g., the collective action problem being solved), the related costs and benefits are not necessarily equally distributed. *Minimal justice* focuses on the victims of climate change and demands that basic needs be met and human rights be respected. Reparative and cooperative justice are, in a sense, more basic than distributive justice, because they do not require a redefinition of legitimate entitlements to economic goods (Van Parijs 2007, 640). Even if one accepts the current distribution of entitlements (e.g., between poor and rich countries), there is still a problem of reparative and cooperative justice. Minimal justice, namely, the requirement that all people can lead a minimally adequate life, is also more basic because it only defines a minimum rather than a fair share. In sum, one does not need to show that climate change implies an unfair distribution of economic goods. It already violates more basic rules of justice, namely, no harm (reparative justice), fair cooperation rules (cooperative justice), and fulfilling basic needs (minimal justice).

Although most people agree that justice is important, the content of justice is often contested. Consider, for instance, cap and trade instruments such as in the Kyoto Protocol. Contrary to what the twofold name suggests, cap and trade instruments consist of three levels: cap, distribution, and trade. The first level defines the cap: what is the collective maximum amount of emissions that will be allowed? The second level distributes the emission rights among the participants. The third level defines whether and how emission rights can be traded. The question of justice primarily concerns the second level, but there is no simple way of defining a fair distribution because many alternative distributions are conceivable. For instance, should we look at past contributions (historical principle), thereby using a kind of reparative justice principle, or at the current distribution (time slice principle) (Singer 2002, 15–50)? If we only look at current contributions, many different options still remain: equal per capita entitlements, priority to the least well-off, equalizing marginal costs, rights to subsistence emissions (compare minimal justice), and so on (Gardiner 2004, 583–89). In summary, the debate on climate justice is complex and the content of a fair solution is not self-evident, even though many proposals might nonetheless point in the same direction. The fact that there is now a lively debate on these issues can itself be seen an expression of the inquiry into justice adapting itself to the phenomenon of climate change.

Temporal Relations and Intergenerational Justice

We are used to thinking within the framework of nation-states when it comes to principles of justice, and our ethical reflections also tend to take place in a limited time frame: we mostly think about relations with people from current generations only. However, the question about a fair relation between generations is at least equally important (De-Shalit 1995; Gosseries 2008b; Gosseries and Meyer 2009). The crucial difference between intra- and intergenerational justice is that reciprocity is impossible between generations. Current generations can expand or limit possibilities for future generations, but the reverse is impossible. Future generations cannot control or influence what current generations will do.[4] The absence of possible reciprocity between distant generations excludes many potential solutions for collective action problems, because most strategies try to ensure cooperation (repeated interaction, communication, creating incentives and costs, making cooperative attitudes visible, etc.). With regard to the global commons (spatial distance), ensuring compliance is certainly difficult *in practice* because there is a lack of global institutions, but ensuring compliance intergenerationally (temporal distance) is impossible *in principle* because there cannot be an agreement with a nonexistent actor (Gardiner 2001).[5] This means that there is only one strategy available: current generations have to abstain voluntarily from the benefits without compensation and without any guarantee that future generations will cooperate. It is no surprise that Gardiner (2001) labels the intergenerational problem as "the *real* tragedy of the commons."

Another key difference follows from the fact that we influence not only the circumstances within which future people will live but also these people themselves. Our (policy) choices will affect the very existence of future people, their number, and their identity (Meyer 2010). This raises a fundamental ethical question: can someone whose existence and identity depend on our actions claim that her rights are violated by us? This issue is known in philosophy as Parfit's nonidentity problem (Parfit 1984). The idea is that a choice of a certain policy cannot be harmful to people affected by it if they owe their existence to it.[6] This problem is heavily discussed in debates on intergenerational justice.[7] One of the most evident ways to deal with this is to refer to the idea of overlapping generations (Gosseries 2008a).

Notwithstanding complexities associated with intergenerational justice, people in general believe that some distributions between generations are more "just" than others. As with the debate surrounding global justice, many approaches exist about what constitutes a just intergenerational distribution: indirect reciprocity, utilitarianism, egalitarianism, sufficientarianism (Gosseries 2008b; Gosseries and Meyer 2009;

---

[4] Some philosophers (O'Neill 1993b) have argued that future generations can have an impact on us – they can give meaning to our lives – but this relation is about nonmaterial benefits and harms.

[5] Not only are cooperative solutions for the intergenerational problem unavailable, but so is the noncooperative way out of the tragedy, namely, privatization. If we place the commons under private ownership, each generation still has an incentive to deplete the resource during its own lifetime. Privatization can prevent the tragedy during one generation but can still occur in time, passing the tragedy on to future generations (O'Neill 1993b, 46).

[6] On the individual level, this idea lies behind the discussion about "wrongful life" (to be distinguished from "wrongful birth"), namely, in bioethics, whether a child, for instance with a severe handicap that was predictable, can claim that his parents made a (moral) mistake by choosing to give birth.

[7] Authors point, for instance, to distinctions between nonexistence and death (Pasek 1993) or between present rights and future rights (for future people), or refer to needs that transcend generations (O'Neill et al. 2008).

Meyer 2010). Ideas such as "sustainable development" seem to suggest that the definition of an intergenerational distribution is self-evident, but different approaches imply different interpretations.

First, sustainable development seems to be in line with the idea of *indirect reciprocity*, namely, that we should pass the same amount of capital (broadly understood) to the next generation as received from the previous generation. The reciprocity is "indirect" because giver and receiver are not the same. Although indirect reciprocity is intuitively plausible (possibly based on our intuitions about parent-child relations), it is not without problems and leaves many questions unanswered. For example, what is its exact justification, and why give the same and not more to future generations? Second, *utilitarianism* would argue that we should give more to future generations. Intergenerational utilitarianism says we should maximize the welfare over generations. If savings in the current generation help to create relatively more welfare in the future (because of the so-called productivity of capital), we have an obligation to do so. Third, some forms of *egalitarianism* would argue against this, for we should worry about inequalities not only between generations but also within generations. We should not sacrifice the worst off today for a little improvement in the future. Fourth, the definition of sustainable development expresses *sufficientarianism* with regard to intergenerational justice. In the Brundtland report, a development is sustainable if it "meets the needs of the present generation without compromising the ability of future generations to meet their own needs" (WCED 1987, 53). Several influential accounts of climate justice focus on basic needs or human rights (Caney 2008, 2010), which is unsurprising, because climate change threatens future basic needs and because such a minimalist account of justice has a strong moral basis: few would argue in favor of a violation of human rights or basic needs. However, from an egalitarian point of view, a focus on basic needs could be seen as too minimal, with regard to both future generations ("should we allow a development in which only future basic needs will be fulfilled?") and current ones ("is investing in the future a problem only if it is threatening current basic needs?").

Our aim here was to show that there is no simple, unequivocal answer to the question of what intergenerational justice is. However, the fact that there is a debate does not imply relativism. In contrast, when discussing policy proposals, many of these theorists tend to agree, especially about the need for mitigation measures.[8]

Politics and Procedural Justice

Political philosophy deals not only with the question of *what* communities should aim for (substantive justice) but also *how* it should be done (procedural justice). In relation to climate change, there are many debates about how environmental democracy

---

[8] Although most of these debates deal with mitigation, some of them also discuss distributive justice and adaptation measures: who will pay for it (Bear 2010)?

should be achieved (Dryzek 1997; Smith 2003). It is a well-known idea within environmental thought that there can be a conflict between ecological problems and democracy (O'Neill 1993a; Dobson 2007). For example, whereas democracy often develops in a slow and piecemeal way, many environmental problems require urgent and wide interventions. Hence it is no surprise that there has been a strand of "green authoritarianism" in environmental thought, both in the past (Jonas 1984, 151–52) and in the present (Shearman and Smith 2007; Lovelock 2009).[9] If democratic regimes are really unable to tackle environmental issues, then we may need stronger states, or a "green leviathan."[10] Looking at the literature, a genuine defense for ecological autocratic regimes seems, however, to be uncommon.[11] Rather than a plea for autocracy, green authoritarianism appears mostly as a rhetorical device, a cautionary tale for what may happen if environmental problems are not taken seriously enough.

There is nonetheless a real tension between democracy and environmental concerns, but rather than exchanging democratic systems for autocratic ones, the threat seems to come from the direction of technocracy (Gorz 1993; Neuteleers and Guillaume 2014). In general, technical expertise becomes a technocracy when experts define not only the means but also the goals of a policy, with their expertise consequently replacing the democratic debate as such. Regarding climate change, this nondemocratic challenge is especially coming from environmental sciences and economics. Technology might play a similar role by confronting democracy with a *fait accompli*, thereby leaving little to decide about important issues (unilateral implementation of geoengineering, for instance). To summarize, the interventions needed to deal with climate change put our democratic systems under pressure. The central idea of democracy is "equal political liberty" (Gutmann 2007), namely, the idea that everyone has an equal part in determining the outcome of political processes. This makes it an adaptive challenge to examine how democratic values (can) contribute to dealing with climate change.

Identity and Cultural Justice

Philosophers have argued that cultural identity is highly valuable to people (Kymlicka 1989; Miller 1995). In the first place, people want to belong to a community, one with a particular history, which is a condition for self-identity and, consequently, for self-respect. Second, the cultural framework provides meaningful options, through which the world is interpreted and from which options for meaningful actions can be chosen (Kymlicka 1995, 89). The fact that cultural identity is so important often leads to political institutions protecting people from externally induced changes (an occupying force that imposes another language, for example).

---

[9] Sometimes this is focused on particular regions such as East Asia (Beeson 2010).
[10] Against this claim, theoretical and empirical research suggests that democracy creates substantial opportunities for environmental-friendly policies (Drosdowski 2006; Li 2006).
[11] Green political theory mostly refers to Heilbroner (1975) and Ophuls (1977).

Climate change not only threatens the fulfillment of basic needs but can threaten people's cultural identity as well. As such, it can also be considered an issue of cultural justice (Heyward 2014). Some aspects of one's culture might be closely related to the natural environment. Environmental change can have an impact on the local culture and threaten its particular way of life (Adger et al. 2013). For instance, the disappearance of traditional food resources may require a radical change in food culture (imagine French food culture without local wine). Such environmental changes may prevent communities from enjoying their traditional ways of life. Should they concern numerous or substantial environmental features, they may threaten the survival of communities as distinct and unique societies. In an extreme form, climate change can even lead to the loss of one's homeland, with, for instance, the submersion of low-lying islands because of sea level rise, thereby leading to the loss of all territorially bound features of a culture.

In the case of such losses, cultural justice would require rectification or compensation. Heyward (2014) argues that two things are required when climate change incurs cultural injustices. In the first place, there is an even stronger demand for mitigation. Second, adapting to climate change also requires measures to rectify or compensate for the loss of culturally nonsubstitutable environmental goods. She identifies four such measures: remembrance (stories about what was and what happened), maintaining a continuous narrative (e.g., by cultural projects), having control over the change process, and acknowledging the injustice that has taken place (e.g., by a statement of regret by those who caused the problem). Moreover, the societies that are subject to the loss of constitutive cultural elements should get the means to deal with this.

These different dimensions of justice – international, intergenerational, procedural, and cultural – point in a similar direction, namely, that anthropogenic climate change has increased our responsibilities. A significant increase of responsibility is not always easy to accept, neither for individuals nor communities. For instance, obtaining new responsibilities when reaching adolescence or adulthood might bring along difficulties: one might be ignorant of these new responsibilities, or not sure how to interpret them, or unwilling to bear them, or have fear of assuming them, or make the wrong decisions. A similar process can take place on the collective level (Thompson 2009). We are somewhat "accustomed" to our ability to exterminate particular species but horrified by apocalyptic climate scenarios such as the end of (human) life or civilization. Yet such future scenarios do not explain the moral discomfort we are experiencing. Our moral unease seems to follow from the significant increase in responsibility, not only for our children or for the least well-off in our society, but for maintaining the preconditions for human life on Earth. As Thompson (2009, 96–97) puts it, "we don't fear the *end of* the natural world; we fear *responsibility for* the natural world."

In a way, both the "end of nature" and the Anthropocene thesis could be seen as an expression of our moral discomfort. This anxiety can also be understood as something good. It is not the fear for an unavoidable negative outcome – which would lead to defeatism – but rather the anxiety that goes along with adaptation and that at the same time motivates us to adapt. The moral dimension of adaptation implies accepting new responsibilities and trying to understand which actions they require.

Values and Virtue Ethics

After having discussed general philosophical topics, it is relevant to end with some more practical reflections about responsibilities by looking at values and moral motivations.

Climate change seems to be one of the most urgent and simultaneously one of the most difficult problems to deal with. Gardiner (2006, 2011a) labels it as a "perfect moral storm": "an event constituted by an unusual convergence of independently harmful factors where this convergence is likely to result in substantial and possibly catastrophic, negative outcomes.... [It] involves the convergence of a number of factors that threaten our ability to behave ethically" (Gardiner 2006, 398). Gardiner discusses the dimensions of this perfect storm. First, our institutions are not well equipped to deal with global and intergenerational relations. Furthermore, as the discussion in previous sections showed, our concepts and theories are still looking for ways to adapt to the new situation, thereby not yet providing us with clear moral guidance. Finally, there is also a problem of moral corruption or motivation. Several factors influence our ability to behave in a morally responsible way. Gardiner (2006) mentions different sources of moral corruption: distraction, complacency, unreasonable doubt, selective attention, delusion, pandering, false witness, and hypocrisy. The issue at stake is not so much that these things exist but rather that they will occur more easily given certain circumstances. Climate change is arguably creating such circumstances (Gardiner 2006). The features of climate change match the conditions for procrastination, namely, pushing decisions forward to an undefined moment in the future. These features include the negligibility of single actions (at the levels of individuals, firms, or countries), the absence of any clear and optimal solution, and high stakes (e.g., the costliness of a solution) (Andreou 2006, 2007). These features evoke procrastination, but they also make us feel comfortable with it because they give us a reason or excuse, such as "my contribution does not make any major change" or "we better wait until we find an optimal solution before we invoke such an enormous financial cost."

Ethics deals primarily with the question of what we ought to do and thus with our responsibilities. One approach within environmental ethics, namely, virtue ethics, deals with the question of character: what kind of person we ought to be. This approach is becoming more popular within environmental ethics (Hill 1983; Sandler 2005). One of the attractions of virtue ethics is that moral motivation plays an important role in it. What we *ought to* do (the primary object of ethical reflection) is one thing, but what we *are doing* is quite another.

Virtue ethics starts from the ideas of excellence and having a flourishing life. It examines stable character traits (or virtues) – for example, temperance, generosity, courage – and the idea is that developing these character treats will lead to a kind of excellence that is beneficial to both the actor and his or her environment, both human and nonhuman. Virtuous people function as role models. If one is wondering how to behave in a certain context, one tends to not think about moral rules but tries instead to imagine how a great person would behave – for instance, a great teacher they have known, that is, someone who represents both how we think we should be and how we *want* to be.

Adapting ourselves is often conceived as being externally forced to adapt to survive or to help others. Virtue ethics possibly inspires a more positive account by distinguishing between adaptation to survive and adaptation to thrive (O'Brien 2013, 309). Rather than just trying to adapt, one can aim for the best way of adapting that goes beyond surviving in a new global climate and aims for flourishing in it (Thompson and Bendik-Keymer 2012). Though attractive, adaptation of values is of course more challenging. Changing my values means also changing who I am, because values are constitutive of my identity. Such fundamental changes will inevitably require a kind of courage (O'Brien 2013).

Changing values to adapt to new circumstances is not a neutral process, because not all value changes are equal. Some adaptations are more appropriate than others. The term "appropriate" should be understood not only in the sense of "being effective" but also in a normative way. For example, climate change will probably entail that particular species become more rare. There are different ways to adapt our values to this. One strategy could be to decrease the value nature has for us. This value adaptation may decrease the anxiety we experience with the loss of nature, thereby making us more adapted to a world with less nature. However, one could question whether this is an appropriate adaptation. Not only do we close down a crucial dimension of human enjoyment but it will also conflict with our moral intuitions about what we should do and about what kind of persons we want to be. Therefore, another strategy might be more apt. In a situation of scarcity of nature, a good adaptation might require, for instance, developing a higher receptivity for the beauty of nature. Although such receptivity was less needed in a situation of abundant natural beauty, it will become an important feature to preserve our enjoyment in a situation of scarcity. Moreover, such a receptivity seems not to conflict with our moral intuitions, which, for instance, require the preservation of value for future generations. If we talk about changing our values, it is impossible to leave the normative dimension out of it, because values cannot simply be treated as instrumental attitudes.

Conclusions

In this chapter we have considered climate change adaptation from a philosophical perspective. Because we expect climate change to require changes in our ethical reflections in particular, this was the primary focus of our inquiry. We proceeded in three steps. Before looking at the field of ethics itself, we examined climate change from epistemological and ontological perspectives. From an epistemological perspective, climate change requires that our concepts and theories adopt a global stance, thereby looking at the Earth as one single global system. The ontological point of view suggests that the relation between humans and nature has fundamentally changed, with humanity now impacting the Earth system on the planetary scale. The ideas of future human extinction or the collapse of civilization challenge us to reflect on the meaningfulness of life in the present.

Such reflections point in the direction of new and increased responsibilities for humans. Although it is quite obvious that we need to extend justice to an international and intergenerational scale, such an extension is complex and not straightforward. First, adapting to climate change seems to require a broadening of the scope of distributive justice to other types of justice (reparative, cooperative, and minimal justice). The lively global justice debate is a sign of ethical theory itself adapting to this wider scope. Subsequently, we pointed to important differences between international and intergenerational justice, namely, nonreciprocity (which excludes cooperative solutions) and the nonidentity problem. As with global justice, our discussion on intergenerational justice showed that the content of a fair intergenerational distribution is far from unequivocal. Nonetheless, current debates on global and intergenerational justice are a manifestation of the adaptation process itself, and though there is much debate about the definition of both global and intergenerational justice, many authors agree about the direction we have to take (e.g., the undertaking of strong mitigation measures). Global and intergenerational justice is the more prominent dimension of justice to think about when it comes to climate change, but two other dimensions are also important: first, the urgency of climate change puts the democratic process under pressure (procedural justice), and second, climate change threatens not only material goods but also cultural goods, which are inevitably connected to our natural environment (cultural justice).

It is one thing to recognize that climate change has increased our responsibilities; it is another to act accordingly. In the context of climate change, many structural conditions seem to impede moral action, creating a "perfect moral storm" (Gardiner 2006). Moreover, pointing to increased responsibilities does not always increase motivation. Some ethical approaches, mostly based on virtue ethics, allow for a more positive account of motivation. Those who are not motivated enough by moral imperatives might find more conviction in the positive idea of building a desirable society likely to persist and flourish in a changing climate. Adapting our values may be part of the process, but new values cannot simply be chosen from a range of options, because they should be in line with prior moral convictions (they are constitutive of who we are). For now, before we perhaps change ourselves through adapting our own values to the reality of a new climate, more challenging still seems adapting ourselves to the idea of climate change, and to the responsibilities this entails.

Acknowledgments

This research was conducted as part of the "DemoEnv" project funded by the French National Research Agency (ANR). We thank Rich Howarth for his support, and we are indebted to Karen O'Brien, Elin Selboe, Paul Knights, Ananka Loubser, and two anonymous referees.

References

Adger, W. N., Barnett, J., Brown, K., Marshall, N. and K. O'Brien 2013. "Cultural dimensions of climate change impacts and adaptation." *Nature Climate Change* 3 (2): 112–117.
Adger, W. N., Lorenzoni, I. and K. L. O'Brien, eds. 2009. *Adapting to Climate Change: Thresholds, Values, Governance*. Cambridge: Cambridge University Press.
Andreou, C. 2006. "Environmental damage and the puzzle of the self-torturer." *Philosophy & Public Affairs* 34 (1): 95–108.
Andreou, C. 2007. "Environmental preservation and second-order procrastination." *Philosophy & Public Affairs* 35 (3): 233–48.
Bear, P. 2010. "Adaptation to climate change: Who pays whom?" In *Climate Ethics: Essential Reading*, edited by Gardiner, S., Caney, S., Jamieson, D. and H. Shue, 247–62. Oxford: Oxford University Press.
Beck, U. 1998. *World Risk Society*. Cambridge: Polity Press.
Beck, U., Giddens, A. and L. Scott. 1994. *Reflexive Modernization: Politics, Tradition and Aesthetics in the Modern Social Order*. Stanford, CA: Stanford University Press.
Beeson, M. 2010. "The coming of environmental authoritarianism." *Environmental Politics* 19 (2): 276–94.
Beitz, C. 1979. *Political Theory and International Relations*. Princeton, NJ: Princeton University Press.
Brooks, T., ed. 2008. *The Global Justice Reader*. Oxford: Blackwell.
Caney, S. 2008. "Human rights, climate change, and discounting." *Environmental Politics* 17 (4): 536–55.
Caney, S. 2010. "Climate change, human rights, and moral thresholds." In *Climate Ethics: Essential Readings*, edited by Gardiner, S. M., Caney, S., Jamieson, D. and H. Shue, 163–77. Oxford: Oxford University Press.
Chakrabarty, D. 2009. "The climate of history: Four theses." *Critical Inquiry* 35: 197–202.
Crist, E. 2013. "On the poverty of our nomenclature." *Environmental Humanities* 3: 129–47.
Crutzen, P. J. 2002. "Geology of mankind." *Nature* 415 (6867): 23–23.
De-Shalit, A. 1995. *Why Posterity Matters: Environmental Policies and Future Generations*. London: Routledge.
Diamond, J. 2005. *Collapse: How Societies Choose to Fail or Succeed*. New York: Penguin.
Dobson, A. 2007. *Green Political Thought*. London: Routledge.
Drosdowski, T. 2006. "On the link between democracy and environment." Diskussionpapier 355. Hannover, Germany: Universität Hannover, Institut für Makroökonomik.
Dryzek, J. S. 1997. *The Politics of the Earth: Environmental Discourses*. Oxford: Oxford University Press.
Dworkin, R. 2000. *Sovereign Virtue: The Theory and Practice of Equality*. Cambridge, MA: Harvard University Press.
Gallopin, G., Hammond A., Raskin, P. and R. Swart. 1997. *Branch Points: Global Scenarios and Human Choice*. Stockholm: Stockholm Environment Institute.
Gardiner, S. 2001. "The real tragedy of the commons." *Philosophy and Public Affairs* 30 (4): 387–416.

Gardiner, S. 2004. "Ethics and global climate change." *Ethics* 114 (3): 555–600.
Gardiner, S. 2006. "A perfect moral storm: Climate change, intergenerational ethics and the problem of moral corruption." *Environmental Values* 15 (3): 397–413.
Gardiner, S. 2011a. *A Perfect Moral Storm: The Ethical Tragedy of Climate Change*. Oxford: Oxford University Press.
Gardiner, S. 2011b. "Is no one responsible for global environmental tragedy? Climate change as a challenge to our ethical concepts." In *The Ethics of Global Climate Change*, edited by D. G. Arnold, 38–59. Cambridge: Cambridge University Press.
Gardiner, S., Caney, S., Jamieson, D. and H. Shue, eds. 2010. *Climate Ethics: Essential Readings*. Oxford: Oxford University Press.
Giddens, A. 1990. *The Consequences of Modernity*. Stanford, CA: Stanford University Press.
Giddens, A. 2002. *Runaway World: How Globalisation Is Reshaping Our Lives*. London: Profile Books.
Gorz, A. 1993. "Political ecology: Expertocracy versus self-limitation." *New Left Review* I (202): 55–67.
Gosseries, A. 2008a. "On future generations' future rights." *Journal of Political Philosophy* 16 (4): 446–74.
Gosseries, A. 2008b. "Theories of intergenerational justice: A synopsis." *SAPIENS* 1 (1): 63–74.
Gosseries, A. and L. Meyer, eds. 2009. *Intergenerational Justice*. Oxford: Oxford University Press.
Guillaume, B. and V. Laramée. 2012. *Scénarios d'avenir. Futurs du climat et de la technologie*. Paris: Armand Colin.
Gutmann, A. 2007. "Democracy." In *Companion to Political Philosophy*, edited by Pettit, P. and R. Goodin, 521–31. Cambridge: Cambridge University Press.
Hansen, J. 2009. *Storms of My Grandchildren: The Truth about the Coming Climate Catastrophe and Our Last Chance to Save Humanity*. New York: Bloomsbury.
Heilbroner, R. 1975. *An Inquiry into the Human Prospect*. New York: W. W. Norton.
Held, D. and A. McGrew. 2000. *The Global Transformations Reader*. Cambridge: Polity Press.
Heyward, C. 2014. "New waves in climate justice: Climate change as cultural injustice." In *New Waves in Global Justice*, edited by T. Brooks, 149–69. Basingstroke, UK: Palgrave Macmillan.
Hill, T. 1983. "Ideals of human excellences and preserving natural environments." *Environmental Ethics* 5 (3): 211–24.
Jamieson, D. 2010. "Climate change, responsibility and justice." *Science and Engineering Ethics* 16: 431–45.
Jonas, H. 1984. *The Imperative of Responsibility: In Search of an Ethics for the Technological Age*. Chicago: University of Chicago Press.
Klein, R. J. T., Huq, S., Denton, F., Downing, T. E., Richels, R. G., Robinson, J. B. and F. L. Toth. 2007. "Inter-relationships between adaptation and mitigation." In *Climate Change 2007: Impacts, Adaptation and Vulnerability. Contribution of Working Group Ii to the Fourth Assessment Report of the Intergovernmental Panel on Climate Change*, edited by Parry, M. L., Canziani, O. F., Palutikof, J. P., van der Linden, P. J. and C. E. Hanson, 745–77. Cambridge: Cambridge University Press.
Kymlicka, W. 1989. *Liberalism, Community, and Culture*. Oxford: Clarendon Press.
Kymlicka, W. 1995. *Multicultural Citizenship: A Liberal Theory of Minority Rights*. Oxford: Clarendon Press.
Leichenko, R. M. and K. L. O'Brien. 2008. *Environmental Change and Globalization: Double Exposures*. Oxford: Oxford University Press.
Li, Q. and R. Reuveny. 2006. "Democracy and environmental degradation." *International Studies* 50 (4): 935–56.

Lovelock, J. 2009. *The Vanishing Face of Gaia: A Final Warning*. New York: Basic Books.
McKibben, B. 1990. *The End of Nature*. New York: Anchor.
Meyer, L. H. 2010. "Intergenerational justice." In *The Stanford Encyclopedia of Philosophy*, Spring 2010 ed., edited by E. N. Zalta. http://plato.stanford.edu/archives/spr2010/entries/justice-intergenerational/.
Miller, D. 1995. *On Nationality*. Oxford: Oxford University Press.
Miller, D. 1999. *Principles of Social Justice*. Cambridge, MA: Harvard University Press.
Miller, D. 2005. "Against global egalitarianism." *Journal of Ethics* 9 (1–2): 55–79.
Nagel, T. 2005. "The problem of global justice." *Philosophy and Public Affairs* 33 (2): 113–47.
Neuteleers, S. and B. Guillaume. 2014. "Introduction: Tendencies towards environmental autocracy and technocracy." *Ethical Perspectives* 21(1): 1–13.
O'Brien, K. 2013. "The courage to change: Adaptation from the inside-out." In *Successful Adaptation to Climate Change: Linking Science and Policy in a Rapidly Changing World*, edited by Moser, S. C. and M. T. Boykoff, 306–19. London: Routledge.
O'Neill, J. 1993a. *Ecology, Policy and Politics*. London: Routledge.
O'Neill, J. 1993b. "Future generations: Present harms." *Philosophy* 68 (263): 35–51.
O'Neill, J., Holland, A. and A. Light. 2008. *Environmental Values*. London: Routledge.
Ophuls, W. 1977. *Ecology and the Politics of Scarcity*. San Francisco: Freeman.
Parfit, D. 1984. *Reasons and Persons*. Oxford: Clarendon Press.
Pasek, J. 1993. *Environmental Policy and "the Identity Problem."* CSERGE Working Papers, University of East Anglia.
Petersen, A. 1985. "The philosophy of Niels Bohr." In *Niels Bohr: A Centenary Volume*, edited by French, A. P. and P. J. Kennedy, 299–310. London: Harvard University Press.
Pogge, T. 1989. *Realizing Rawls*. Ithaca, NY: Cornell University Press.
Ponting, C. 1992. *Een Groene Geschiedenis Van De Wereld*. Amsterdam: Amber. Translated from A Green History of the World (London: Sinclair-Stevenson, 1991).
Rawls, J. 1999a. *The Law of Peoples; with, the Idea of Public Reason Revisited*. Cambridge, MA: Harvard University Press.
Rawls, J. 1999b. *A Theory of Justice*. Cambridge, MA: Belknap Press of Harvard University Press.
Rees, M. 2003. *Our Final Hour: A Scientist's Warning: How Terror, Error, and Environmental Disaster Threaten Humankind's Future in This Century – on Earth and Beyond (2003)*. New York: Basic Books.
Ricoeur, P. 1990. *Soi-Meme Comme Un Autre*. Paris: Editions du Seuil.
Rockstrom, J., Steffen, W., Noone, K., Persson, A., Chapin, F. S., Lambin, E. F., Lenton, T. M. et al. 2009. "A safe operating space for humanity." *Nature* 461 (7263): 472–75.
Sandel, M. J. 2005. *Public Philosophy: Essays on Morality in Politics*. Cambridge, MA: Harvard University Press.
Sandler, R. D. 2005. "Introduction: Environmental virtue ethics." In *Environmental Virtue Ethics*, edited by R. D. Sandler and P. Cafaro, 1–12. Lanham, MD: Rowman and Littlefield.
Schellnhuber, H. J. 1999. "'Earth system' analysis and the second Copernican revolution." *Nature* 402: C19–C23.
Shearman, D. and J. W. Smith. 2007. *The Climate Challenge and the Failure of Democracy*. Westport, CT: Praeger.
Shepherd, J. G. 2012. "Geoengineering the climate: An overview and update." *Philosophical Transactions of the Royal Society A: Mathematical, Physical and Engineering Sciences* 370 (1974): 4166–75.
Simonet, G. 2010. "The concept of adaptation: Interdisciplinary scope and involvement in climate change." *S.A.P.I.EN.S* 3 (1): 1–9.

Singer, P. 2002. *One World: The Ethics of Globalization*. New York: Yale University Press.
Smith, G. 2003. *Deliberative Democracy and the Environment*. London: Routledge.
Steffen, W., A. Sanderson, P. Tyson, J. Jäger, P. Matson, B. Moore III, F. Oldfield, K. Richardson et al., eds. 2004. *Global Change and the Earth System: A Planet under Pressure*. IGBP Global Change Series. Berlin: Springer.
Stiglitz, J. E. 2002. *Globalization and Its Discontents*. New York: W. W. Norton.
Sylvan, R. 2003. "Is there a need for a new, an environmental, ethic?" In *Environmental Ethics: An Anthology*, edited by Light, A. and H. Rolston III, 47–52. Oxford: Blackwell.
Thompson, A. 2009. "Responsibility for the end of nature: Or, how i learned to stop worrying and love global warming." *Ethics and the Environment* 14 (1): 79–99.
Thompson, A. and J. Bendik-Keymer. 2012. "Introduction: Adapting humanity." In *Ethical Adaptation to Climate Change*, edited by Thompson, A. and J. Bendik-Keymer, 1–23. Cambridge, MA: MIT Press.
UNFCC. 1992. *United Nations Framework Convention on Climate Change*. http://unfccc.int/resource/docs/convkp/conveng.pdf.
Van Parijs, P. 2007. "International distributive justice." In *A Companion to Contemporary Political Philosophy*, edited by Goodin, R. E., Pettit, P. and T. Pogge, 638–52. Oxford: Blackwell.
Ward, P. and D. Brownlee. 2000. *Rare Earth: Why Complex Life Is Uncommon in the Universe*. New York: Springer.
WCED. 1987. *Report of the World Commission on Environment and Development: Our Common Future*. Transmitted to the General Assembly as an Annex to Document a/42/427. http://worldinbalance.net/pdf/1987-brundtland.pdf.
Zalasiewicz, J., Williams, M., Steffen, W. and P. Crutzen. 2010. "The new world of the Anthropocene." *Environmental Science and Technology* 44 (7): 2228–31.

3

Urban Climate Change Policy Transitions

Views from New York City and London

WILLIAM SOLECKI, LESLEY PATRICK, AND ZOE SPRIGINGS

The objective of this chapter is to detail how climate change adaptation emerged as a public policy issue for the cities of New York and London during the past decade (approximately 2003–13). The chapter argues that climate change adaptation strategies developed in each city through a set of complex interactions among and between policy entrepreneurs, government officials, and members of the scientific-expert community. While benchmarks typically associated with issue development (e.g., problem identification, risk assessment, prioritization, policy entrepreneurs) were present as climate change adaptation was brought into the public policy arena, a set of issues, events, and activities specific to the climate change, including scientific uncertainty and interactions between local officials and climate change decision makers in other large cities, helped push the agenda forward. In this chapter, the narrative of climate change adaptation policy development in the two cities is detailed and investigated through the use of an urban policy transition frame that highlights the relative role of system tipping points and policy triggers in public policy formation. As part of the analysis, we also examine significant climate change adaptation policy shifts in the two cities and attempt to explain why they occurred.

The centerpiece of New York City's action on climate change has been the PlaNYC effort and the associated long-term sustainability plan, PlaNYC 2030, originally released in April 2007 (updated in 2011 [New York City Office of the Mayor 2011] and further enhanced in a post–Hurricane Sandy effort in 2013 [New York City Office of the Mayor 2013]). These plans heralded many climate-related, sector-level action efforts focused on different aspects of city operation and function. In the case of London, the London Plans of 2008 and 2011 served as a primary consolidation of climate change action. Each of the city's activities included elements designed to manage climate change risk and increase the resilience of infrastructure and systems operations and communities to climate change impacts. We set climate action policy shifts in both cities within the larger history of response to environmental challenges and crises. The analysis of the policy transition for London and New York is based on analysis of primary documents and reports and associated secondary data. We draw on the key policy statements and reports during the period 1997 to 2013. The selection of specific documents and associated analysis was guided by professional experience in the field.

The chapter hopes to contribute to the growing literature on urban climate change governance and policy analysis (for recent reviews, see Bulkeley 2013, 2010; Bulkeley and Newell 2010; Corfee-Morlot et al. 2009; Hoornweg et al. 2011; McCarney et al. 2011; Satterthwaite et al. 2008). This literature has begun to document different types of city government responses to climate change risks (Betsill and Bulkeley 2006), the relative role of local decision makers, stakeholders, and planning approaches (Anguelovski and Carmin 2011; Carmin et al. 2012), and how cities' climate policies reflect underlying questions of equity and justice (Broto and Bulkeley 2013; Bulkeley 2010). Much of the research attention has been directed toward climate change mitigation issues, including governing low-carbon cities and greenhouse gas (GHG) reduction policy. However, the focus in cities on climate change adaptation has increased in the past several years. In this work, cities are often presented as sites of significant discussion on appropriate climate policy and in many cases genuine climate action as opposed to the political stalemates that characterize discussions among nation-states at the international level (Kousky and Schneider 2003; Rosenzweig et al. 2010). With respect to policy shifts, meaningful climate change action was first identified in early adopter cities (e.g., particularly western European cities) in the latter half of the 1990s, and a second wave, according to Bulkeley (2010), emerged in the early part of the 2000s decade.

The chapter is organized into the following sections. In section 2 we present a brief outline of policy transition theory relevant to understanding policy transitions with respect to climate change action in cities. The third section discusses the emergence of climate change adaptation within New York since the late 1990s with two significant policy transitions noted. Next, in section 4, we introduce some of the parallel steps of development in London and describe how they informed and influenced activities in New York. Section 5 reviews the case study material and presents a broad conceptual model on urban climate policy development. A conclusion statement is presented that connects climate change action to the wider literature of sustainability transitions in cities.

Urban Environmental Policy Transitions

An extensive body of research literature exists on how policy issues develop, are implemented, and shift over time. The policy development process involves several broad steps, including problem/issue identification, policy formulation, implementation, and evaluation. Often this sequence of steps occurs over a lengthy period as a general understanding of the issue emerges and the capacity of government to respond is assessed and engaged. One important analytical concern is to determine which factors mediate how and why policy transitions (i.e., significant shifts in policy focus, direction, and/or structure) take place. For example, some issues emerge very rapidly within the public policy agenda, whereas others might emerge only after decades of discussion. The central questions addressed in this chapter are, in what context does climate change adaptation action emerge as an urban environmental policy issue, what events motivate its development, and how do these activities come to be seen as a policy transition?

Past environmental change, crisis, and policy transitions provide useful examples for how to define and interpret the development of climate change as a local public policy issue and how associated transitions took place. For example, as New York City grew during the nineteenth and twentieth centuries, many environmental challenges emerged that were characterized by lead-up events and attempts at solutions, widespread heightened levels of concern, and eventual policy transition resolution. The crises and transitions involved important public health and quality of life issues that were seen as impinging on the city's continued economic development (Solecki 2012). Some of the major policy shifts and associated time frame included (1) acquisition of a steady and copious supply of fresh drinking water (early 1840s), (2) creation of urban open green space (1850s), (3) professionalization of waste management and sanitation (1880s), (4) promotion of mobility and transit (1910s), and (5) reduction of air and water pollution (1960s).

A variety of conceptual policy transition models have been formulated that highlight and explain the importance of varying factors and drivers in policy shifts. Some of the factors and drivers most noted are extreme disaster events as a motivating condition, policy entrepreneurs (i.e., those focused on promoting a new policy initiative), available resources, and well-defined and achievable solution opportunities. The policy window conceptual model (Kingdon 1995; Solecki and Michaels 1994) focuses on how a sudden shift or change (e.g., natural disaster or fiscal crisis) creates the conditions for a new policy environment or regime to emerge. Extensive research has been done on how extreme weather events such as hurricanes and floods provide conditions for new policy, management, and operational regimes to be put into place (Birkmann et al. 2010). Important factors associated with policy shifts emerging from an extreme event include whether the threat for another similar or greater event is perceived as immediate and how well the threat is understood (Alam and Rabbani 2007; Zahran et al. 2008). Often policy entrepreneurs are eager to implement new plans for action, and they recognize the policy window created by extreme events as an opening for pushing forward their agenda (Lambright et al. 1996).

An important evolution in policy transition analysis is the conceptualization of policy development and policy shifts within the context of a complex systems framework (Holling 1978; Holling et al. 1995; Repetto 2006) and associated nonlinear behavior (Brown 1994; Kline 2001). This literature emerged from at least two disciplinary perspectives. One thread is from more traditional social science frames such as political science, economics, and geography. Research from these perspectives defines policy development and shifts as embedded within decision-making processes, constraints, and social structures. Policy regime shifts in this context occur at moments when events or other circumstances change public opinion or other factors that result in increased potential for successful changes in public policy followed by periods of little or no policy change. This process is described as punctuated equilibrium (Baumgartner 2006; Repetto 2006), and it is where policy tipping points can be identified (Brock 2006). The speed of public opinion shifts is seen as a significant determinant of policy changes.

A parallel thread of research has emerged from the fields of ecology and conservation management and is best expressed by the work of Holling and Gunderson (Holling 1978; Holling et al. 1995) on adaptive cycles. In this context, transitions result from internal and external pressures stressing the resilience of a system. Resilience provides systems the capacity to recover from shocks and to incorporate adaptive actions in response to new and emerging realities. This approach has facilitated extensive research into adaptive management strategies and system function and their application to ecosystems and to a broad set of socioecological systems (see Walker and Salt 2012 as a recent example of this work). Another dimension of this discussion is the size, speed, and significance of the transition and the question of whether the policy change involves a subtle adjustment to an existing management regime or a fundamental and potentially irreversible change. The work of Scheffer (2009; Scheffer et al. 2003, 2009) and Walker and Myers (2004) points to conditions that are associated with critical transitions that involve complete reorganization of the policy regime system. Together, the adaptive cycle model and critical transition model provide a robust understanding of how policy shifts occur because they conceptualize the role of both internal and external system-level drivers and stressors.

Many of the policy transitions in cities also were associated with the coupling of new scientific data and knowledge to the practical everyday experiences of city businesses residents. For instance, the open space movement in New York during the 1850s flowered with the design and construction of the Central Park, partially as response to expert assertions that densely settled cities like New York needed open spaces to serve as pressure release valves for the working classes and the stress of urban living (Cranz 1982). Similarly, science advancements in public health sanitation engineering in the last third of the nineteenth century were integrated into building design and plumbing and waste management to reduce the likelihood of disease outbreaks. Science also moved to the forefront in the middle of the twentieth century as advances in atmospheric chemistry and meteorology were applied to air pollution policies and actions to lessen the social and health burden of smog and deadly thermal inversions (Solecki and Shelley 1996).

The character of these policy transitions also reflects larger social transformations occurring within society at the time. Modernity, like the concepts of sustainability and resilience today, became a powerful meta-narrative for the reimagining and reconstructing of urban life during the early decades of the twentieth century. New York City officials and regional planners, including the Regional Plan Association, pushed forward an agenda to relieve the chronic traffic congestion of city streets and the chaotic construction patterns with a set of policies focused on land use zoning, planning, and subway and highway construction designed to (re)build a more rational and modern New York (Ward and Zunz 1992). In the contemporary era, the concept of sustainability is similarly being used as a device to promote and validate a set of municipal-, neighborhood-, and household-level actions in cities. The broad-scale sustainability transition connects to a variety of urban sectors, including water/wastewater, solid waste, energy, food, and building (Markard et al. 2012; Meadowcroft 2011). The linkages between urban sustainability policy agenda and the urban climate change policy agenda are complex, as highlighted in the case studies presented in this chapter.

In the following two sections, the details of how climate change action emerged as a public policy issue in New York and London are presented. The objective is to illustrate the chronology of events and identify specific policy drivers and determine whether they reflect the traditional types of drivers defined within existing literature and what elements, if any, are unique or of particular importance within the two city case studies.

Climate Change Action and Policy Shifts in New York City

Several conditions set the stage for climate change action in New York City over the past several decades. Central to the process was an increased knowledge about climate change impacts and capacity for mitigation and adaptation. In addition, high precipitation and storm surge events caused coastal and street flooding that impacted infrastructure and operations throughout the city, creating a sense of urgency about the potential for catastrophic impacts and highlighting the need for solutions. The policy entrepreneurship of Mayor Michael Bloomberg and the development of national and global networks worked in tandem to identify policy windows and present appropriate responses.

Foundation for Climate Change Action in New York City

Concern about climate change impacts on New York City precipitated the 1994 conference "Metropolitan New York in the Greenhouse: The Baked Apple?," which was convened as a first step in preparing the New York City region for the possibility of climate change and its associated impacts and to encourage direct action toward mitigation or adaptation (Hill 1996). However, the first major, scenario-based scientific assessment of climate impacts across the New York metropolitan region wasn't developed until the late 1990s, as part of the first comprehensive national climate assessment. In 2000, the *Metropolitan East Coast (MEC) Assessment of Impacts of Potential Climate Variability and Change* report (Rosenzweig and Solecki 2001) was produced. Report recommendations included several potential adaptations at both conceptual and operational levels, the creation of an Inter-Agency Climate Task Force, and the creation of a Climate Awareness Program to inform decision makers and the general public about climate change and potential responses. Overall, the report was widely cited but not immediately translated into significant policy changes or immediate management actions. The report could be seen as largely alerting stakeholders to the issue of climate change, but there was not sufficient momentum to initiate climate actions.

Michael Bloomberg was elected mayor of New York in November 2001, while the city was still very much in the immediate throes of the September 11 attack. Bloomberg's attention to climate change issues first became publicly present in 2005, when he joined with 131 other mayors to embrace the Kyoto Protocol for GHG reduction (Sanders 2005). The following year, in an Earth Day address, Bloomberg stated that the city's "commitment to the environment also extends to improving air and water quality, . . . [and] to substantially reduce the release of greenhouse gas emissions into the air" (New York City Office of the Mayor 2006a).

Just one month later in 2006, Bloomberg formally announced the creation of a new mayoral office – the Office of Long-Term Planning and Sustainability (OLTPS) – the goal of which was to develop and implement a comprehensive plan to create a greener and more sustainable city. The aim of the plan was to "prepare the city for one million more residents, strengthen our economy, enhance the quality of life for all New Yorkers, and deal with climate change," suggesting that the comingled challenges of climate change, population growth, and continued economic viability were equally important (New York City Office of the Mayor 2006b). GHG mitigation became a central goal of PlaNYC 2030 climate activities, and then an ambitious goal of a 30 percent reduction of GHG emissions from 2005 levels by 2030 was put forward (New York City Office of the Mayor 2007).

Initiating Climate Change Adaptation Action in New York City

Climate change adaptation also was considered in the final draft PlaNYC document released in April 2007. It called for the creation of a climate change adaptation task force, the development of adaptation plans, and the need to consider highly vulnerable communities in the city. The discourse of adaptation planning and action, however, changed later that year. On August 8, 2007, a severe and largely unpredicted thunderstorm swept through the city, resulting in major and in some areas prolonged service disruptions of the Metropolitan Transportation Authority's (MTA) transit system. The flash flooding rendered almost the entire subway system inoperable, causing significant economic losses that day because employees and customers could not get into the city's central business districts. Suddenly the prospect of climate change impacts seemed more immediate and relevant to the everyday. The event became a policy window for the initiation of climate change adaptation policy in New York City and marks the first of two significant transitions in the city's climate action.

In the immediate aftermath of the August storm, the New York State governor directed the MTA to conduct an assessment of the system's vulnerability to future storms. Specific recommendations for improving the MTA's operations, engineering and regional interagency issues, and communications were put forth, including the creation of an Emergency Response Center and Inter/Intra-Agency Flooding Task Force. These adaptation measures, developed in response to crisis, increased the capacity of the MTA in the face of future storm events.

Connecting PlaNYC with Climate Change Adaptation

Approximately one year after the August storm, the mayor formally launched twin PlaNYC initiatives directly connected with climate change adaptation. In August 2008 Bloomberg announced the creation of the New York City Climate Change Adaptation Task Force, a group made up of representatives from the city's Departments of Environmental Protection, Planning, Public Health, and Transportation, among others, as well as state and regional transportation agencies, including the MTA and private railroad and telecommunication companies. The primary focus was on what steps could or should be taken to ensure the continued function of the city's critical infrastructure (i.e., roads, subways, communication services, energy, water, and public parks and open space) (New York City Office of the Mayor 2008).

At the same time, the mayor brought together the New York City Panel on Climate Change (NPCC). The panel was composed of academic and research experts and was convened to advise the Climate Change Adaptation Task Force on the development of adaptation strategies to secure the city's infrastructure from the effects of climate change. Central to the panel's charge was the development of a climate change science base, including past and current temperature and precipitation trends as well as expected future trends and scenarios developed from downscaled global climate models. The city was particularly interested in understanding how the frequency and intensity of heat waves, extreme wind events, inland and street-level flooding, and coastal flooding and storm surge would change over time.

The recommended climate action in the 2010 NPCC report focused on the concept of Flexible Adaptation Pathways (derived from a policy frame used by London climate stakeholders), which involves an understanding that adaptation strategies should provide opportunities for decision makers and stakeholders to adjust their plans and activities as new and assumedly more sophisticated science information emerges in the future. Other important elements of the report included developing potential adaptation decision tools and processes to evaluate existing codes and standards and discern whether they need to be adjusted in the face of climate change.

An important contributor to the development of the Adaptation Task Force and the NPCC was the professional and personal connections between the New York City participants and their counterparts in other cities (e.g., Chicago, San Francisco, Boston), who were also developing their own climate action efforts. The role of intercity networks was recognized relatively early as an important component of individual city climate planning and action. Networking played an important role in promoting climate change mitigation and adaptation in European cities in the early 2000s (Kern and Bulkeley 2009). In the latter part of the decade, networks began to emerge across North American cities as well within international organizations such as C20 (later renamed C40) and the World Mayors Council on Climate Change (Gore and Robinson 2009; Rosenzweig et al. 2010; Rutland and Aylett 2008; Toly 2008). In the case of New York City, very significant connections came with the city of London and the work of the Greater London Authority. Close collaborations resulted from the ongoing discussions between the groups in both cities.

Shifting from Climate Change Adaptation to Climate Resilience

During 2010 and into the early part of 2011 the second significant shift in the city's policy approach to climate change adaptation occurred. The shift involved mainstreaming climate adaptation into everyday climate risk practice and management around the concept of climate resilience and opportunities for broader societal resilience. Resilience in this context includes increased preparedness for future climate changes; improved communication of climate impacts at the community level; fortification of critical infrastructure, including buildings and coastlines; and opportunities to enhance existing resources with emerging climate change information.

This switch seemed to have occurred for several reasons. Most importantly, the concept of resilience was increasingly being used and debated within local, state, and national government circles as an effective mechanism to connect ongoing climate risk management strategies and related disaster risk reduction efforts and public health concerns. Besides the Greater London Authority effort mentioned earlier, the *America's Climate Choices* report produced by the National Academy of Sciences of the United States of America and published in May 2011 prominently presented the need to "increase the *resilience* of human and natural systems to climate change." Later, in October 2011, the Progress Report of the Interagency Climate Change Adaptation

Task Force released its report *Federal Actions for a Climate Resilient Nation*, which significantly advanced the focus on broader climate resilience as opposed to more narrow climate change adaptation.[1]

The update and reauthorization of the PlaNYC document in April 2011 illustrates the shift from climate change adaptation to climate resilience and how resilience was to be more fully integrated into a wider set of task-specific climate action initiatives. In PlaNYC 2007 the term *resilience* is not present once, but in the 2011 report *resilience* was used fifty-seven times (and the term *adaptation* declined from twenty-one to five occurrences). Furthermore, the 2007 PlaNYC report offered just three climate action initiatives focused on developing processes, personnel, and strategies for response. The 2011 plaNYC included more numerous and specific new initiatives and outcome goals that reflected the ambition of enhancing climate resilience by providing better information and science, tools for decision making, and integration across government agencies and between government decision makers and private stakeholders.

The city's efforts at promoting climate resilience also have recently expanded beyond the PlaNYC effort. For instance, climate change projections have been integrated into emergency management and preparedness via the city's recently established Office of Recovery and Resiliency and community-level disaster response via the city's Department of Planning.[2] The NPCC report conclusions and recommendations have been connected to a wide variety of climate resilience–focused actions, including (1) decision support protocols such as risk prioritization; (2) a reevaluation of existing plans, codes, standards, and regulations; and (3) urban design review and competitions including public art displays (e.g., *Rising Currents* exhibit at New York's Museum of Modern Art[3]). Climate resilience also has been evident within high-profile city planning documents, including the NYC *Green Infrastructure Plan* (2011 Update), the *Sustainable Stormwater Management Plan Progress Report* (2010), *Vision 2020: New York City Comprehensive Waterfront Plan* (2011), and the *Greener Greater Buildings Plan* (2009).

Hurricane Irene and Hurricane Sandy: Putting Resilience into Practice

Soon after New York released its update of the PlaNYC document, Hurricane Irene struck the metropolitan region in late August 2011. Local officials and other stakeholders broadly described it as a focusing event that could be utilized to put extreme event resiliency planning proposals into practice. Hurricane Irene was a significant rain event that resulted in extensive flooding into distant suburban and exurban areas north

---

[1] In the previous year, the same body produced a report titled *Recommended Action in Support of a National Climate Change Adaptation Strategy*.
[2] See, for more discussion, http://www.nyc.gov/html/dcp/html/climate_resilience/index.shtml.
[3] http://www.moma.org/visit/calendar/exhibitions/1031.

and west of the NYC and resulted in only slight storm surge flooding in the city itself. In October 2012, fourteen months after Irene, Hurricane Sandy hit the metropolitan region, causing catastrophic damage, the most significant of which came from record storm surge and coastal flood. In the aftermath of Hurricane Sandy, Mayor Bloomberg created the Special Initiative of Rebuilding and Resilience (SIRR) and reconvened New York City on Climate Change.

The SIRR focused on assessing the damage from Sandy, understanding how future climate change might influence the level of coastal risk and promoting resiliency efforts in the city's neighborhoods most at risk of current and future flooding. The SIRR released its report in June 2013 (New York City Office of the Mayor 2013), and the NPCC released its climate projection updates at the same time (New York City Panel on Climate Change 2013). Similar to the two previous PlaNYC documents, the SIRR report highlighted dozens of new initiatives and actions designed to reduce vulnerabilities, add in rebuilding, and institutionalize resiliency practice. In this regard, Sandy and SIRR did not signal a change in policy but presented an opportunity and policy window to catalyze new and larger-scale action.

As described, New York looked to London, as much as to other U.S. cities, as it developed its climate change adaptation plans. Meanwhile, London consciously positioned its adaptation policies in relation to international cities such as New York rather than cities in the United Kingdom. They were both founder members of the organization now known as the C40 Cities Climate Leadership Group, a global network of large cities taking action to address climate change. Both cities realized early on the benefits of transnational cooperation with other cities, and this way of working played to their identity as global cities.

Climate Change Adaptation and Policy Shifts in London, United Kingdom

The climate impacts evidence base for London was relatively developed by the turn of the twenty-first century, due to U.K. national research, and this continually emerging scientific knowledge enabled policy making and local–national cooperation. The U.K. Climate Impacts Programme (UKCIP) was established by the new U.K. government in 1997 and generated scenarios in 1998, 2002, and 2009.[4] The London Climate Change Partnership (LCCP) was formed in 2001 to commission a tailored report on the specific consequences for London, *London's Warming*, produced by the UKCIP using its 2002 national scenarios (U.K. Climate Impacts Programme 2002). The United Kingdom's compact size meant that national scenarios, with regional

---

[4] The UKCIP was established with the aim of providing a framework for an integrated national assessment of climate change impacts, and subsequently to help organizations assess how they might be affected by climate change, so they can prepare.

components, were meaningful for London. The city therefore directly benefited from national resources directed toward identifying specific problems and their solutions. The LCCP was funded by the Greater London Authority[5] (GLA), the U.K. Department of Environment, Food, and Rural Affairs (Defra), and local utility Thames Water. A Defra minister attended the launch of the London report, and a Defra official sat on the steering board of the LCCP. It is evident that London and the national administration collaborated in their approach to addressing climate impacts from early on, building on a shared evidence base.

The U.K. government was positioning itself at this time as an international thought leader on climate change mitigation and adaptation, a position it has maintained into the present. This created opportunities for London. In September 2004, Prime Minister Tony Blair confirmed that "climate change will be a top priority for our G8 Presidency next year" (Blair 2004). He also announced that the United Kingdom would host an international symposium of climate change scientists prior to the July 2005 summit to inform its discussions (Blair 2004). Impacts were put top of the symposium agenda (Defra 2006). That summer, Chancellor Gordon Brown announced he had commissioned Nicholas Stern to report to him and the prime minister on the economics of climate change. One of the three objectives was to review "the potential of different approaches for adaptation to changes in the climate" (Stern 2006). The influential *Stern Review* (Stern 2006) took an international perspective and was reported around the world. It provided a new political lexicon and evidence base for addressing climate change from a costs and benefits perspective. Its headline finding was that the benefits of strong, early action on climate change outweighed the costs. In November 2006 the government announced the Climate Change Bill, as Tony Blair acted on his assertion that "to acquire global leadership, on this issue Britain must demonstrate it first at home" (Blair 2004).

Initiating Climate Change Adaptation in London

National priorities had created a window of policy opportunity for London and its mayor to step forward and win support at the U.K. level for pioneering climate change action and at an international level as the capital city of a country leading global action. Mayor Livingstone is easily identifiable as the London policy entrepreneur, acting in parallel with the policy entrepreneurship of Prime Minister Blair. The enmeshing of Livingstone's local, national, and global leadership was evident in the frenetic activity of autumn 2005. September saw the publication of *The Impacts of Climate Change on London's Transport Systems* (Greater London Authority 2005a). In October,

---

[5] The Greater London authority, known as the GLA, is the permanent staff supporting the Mayor of London and London Assembly.

Livingstone brought together twenty-five global megacities to launch the Large Cities Climate Leadership Group in London.[6] In November the GLA published *Adapting to Climate Change: A Checklist for Development* (Greater London Authority 2005b), designed for national use. In his address at World Environment Day and at the Large Cities conference, Livingstone argued that cities were key influencers on climate change policy, because of their large populations and because they are centers of innovation, and merited specific attention (Livingstone 2005).

From the beginning, London adaptation policy had a strong consciousness of its identity as a global city, at the vanguard of climate action. The very first report asserted that London was different to the rest of the United Kingdom's cities and instead drew its parallels with other global cities (U.K. Climate Impacts Programme 2002). It is true that the density of population in London (twice that of most other U.K. cities) and its density of capital as a leading global financial center did make cities like New York more natural comparators, but also competitors. London's mayors have therefore always sought to put their city ahead of the game. They have consistently positioned London internationally in their adaptation policy announcements, with Mayor Livingstone describing "my vision, which guides all my strategies, is to develop London as an exemplary, sustainable world city" (Greater London Authority 2008) and Mayor Johnson asserting that to "maintain our status as a leading global city, we must adapt" (Greater London Authority 2010). The most recent London Plan of 2011, designed to apply until 2031, placed adaptation squarely in the foreground of London's local and global identity. It was presented as a continuation of London's "proud tradition" of planning innovation and also a means of profiting from being "a global example" (Greater London Authority 2011a, 6–7). With adaptation policy, London actively promoted its local knowledge as something to share internationally, and the consequences can be seen in the New York case study.

Connecting the London Plan with Climate Change Adaptation

Alongside this opportunity to be heard, there was a pressing need for London to consider adaptation to a growing urban population. This was set out in London's first official spatial development strategy, the London Plan of 2004, which described the city's recent and anticipated population growth and spatial expansion and the consequent need for new development. The expanding economy meant that significant funds were available for investment. Both the mayor and the U.K. government were supportive of directing investment toward the east of London and beyond into the Thames Estuary, historically some of the most deprived areas of London and the United Kingdom (Department for Communities and Local Government 2007; U.K. Government 2004). The area, known as the Thames Gateway, was the proposed location for at

---

[6] The Large Cities Climate Leadership Group later became the C20 and then the C40 as its membership expanded.

least 120,000 new homes and up to 250,000 new jobs (London Assembly Environment Committee 2005), as well as the London 2012 Summer Olympics. However, it was situated in the floodplain of the River Thames, a high-risk area (McRobie et al. 2005). Moreover, its anticipated life-span extended into the second half of the twenty-first century, beyond the proposed 2030 expiry date of London's key flood defense: the Thames Barrier. The London Assembly Environment Committee expressed severe concerns regarding climate-exacerbated flood risk in their assessment, *London under Threat?* (2005), and it was a matter of considerable debate whether this development would be unacceptably exposed.

Assessing and addressing the future flood risk became a practical and political necessity. London and the U.K. government pursued an approach that offered future commitments and strong words while pushing ahead with the proposed development. The mantra of sustainable development to manage flood risk was promoted in London and at the national level (Department for Communities and Local Government 2007; Greater London Authority 2008). The GLA Act 2007 imposed mitigation and adaptation obligations on the mayor of London, and the London Plan 2004 was explicitly revised in 2008 to take account of mitigating and adapting to climate change, including an objective to make London an exemplary world city in this regard. It fully endorsed the Thames Gateway development but committed to take account of findings from the Environment Agency, an environmental protection body funded by the U.K. government. The Environment Agency had been commissioned by the government to work with the U.K. Meteorological Office, also government funded, to advise on a flood risk management plan for the Thames Estuary in the light of climate change. Thus final decisions were postponed, but the issue of adaptation had been accepted onto the political agenda as a dimension of growth and development.

This analysis suggests that extreme weather was not a significant driver of London adaptation policy. The last fatal flooding in London was in 1928, and it is protected by substantial flood defenses on the Thames, so it is unsurprising that flooding has drifted out of popular consciousness (Lavery and Donovan 2005). London's adaptation policy documents make only brief reference to recent weather events, focusing instead on foreign events, particularly the catastrophic damage to New Orleans by Hurricane Katrina in 2005 (London Assembly Environment Committee 2005). Nonetheless, London planners remain alive to flood risk, and the precautionary principle which underpinned the construction of the Thames Barrier is still evident (Greater London Authority 2011a; Baxter 2005). Policy making appears to be driven by concern about the potential for catastrophic impacts rather than reaction to recent extremes.

The fact that recent experience was not needed to persuade voters, politicians, and bureaucrats may be attributed to the collective trust in the advisory bodies that foretold future extreme weather. Cross-party agreement underpinned the U.K. Climate Change Act 2008 and the GLA Act 2007, which imposed mitigation and adaptation obligations on the mayor of London. This consensus created the policy window for climate action through sustained local and national funding for scientific research and sustained political support for planning. Although there are signs of the consensus now being unpicked at a national level, London continues to prioritize climate change in its policy statements, and adaptation is now embedded in its modus operandi. The question for the future is what form this adaptation will take.

Shifting from Climate Change Adaptation to Climate Resilience

As in the case of New York, the term *resilience* has recently emerged in London's adaptation policy, swiftly moving from obscurity in 2002 to become a synonym for adaptation by 2011. However, as with the national and international adaptation discourse, the popularity of resilience far outstrips its clarity. When *resilience* materialized in adaptation policy, it dealt with content usually found within a sustainable development paradigm but used language with overtones of the security paradigms. Resilience first appeared as a synonym of adaptation in 2005 but did not become dominant until 2011. Their coexistence and interchangeability are epitomized by the title of the first adaptation strategy, *Managing Risks and Increasing Resilience: The Mayor's Climate Change Adaptation Strategy* (Greater London Authority 2011b). Adaptation has been formally defined by the London authorities as "a process of identifying climate risks and opportunities," but resilience remains undefined, although there are some contextual clues (Greater London Authority 2011b). It came to prominence as part of a revised adaptation narrative consciously deploying the terminology of business economics and risk management.

With reference to extreme weather events, the concept of resilience currently occupies a liminal position between security and sustainable development paradigms. The history of its usage in London offers some clues about how adaptation as resilience might develop and be implemented in the future. In London, *resilience* appeared most explicitly in the form of the London Resilience Forum, which sprang from the U.K. Civil Contingencies Act (2004). Here the security paradigm was dominant. This act established a framework for emergency planning and response, specifically accounting for terrorism, but also including extreme weather. In this context, the U.K. government's official definition of resilience has been the "ability of the community, services, area or infrastructure to detect, prevent, and, if necessary to withstand, handle and recover from disruptive challenges" (U.K. Cabinet Office 2013). The local forum had the responsibility for bringing together local emergency responders, including the police. It has been argued that the act represented a significant step forward by promoting planning, not just reactive responses (Bosher and Dainty 2011). Still, it remained primarily concerned with technological and state security solutions and protecting the status quo.

This security approach became particularly evident with the election of Boris Johnson in 2008, reelected in 2012, in the wake of the economic crisis. Adaptation qua resilience was framed as primarily an economic necessity for London to be a secure place for local people to live and for international financiers to invest (Greater London Authority 2010, 2011b). London's interest in international adaptation was presented as a desire to have stable trading partners. Distance was clearly put between these new arguments and previous environmental arguments, saying adaptation makes sense "even from a purely economic standpoint" (Greater London Authority 2011b). While the terminology resonates with that of the security paradigms, the content looks very different. The resilience actually described in the London Adaptation Strategy (Greater London Authority 2011b) is one of renewed "village" spaces and tackling social inequality, quality of life, and a new green economy. This usage has real potential to be used as a transformative approach to adaptation, which considers the social determinants of vulnerability, worlds apart from emergency response (Pelling 2011).

Adaptation as resilience in London is still in this liminal state, with the potential to develop within either the sustainable development or the security paradigms. The London Strategic Flood Framework (Greater London Authority 2012) by the London Resilience Partnership partly indicates a move toward long-term adaptation to environmental change. Conversely, climate change adaptation is not explicitly mentioned, and the focus is largely on emergency response. In the most recent London Plan (Greater London Authority 2011a), resilience was used in both a traditional security context and in a climate change adaptation context. Clarity has not yet been found. Perhaps if London continues to avoid further terrorist attacks or severe weather, it may be the more transformational paradigms of sustainable development that will dominate, and adaptation will remain on a separate track to emergency concerns. However, a terrorist attack or a severe weather event might provoke a more securitized response, a battening down of the hatches to protect the status quo, with adaptation brought under the umbrella of emergency response. Future research will be needed in this area.

Discussion and Conclusions: Toward an Urban Climate Change Policy Transitions Framework

Since the late 1990s, cities throughout the world have been considering the implications of climate change and the potential response of local policy, planning, and action. A fundamental issue is that climate change is an issue that in some ways is similar to those that cities have faced before (e.g., the issue presents a recognized and growing threat to everyday life in the city), and in other ways it represents something different in the sense that the hazards associated with climate change could lead to locally catastrophic impacts, while the solution spaces for climate change (both adaptation and mitigation) have local as well as global scale dimensions.

The case study reviews of New York and London illustrated that many traditional drivers (i.e., activities which foster the demand and capacity for policy action) and

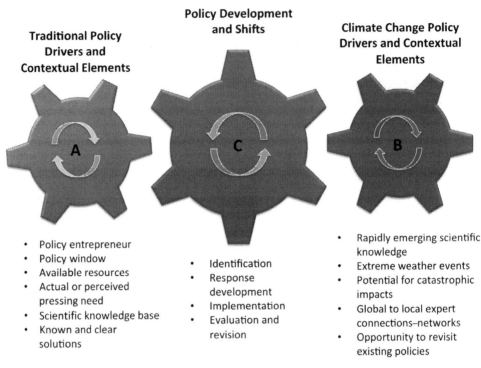

Figure 3.1. A diagram of key drivers and contextual elements of traditional and urban climate change policy transitions.

dynamic contextual elements (i.e., properties which influence the speed and rate of policy action) of the urban policy development and transition process were observed as climate change action emerged as a public policy issue in each city. On the basis of the two city cases, it is also possible to add to the framework several drivers and elements specific to increased climate risk and climate nonstationarity. The traditional policy- and climate change–relevant drivers and contextual elements are illustrated in Figure 3.1. Together these drivers and elements (cogs A and B) help to push forward policy development and shifts (noted in cog C). Conceptually one could argue that the drivers listed in cog A, "Traditional Policy Drivers and Contextual Elements," and cog B, "Climate Change Policy Drivers and Contextual Elements," together provide the push for policy development and implementation.

In the case of New York City, a disaster event coupled with available human and financial resources helped make an underlying issue a policy priority, resulting in a surge of momentum toward a policy shift or change. Cog B includes the dynamic quality of the scientific knowledge base, the awareness of the extreme weather events, unprecedented impacts of inaction, global-to-local expert connections, and the opportunity to modify existing policies.

The proposed model fits well with the London experience. All the elements of cog A were evident in London: the policy entrepreneurs of Blair and Livingstone, the window of local opportunity offered to London by national priorities, the resources of a growing economy, the solutions available from well-funded scientific institutions, and the pressing need of population growth into the floodplain. The elements of cog B were also present, with one important exception. The scientific basis for climate change was well established, including evidence of potentially catastrophic climate impacts on London; opportunities existed for the policy to be revised as the science developed; and there was a strong global–local link between London and the U.K. government as well cities around the world. Extreme weather, however, is notable by its absence.

New climate science and knowledge are foundational to climate change policy not only because it is appropriate to integrate the best and most up-to-date science in planning and policy but also because decision makers and stakeholders recognize that the climate science is rapidly advancing and that new policy needs to be flexible to incorporate new knowledge emerging from the scientific community. As a result, defining an efficient and effective science-to-policy information transfer mechanism is important (Horton et al. 2011). The local policy community understands that the science underlying the climate policy needs to be updated on a regular basis.

Emerging scientific knowledge on the magnitude and extent of climate change and its impacts has become another important motivator for public action (Scheffer et al. 2003). While other public concerns, such as crime, education, and public health, are typically seen as more urgent, climate change has been recognized as having the potential for significant societal impacts, particularly with respect to extreme weather events such as severe storms and floods and heat waves. Recent events such as Hurricane Sandy become benchmarks through which future climate risks are interpreted. Climate change science presents scenarios where disastrous weather events will become more frequent, intense, and damaging. As a result, climate change has increasingly become an issue that mayors and other city managers cannot ignore (Rosenzweig et al. 2010).

Urban climate policy development and transition have also been associated with the meaningful communication networks that have linked the decision makers and managers in different cities. Examples of climate action strategies in comparable cities offer templates for action, as do discussions of lessons learned and best practices. For example, climate change action in Chicago and London influenced the framework of policy development in New York. New York policy action has in turn influenced other cities within the United States and globally. Formal and informal networks of cities are likely to continue to function and strengthen as city leaders and agency staff personnel develop and maintain connections. London and New York are key members of C40,[7] which brings together representatives and decision makers from the world's largest cities to exchange advances and best practices regarding climate change mitigation and adaptation.

---

[7] http://www.c40cities.org/.

Climate change drivers and contextual elements (cog B) can enable policy change while working in parallel with more traditional drivers in cog A. As an example, emerging scientific information about climate change projections and impacts coupled with the experience of extreme weather events can create a pressing need for a policy shift while also providing context to evaluate and revise existing policies. These cross-connections between traditional and climate change policy associated drivers and elements can help policy entrepreneurs to promote opportunities for reviewing and reevaluating wide swaths of current public policy regulations, codes, and standards. For example, in New York City, climate change has pushed forward comprehensive reviews of emergency response preparedness planning, flood protection policies, and building codes and standards for wind damage prevention, among other shifts (Solecki et al. 2013). In this way, climate change is not only altering the way we think about climate risk, it is also transforming the way we manage cities and, inevitably, how we think about cities. For example, cities and their infrastructure and residents, as a result of climate change, are increasingly under pressure to be flexible and adaptive to more dynamic environmental conditions and the growing probability of extreme weather events.

The urban policy response to climate change has become profoundly connected to the larger-scale question of how to promote urban sustainability in a reflexive manner. The meta-narrative of sustainability has helped accelerate action on climate change in cities, and in turn, urban climate policy actions define the character and opportunity for future sustainability efforts. The recent rise of the concept of resilience and its role as a narrative intertwined with the goals of urban climate policies and sustainability policies has brought new questions regarding the rate and nature of future urban environmental transitions. Will policy advancement come through small, subtle adjustments to at-risk infrastructures and communities or through larger-scale transformations involving a change of systems? Climate change policies are now very much on the agenda of large cities like New York and London. Although the future will tell to what extent the planning and action now under way will lead to meaningful risk reduction and protection of the lives and property of the two cities' residents, a better understanding of urban policy development in an interconnected world can provide important insights into how to achieve sustainability.

References

Alam, M. and M. D. G. Rabbani. 2007. "Vulnerabilities and responses to climate change for Dhaka." *Environment and Urbanization* 19 (1): 81–97.

Anguelovski, I. and J. Carmin. 2011. "Something borrowed, everything new: Innovation and institutionalization in urban climate governance." *Current Opinion in Environmental Sustainability* 3 (3): 169–75.

Baumgartner, F. R. 2006. "Punctuated equilibrium theory and environment policy." In *Punctuated Equilibrium and the Dynamics of U.S. Environmental Policy*, edited by R. Repetto, 24–46. New Haven, CT: Yale University Press.

Baxter, P. J. 2005. "The east coast Big Flood, 31 January–1 February 1953: A summary of the human disaster." *Philosophical Transactions of the Royal Society A: Mathematical, Physical and Engineering Sciences* 363 (1831): 1293–312.

Betsill M. M. and H. Bulkeley. 2006. "Cities and the multilevel governance of global climate change." *Global Governance* 12 (2): 141–59.

Birkmann, J. P. Buckle, P., Jaeger, J., Pelling, M., Setiadi, N., Garschagen, M., Fernando N. and J. Kropp. 2010. "Extreme events and disasters: A window of opportunity for change? Analysis of organizational, institutional and political changes, formal and informal responses after mega-disasters." *Hazards* 55 (3): 637–55.

Blair, T. (2004, September 15). Full text: Blair's climate change speech. Speech given by the prime minister on the environment and the "urgent issue" of climate change. *The Guardian*, September 15. http://www.guardian.co.uk/politics/2004/sep/15/greenpolitics.uk.

Bosher, L. and A. Dainty. 2011. "Disaster risk reduction and 'built-in' resilience: Towards overarching principles for construction practice." *Disasters* 35 (1): 1–18.

Brock, W. A. 2006. "Tipping points, abrupt opinion changes, and punctuated policy change." In *Punctuated Equilibrium and the Dynamics of U.S. Environmental Policy*, edited by R. Repetto, 47–77. New Haven, CT: Yale University Press.

Broto, V. C. and H. Bulkeley. 2013. "A survey of urban climate change experiments in 100 cities." *Global Environmental Change* 23 (1): 92–102.

Brown, C. 1994. "Politics and the environment: Nonlinear instabilities dominate." *American Political Science Review* 88 (2): 293–303.

Bulkeley, H. 2010. "Cities and the governing of climate change." *Annual Review of Environment and Resources* 35: 229–253.

Bulkeley, H. 2013. *Cities and Climate Change*. Routledge Critical Introductions to Urbanism and the City. London: Routledge Press.

Bulkeley, H. and P. Newell. 2010. *Governing Climate Change*. Routledge: London.

Carmin, J., Anguelovski, I. and D. Roberts. 2012. "Urban climate adaptation in the Global South: Planning in an emerging policy domain." *Journal of Planning Education and Research* 32 (1): 18–32.

Corfee-Morlot, J., Kamal-Chaoui, L., Donovan, M. G., Cochran, I., Robert A. and P. J. Teasdale. 2009. *Cities, Climate Change and Multilevel Governance*. Organ. Econ. Co-Op. Dev. Environ. Working Paper 14. Paris: OECD Publications.

Cranz, G. 1982. *The Politics of Park Design: A History of Urban Parks in America*. Cambridge, MA: MIT Press.

Defra. (2006). *Executive Summary of the Conference Report. Avoiding Dangerous Climate Change: Scientific Symposium on Stabilisation of Greenhouse Gases February 1st to 3rd, 2005*. January. Exeter, U.K.: Met Office.

Department for Communities and Local Government. 2007. *Thames Gateway: The Delivery Plan*, London: DCLG.

Gore, C. and P. Robinson. 2009. "Local government response to climate change: Our last, best hope?" In *Changing Climates in North American Politics: Institutions, Policymaking and Multilevel Governance*, edited by H. Selin and S. D. VanDeveer, 138–58. Cambridge, MA: MIT Press.

Greater London Authority. 2005a. *The Impacts of Climate Change on London's Transport Systems*. London: GLA.

Greater London Authority. 2005b. *Adapting to Climate Change: A Checklist for Development*. London: GLA.

Greater London Authority. 2008. *The London Plan: Spatial Development Strategy for Greater London. Consolidated with Alterations since 2004*. London: GLA.

Greater London Authority. 2010. *The Draft Climate Change Adaptation Strategy for London Consultation Draft*. London: GLA.

Greater London Authority. 2011a. *The London Plan: Spatial Development Strategy for Greater London.* London: GLA.
Greater London Authority. 2011b. *Managing Risks and Increasing Resilience: The Mayor's Climate Change Adaptation Strategy.* London: GLA.
Greater London Authority. 2012. *London Strategic Flood Framework Version 2.0.* London: GLA.
Hill, D. 1996. *The Baked Apple: Metropolitan New York in the Greenhouse.* New York: Annals of the New York Academy of Sciences.
Holling, C. S., ed. 1978. *Adaptive Environmental Assessment and Management.* Chichester, U.K.: Wiley.
Holling, C. S., Gunderson, L. and S. Light, eds. 1995. *Barriers and Bridges to the Renewal of Ecosystems and Institutions.* New York: Columbia University Press.
Hoornweg, D., Freire, M., Lee, M. J., Bhada-Tata, P. and B. Yuen. 2011. *Cities and Climate Change: Responding to an Urgent Agenda.* Washington, DC: The World Bank.
Horton, R. M., Gornitz, V., Bader, D. A., Ruane, A. C., Goldberg, R. and C. Rosenzweig. 2011. "Climate hazard assessment for stakeholder adaptation planning in New York City." *Journal of Applied Meteorology and Climatology* 50 (11): 2247–66.
Kern, K. and H. Bulkeley. 2009. "Cities, Europeanization and multi-level governance: Governing climate change through transnational municipal networks." *Journal of Common Market Studies* 47 (2): 309–32.
Kingdon, J. 1995. *Agendas, Alternatives, and Public Policies.* 2nd ed. New York: Addison-Wesley.
Kline, D. 2001. "Positive feedback, lock-in and environmental policy." *Policy Science* 34 (1): 95–107.
Kousky, C. and S. Schneider. 2003. "Global climate policy: Will cities lead the way?" *Climate Policy* 3 (4): 359–72.
Lambright, W. H., Chagnon, S. A. and L. D. D. Harvey. 1996. "Urban reactions to the global warming issue: Agenda setting in Toronto and Chicago." *Climatic Change* 34 (3–4): 463–78.
Lavery, S. and B. Donovan. 2005. "Flood risk management in the Thames Estuary looking ahead 100 years." *Philosophical Transactions of the Royal Society A: Mathematical, Physical and Engineering Sciences* 363 (1831): 1455–74.
Livingstone, K. 2005. "Mayor brings together major cities to take lead on climate change." Press release. October 5. http://www.london.gov.uk/media/press_releases_mayoral/mayor-brings-together-major-cities-take-lead-climate-change.
London Assembly Environment Committee. 2005. *London under Threat? Flooding Risk in the Thames Gateway.* London: GLA.
Markard, J., Raven, R. and B. Truffer. 2012. "Sustainability transitions: An emerging field of research and its prospects." *Research Policy* 41 (6): 955–67.
McCarney, P., Blanco, H., Carmin, J. and M. Colley. 2011. "Cities and climate change." In *Climate Change and Cities: First Assessment Report of the Urban Climate Change Research Network*, edited by Rosenzweig, C., Solecki, W. D., Hammer, S. A. and S. Mehrotra, 249–69. Cambridge: Cambridge University Press.
McRobie, A., Spencer, T. and H. Gerritsen. 2005. "The big flood: North Sea storm surge." *Philosophical Transactions of the Royal Society A: Mathematical, Physical and Engineering Sciences* 363 (1831): 1263–70.
Meadowcroft, J. 2011. "Engaging with the politics of sustainability transitions." *Environmental Innovation and Societal Transitions* 1 (1): 70–75.
New York City Department of City Planning. 2011. *Vision 2020: New York City Comprehensive Waterfront Plan.* http://www.nyc.gov/html/dcp/pdf/cwp/vision2020_nyc_cwp.pdf.

New York City Office of the Mayor. 2006a. "Mayor Bloomberg discusses the importance of protecting the air New Yorkers breathe, the water New Yorkers drink and the land New Yorkers share." Press Release 125-06.

New York City Office of the Mayor. 2006b. "Mayor Bloomberg delivers sustainability challenges and goals for New York City through 2030." Press Release 432-06.

New York City Office of the Mayor. 2007. "Energy savings and reduction of greenhouse gas emissions in city buildings and operations." Executive Order 109.

New York City Office of the Mayor. 2008. "Mayor Bloomberg launches task force to adapt critical infrastructure to environmental effects of climate change." Press Release 308-08.

New York City Office of the Mayor. 2009. *Greener, Greater Buildings Plan (GGBP)*. http://www.nyc.gov/html/gbee/html/plan/plan.shtml.

New York City Office of the Mayor. 2010. "PlaNYC sustainable stormwater management plan progress report." http://www.nyc.gov/html/rabrc/downloads/pdf/sustainable_stormwater_mgmt_plan_progress_report_october_2010_final.pdf.

New York City Office of the Mayor. 2011. *PlaNYC 2030, 2011 update*. http://www.nyc.gov/html/planyc2030/html/theplan/the-plan.shtml.

New York City Office of the Mayor. 2013. *A Stronger, More Resilient New York*. http://www.nyc.gov/html/sirr/html/report/report.shtml.

New York City Panel on Climate Change. 2013. *Climate Risk Information 2013: Observations, Climate Change Projections, and Maps*. Edited by Rosenzweig, C. and W. Solecki. New York: Prepared for use by the City of New York Special Initiative on Rebuilding and Resiliency.

Pelling, M. 2011. *Adaptation to Climate Change: From Resilience to Transformation*. London: Routledge.

Repetto, R., ed. 2006. *Punctuated Equilibrium and the Dynamics of U.S. Environmental Policy*. New Haven, CT: Yale University Press.

Rosenzweig, C. and W. Solecki, eds. 2001. *Climate Change and a Global City: The Potential Consequences of Climate Variability and Change*. Metropolitan East Coast, Report for the U.S. Global Change Research Program. New York: Columbia Earth Institute.

Rosenzweig, C., Solecki, W., Hammer, S. A. and S. Mehrotra. 2010. "Cities lead the way in climate-change action." *Nature* 467 (7318): 909–11.

Rutland, T. and A. Aylett. 2008. "The work of policy: Actor networks, governmentality, and local action on climate change in Portland, Oregon." *Environment and Planning D* 26 (4): 627–46.

Sanders, E. 2005. "Rebuffing Bush, 132 mayors embrace Kyoto rules." *New York Times*, May 14.

Satterthwaite, D., Huq, S., Reid, H., Pelling, M. and P. R. Romero Lankao. 2008. *Adapting to Climate Change in Urban Areas: The Possibilities and Constraints in Low- and Middle-Income Nations*. London: IIED.

Scheffer, M. F. 2009. *Critical Transitions in Nature and Society*. Princeton, NJ: Princeton University Press.

Scheffer, M. F., Bascompte, J., Brock, W. A., Brovkin, V. S., Carpenter, S. R., Dakos, V., Held, H., et al. 2009. "Early-warning signals for critical transitions." *Nature* 461 (7260): 53–59.

Scheffer, M., Westley, F. and W. Brock. 2003. "Slow responses of societies to new problems: Causes and costs." *Ecosystems* 6: 493–502.

Solecki, W. 2012. "Moving toward urban sustainability: using lessons and legacies of the past." In *Metropolitan Sustainability: Understanding and Improving the Urban Environment*, edited by F. Zeman, 680–96. Woodhead Publishing Series in Energy 34. Sawston, Cambridge, UK: Woodhead Publishing.

Solecki, W. D. and S. Michaels. 1994. "Looking through the post-disaster policy window." *Environmental Management* 18 (4): 587–95.
Solecki W. D. and F. M. Shelley. 1996. "Pollution, political agendas, and policy windows: environmental policy on the eve of Silent Spring." *Environment and Planning C: Government and Policy* 14 (4): 451–68.
Solecki, W., Rosenzweig, C., Hammer, S. and S. Mehrotra. 2013. "The urbanization of climate change: Responding to a new global challenge." In *The Urban Transformation: Health, Shelter, and Climate Change*, edited by Sclar, E., Volvavka-Close, N. and P. Brown, 197–220. Abingdon, UK: Routledge.
Stern, N. 2006. *The Economics of Climate Change: The Stern Review*. Cambridge: Cambridge University Press.
Toly, N. J. 2008. "Transnational municipal networks in climate politics: From global governance to global politics." *Globalizations* 5 (3): 341–56.
U.K. Cabinet Office. 2013. *UK Civil Protection Lexicon*. Version 2.1.1 February 2013. https://www.gov.uk/government/publications/emergency-responder-interoperability-lexicon.
U.K. Climate Impacts Programme. 2002. *London's Warming: The Impacts of Climate Change for London: Summary Report*. London: London Climate Change Partnership.
U.K. Government. 2004. *London Thames Gateway Development Corporation*. Order 1642. http://www.legislation.gov.uk/uksi/2004/1642/contents/made.
Walker, B. and J. A. Meyers. 2004. "Thresholds in ecological and social–ecological systems: A developing database." *Ecology and Society* 9 (2): 3. http://www.ecologyandsociety.org/vol9/.
Walker, B. and D. Salt. 2012. *Resilience Practice: Building Capacity to Absorb Disturbance and Maintain Function*. Washington, DC: Island Press.
Ward, D. and O. Zunz. 1992. *The Landscape of Modernity: Essays on New York City, 1900–1940*. New York: Russell Sage Foundation.
Zahran, S., Brody, S. D., Vedlitz, A., Grover, H. and C. Miller. 2008. "Vulnerability and capacity: Explaining local commitment to climate-change policy." *Environment and Planning C: Government and Policy* 26 (3): 544–62.

4

Planning for Climate Change Adaptation in Urban Areas

JAN ERLING KLAUSEN, INGER-LISE SAGLIE, KNUT-BJØRN STOKKE,
AND MARTE WINSVOLD

Adaptation and the Pressure of Urbanization

Climate change adaptation is an emerging issue on urban policy agendas, in Norway as in other countries. The changing climate introduces a range of new factors to be taken into consideration in urban planning and development, by governments, developers, and others. Yet this issue is still incipient and seems to gain prominence only gradually and unevenly. This chapter provides a momentary impression of how the matter currently stands in five Norwegian cities. To what extent has climate change considerations made their way into public and private planning and development decisions?

In Norway, as in many other countries, a long-standing urbanization trend puts pressure on limited urban areas. The pressure is further increased by the current planning paradigm of "the compact city," evident in national planning guidelines for integrated land use and transport as well as in the adoption of local densification policies (Næss et al. 2009). This paradigm is a response to the challenge of creating a sustainable urban form, ensuring, among other things, reduction of carbon emissions from the transport sector (St. meld. nr. 31 1992–93). Municipal densification policy has resulted in the development of centrally located brown field areas, including former harbor areas and abandoned industrial areas, often situated along rivers and the sea. Such locations seem to be particularly popular among house buyers and profit-seeking developers. According to scenarios, however, Norwegian cities will be exposed to more frequent incidents of heavy rainfall and to sea level rise (Hanssen-Bauer 2009). This is in turn expected to cause landslides and flooding, in the form of storm floods as well as pluvial flooding in densely built areas with sealed surfaces. The most popular areas of development may well turn out to be among those that are most vulnerable to climate change effects.

A key concern of this chapter is how urban planning and development – and, as a consequence, climate change adaptation – takes place in a context marked by extensive interaction between public governments and the private sector. In Norway,

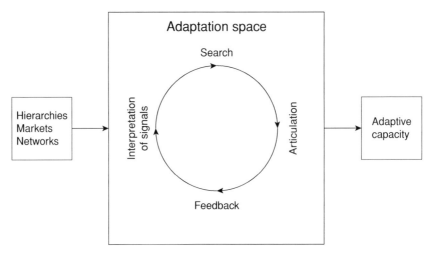

Figure 4.1. A conceptual model for adaptation, combining collective learning and governance mechanisms. *Source*: Winsvold et al. (2009).

planning authority pursuant to the Planning and Building Act (PBA) is delegated to the municipalities. The national government still exercises some authority, for instance, by issuing national guidelines for local planning. In later years, however, investors and private developers have gained an increasingly prominent position in urban planning and development. 80 to 90 percent of approved and legally binding plans are now developed by private actors. The public role is increasingly reactive instead of proactive. It is more often than not characterized by negotiations between private actors and public authorities (Falleth et al. 2010; Hansen and Guttu 1998; Holsen 2007). Private actors provide funding for much wanted urban development, which in turn gives these actors a strong bargain position vis-à-vis public authorities (Mäntysalo and Saglie 2010). The situation has been described as governance "in the shadow of hierarchy" (Sehested 2005). The local planning administration is not just a level in a hierarchical government system; it is a partner in local private–public negotiations in networks as well.

The Governance of Collective Learning

In the present context, we choose to conceptualize climate change adaptation as a collective learning process. This approach allows us to decompose the adaptation process into separate stages, each of which can be investigated empirically. For this purpose, we use the simplified model presented in Figure 4.1. It is based partially on the work of Berkhout et al. (2006). There is, of course, a vast literature that problematizes the linear, rationalistic assumptions that underlie this model.[1] Collective learning is

---

[1] It can indeed be contended that critique of rational models has been a major issue driving the development of the organizations literature through most of the twentieth century and to this day. For instance, some would contend

probably a lot less linear and well ordered than this model suggests, yet for our present purpose a simplified model will suffice.

The model is not particular to climate change adaptation. It is a generalized approach to describing how organizations respond to changes in their environments – in this case, information about climate change. In the model, four stages are identified. Ideally, the learning process should proceed through all four stages. First, it is assumed that the learning process starts when organizations receive and interpret *signals* about incipient changes in their environments. Second, the organization initiates a *search* for viable solutions as a response to the problems posed by the signals. Third, these solutions are *articulated* into specific, implementable measures. Fourth, there is a *feedback* stage. A precondition for learning is that the organization tries to assess how suitable the measures have been in terms of dealing with the problem at hand. Such insights should ideally affect the way the problem is approached in the future. In other words, learning should be conceived as a phenomenon of some endurance, as something that goes beyond the organization's ongoing efforts to resolve problems as they emerge. Levitt and March (1988), in this vein, suggest that learning can be described as *organizational* in a meaningful way only when the insights lead to changes in the organization's routines:

> Organizations are seen as learning by encoding inferences from history into routines that guide behavior.... Routines are independent of the individual actors who execute them and are capable of surviving considerable turnover in individual actors.
>
> (Levitt and March 1988, 320)

"Routines" can include forms, rules, procedures, conventions, strategies, and technologies through which the organization operates (Levitt and March 1988). This, then, is what distinguishes organizational learning from individual learning as well as from day-to-day problem solving. When insights change procedures, a degree of permanence and impersonality is established.

Our ultimate interest is, however, not in learning processes in single organizations but how learning processes contribute differently to climate change adaptation in a broader urban context. We are interested in how climate change adaptation is "learned" by the broad variety of actors who are in some way or another involved in urban planning and development – including developers, investors, planers, city officials, and so forth. These actors obviously operate in an environment that is organized and coordinated in ways that are a lot more complex and differentiated than what is the case inside of one single organization. Hierarchically organized governmental agencies interact with market operators such as developers and investors, sometimes in the context of self-regulating networks. Our research question is the following:

---

that organizations are predominantly social systems in which shared conceptions of reality are shaped through social interaction (see, e.g., Berger and Luckmann 1966). Other theorists see organizations as "loosely coupled" systems in which there is no pregiven causality between problem formulation and decisions (see, e.g., March and Olsen 1976).

Table 4.1. *Key features of hierarchies, markets, and networks*

	Hierarchy	Market	Network
Base of interaction	Authority and dominance	Exchange and competition	Cooperation and solidarity
Purpose	Consciously designed and controlled goals	Spontaneously created results	Consciously designed purposes or spontaneously created results
Guidance, control and evalutation	Top-down norms and standards, routines, supervision, inspection, intervention	Supply and demand, price mechanism, self-interest, profits and losses as evaluation, courts, invisible hand	Shared values, common problem analyses, consensus, loyalty, reciprocity, trust, informal evaluation – reputation
Theoretical basis	Weberian bureaucracy	Neo-institutional economics	Network theory

Source: Based on Bouckaert et al. (2010).

How do different modes and mechanisms of social coordination affect collective learning for urban climate change adaptation?

This analysis can provide a more nuanced understanding of the various coordinative processes that go into a city's adaptive capacity as a whole. Following a brief theoretical discussion of these issues, we turn to the empirical findings to establish whether the hierarchies, markets, and networks observed in our case study cities can indeed be said to have provided collective learning on climate change adaptation.

In the following, we draw on the long-standing distinction between hierarchies, markets, and networks as the three fundamental mechanisms of social coordination (Bouckaert et al. 2010; Jessop 1998, 2002; Williamson 1975). Main features of these mechanisms are presented and related to the model of collective learning. Key features of the three coordinative mechanisms are presented in Table 4.1.

Hierarchy is traditionally the predominant form of coordination in the public sector and formed a basic feature of Max Weber's ideal type model of public administration (Weber [1924] 1947). Hierarchies provide coordination through formalized structures of authority distribution and specialization (Simon 1957). Decisions are made on the top level of the organization and communicated downward to the appropriate subordinate levels for implementation. Depending on the principle of specialization, the subordinate levels are divided into departments or units with clearly defined tasks (Gulick 1937). The coordinative powers of the hierarchical model in a general sense rely on the exertion of authority by means of transforming of top-level decisions into unambiguous, codified directives that govern the activities of subordinate levels. The ability to do so is, logically speaking, a precondition for majority rule and democratic accountability (Dahl 1990; Hyland 1995; Olsen 1978).

Related to the model of collective learning, information (signals) on climate change impacts must be received and interpreted by the top level of the system. These decisions would need to be made in the form of binding directives to the appropriate subordinate levels. This could for instance be a decision made by a municipal council on the minimal height above sea level for new builds, which would have binding powers over the town's planning department. Moreover, the subordinate units would be expected to conduct reporting of their activities to the superior levels. In this sense, the feedback mechanism is identical with the fulfillment of the control function, which is a basic prerequisite for hierarchical systems.

Markets coordinate collective action based on the invisible hand of supply and demand (Smith [1776] 1957). If there is demand for a product or a service, it is assumed that the ensuing profit potential will ensure supply (Arrow 1974; Ouchi 1979, 1980). The market price is determined by the balance between supply and demand – increasing demand will hitch up prices, triggering an increasing supply that will push prices downward until a new equilibrium is reached (Samuelson and Nordhaus 1989). Price formation takes place independently of the actions of any single supplier or demander, and there are no formal links between the member organizations – rather, the resulting degree of social coordination is "the systemic result of partisan mutual adjustment of each unit in the market to its perceived environment" (Alexander 1995, 57, quoted by Bouckaert et al. 2010, 41).

In this approach, signals about climate change would have to be communicated through the price mechanism. For instance, if home buyers become aware of the risks associated with climate change, there would be an increasing demand for climate-resilient houses. Increasing demand would increase the prices of such houses, and this signal would be interpreted by private developers. The search for measures would probably result in the "articulation" of an increasing supply of climate-resilient houses, to ensure continued profitability. Sales figures and price changes in the housing market would provide the necessary feedback for determining a lasting change in company policy along these lines. Through markets, adaptation would take place even in the total absence of governmental policy, indeed without any direct communication between actors at all.

Networks coordinate actors who are operationally autonomous yet aware of interdependencies caused by shared challenges and opportunities (Rhodes 1997; Schmitter 2002). Networks can be more or less formalized and are, accordingly, often conceptualized as "self-regulating" (Osborne and Gaebler 1992; Van Kersbergen and Van Waarden 2004). Like markets, networks lack fixed authority structures, and their coordinating powers depend on voluntary agreements (Klijn and Koppenjan 2000; Kooiman 1993) based on trust and shared visions (Peters 2013). The necessity of relying on arguing and bargaining as basic forms of interaction reflects the nonhierarchical nature of networks, in the sense that there is no basis for centrist command and control. Yet networks at the same time allow direct interaction that is largely absent in the "invisible hand" model of market coordination.

As for the model of collective learning, the density of linkages between network actors, as well as the reliance on direct and horizontal modes of interaction between them, could potentially serve to increase the chances of receiving and interpreting signals about climate change impacts. Furthermore, networks may serve to disseminate information about climate change, thereby alleviating market failures due to imperfect information.

In the urban context, all three mechanisms of coordination are abundantly present. The implication is that cities are marked by highly varied preconditions for collective learning processes related to climate change adaptation. Hierarchies, markets, and networks are, however, generally speaking associated with particular strengths as well as weaknesses – in the words of Bop Jessop (1998, 32), "*markets, states and governance fail in different ways.*" Hierarchies provide the only opportunity for enforced coordination and may in combination with a specialized and competent administrative apparatus provide decision makers with extensive powers to ensure that specific concerns are taken into consideration (March and Simon 1958; Weber [1924] 1947). But the literature on "government failures" warns that the reliance on codified top-down steering may impede on the system's overall flexibility in terms of adjusting to local contingencies (Jessop 1998, 41) and may also be marked by problems related to overconformity, unresponsiveness, and inefficiency (Scott 1987, 298–317). Markets can be expected to provide a degree of simultaneous adaptation, due not least to investors' fears of capital depreciation. On the flip side of the coin, however, is the specter of market failure, in this context, notably, the way that market forces fail to address positive and negative externalities (Jessop 2002, 44) – in this case, potential losses associated with climate change. Potential governance failures, finally, include problems of accountability and biased power structures (Haus and Heinelt 2005, 21; Pierre and Peters 2000, 67–68) as well as political apathy, high transaction costs, or low expectations of joint benefits (Torfing 2007).

All in all, we expect the strengths and weaknesses of each mechanism of coordination to affect the cities' capabilities for collective learning correspondingly. We now turn to the empirical evidence.

The Five Case Studies

The empirical studies presented in this chapter were carried out by the authors. Five cities were selected for the study: Fredrikstad, Sarpsborg, Ålesund, Bodø, and Hammerfest (see Figure 4.2). All five cities are municipalities with elected city councils. The towns were selected based on a number of criteria. First, these towns

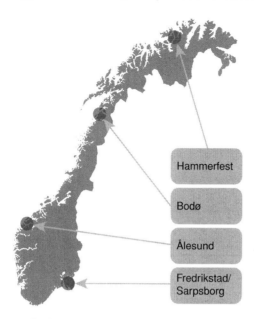

Figure 4.2. Geographical location of the five case study towns. *Source*: Authors.

have experienced a significant increase in population (between 0.5 and 1.3 percent per year since 1990). The ensuing demand for housing development was expected to provide fertile ground for identifying relevant projects for empirical study. Second, the five towns represent a variety of climate change vulnerabilities. Although all cities will be affected by increased precipitation, Fredrikstad and Sarpsborg are also especially prone to river flooding. With temperature increase, Hammerfest and Bodø are predicted to experience more winter flooding due to rain falling on frozen ground. Ålesund is exposed to landslides, and all cities except Sarpsborg will be affected by sea-level rise. Third, to avoid specific regional effects, the selected towns are located throughout the country. Fourth, case study towns were selected partially based on size, a factor that could potentially affect access to resources relevant to adaptation. The five towns vary substantially in this sense, from 73,000 inhabitants (Fredrikstad), to 52,000 (Sarpsborg), 46,000 (Bodø), 42,000 (Ålesund), and 10,000 (Hammerfest). Finally, it was seen as appropriate to include some towns that were involved in networking activities concerning climate change awareness, and some that were not. Fredrikstad and Sarpsborg are members of Cities of the Future, while Fredrikstad and Hammerfest participate in NORADAPT.[2] Ålesund and Bodø have not been involved in such networks.

---

[2] The Cities of the Future network is a collaboration between the government and the thirteen largest cities and towns in Norway to reduce greenhouse gas emissions and improve quality of life in cities. NORADAPT – Community Adaptation and Vulnerability in Norway – is a collaborative research project funded by the Research Council of Norway.

In-depth interviews with stakeholders from different groups involved in urban development were conducted in all five towns, including individual as well as group interviews. Altogether, twenty-three interviews with forty-four different persons were conducted in the period between September 2008 and February 2010. Owing to differences in municipal organization and different types of vulnerabilities to climate change, the informants in the five towns did not hold identical positions. In all towns, however, representatives from the municipal planning office and a sample of developers were interviewed. Other interviewees included various professions serving as consultants to the developers (including architects, landscape architects, and engineers), local council members, representatives of the port authority, representatives of the environment section in the municipality, representatives of the Housing Bank, and business investors. Apart from the municipal planners and developers, informants were chosen based on position or on tips from informants that had already been interviewed. Representatives of development projects were chosen based on the public lists of new regulation plans and in discussion with members of the municipal planning section, who identified especially vulnerable geographical areas.

Since climate change adaptation still was an emerging issue, we expected the issue to be scarcely mentioned in policy documents and plans. Interviews were therefore our primary data source. In addition, we conducted a document study of the most relevant municipal documents: the municipalities' General Plans, other municipal plans with reference to climate change (e.g., energy plans), and all regulation plans of a certain size submitted to the city council in the interview period. The analysis is based on the informants' account of how they perceived and dealt with the issue of climate change adaptation, as well as the references of climate change adaptation in the municipal documents.

Collective Learning on Adaptation in Practice

We now turn to the empirical observations in the five case study towns. The purpose is to analyze how the different mechanisms of social coordination affect collective learning for urban climate change adaptation – if at all. Looking at these three mechanisms of social coordination, we consider the four stages of adaptation response described earlier: signals, solutions, articulation, and feedback.

Hierarchy

In all the five case towns, we observed three hierarchical structures. First, although decentralized, the relationship between the state and the five municipalities clearly has many hierarchical features. The Ministry of the Environment is the highest planning authority, and national legislation and regulations structure the exertion of authority on the local level. Second, the relationship between different layers of the five municipal administrations is hierarchically organized, as a pyramid-shaped structure of command and control inside the administrative departments and with the elected politicians holding the top position. Third, the relationship between the municipality and developers could be described as hierarchical, because urban development takes place within a hierarchically structured planning procedure.

Interpretation of Signals

In general, few and weak signals indicating a need to adapt were sent or received within the hierarchical structures. The PBA of 2009 includes a paragraph that requires an assessment of vulnerabilities in all new building projects. Climate change is not, however, explicitly mentioned as a reason for increased vulnerability or risk. Governmental flood inundation maps based on estimates include historical data only, not projections. There is also an expert report on sea-level rise published by the government. This report, however, contains no specific instructions, only information on estimated sea levels at different intervals in fifty- and one-hundred-year time spans.

National and regional government bodies were overall found to send few if any signals to the municipalities. No signals were sent from local politicians or from higher levels in the municipal administration to the planners-in-charge either. Finally, the municipalities apparently sent few signals to the developers. Occasionally the local planning authorities asked developers to take steps, but these requests seemed irregular and coincidental. The municipalities differed in how consistently they demanded adaptation measures. In Ålesund, only one out of two neighboring waterfront projects was required to elevate the ground floor. Apparently, this had to do partially with the fact that the other project was a private development. For lack of national standards and certain, certifiable knowledge, the town authorities hesitated to impose such regulations on a private development.

The signals received by municipal officials are not transmitted through the planning hierarchy. Mostly they are picked up from the general public debate, from media coverage, or from professionals with whom the municipality cooperate – from the public and the private sectors. The municipalities do not receive signals by virtue of being hierarchies. There are no specific positions responsible for being attentive to such signals in particular or to accumulate information on the subject. In the absence of signal transmission within the hierarchical structure, climate change adaptation largely relies on the presence of dedicated individuals within the organizations. However, if these individuals succeed in institutionalizing their concern and, for example, commit the municipality to cooperation projects or networks, climate awareness may outlast these individuals' tenures. This had happened in one of our case municipalities: Hammerfest.

Search

Experiences with climate-related incidents in the past normally guide the search for solutions. One can differentiate between measures that aim at preserving and protecting existing ways of living and doing things and measures that involve new ways of thinking and living. The first category was predominant among our observations. "There is always a technical solution" was a common saying among stakeholders from all types of organizations, as exemplified by one developer in Bodø:

> Sea-level rise... that is something those guys that know something about the technical matters must fix.... The technical guys have to make sure that everything is correct and that this will stand for several hundred years.

In this case, the "technical guy" – the building engineer – had made sure the building was waterproof and able to withstand water pressed upward in the facade, a not uncommon procedure in exposed locations. He was also worried about the current sea level and made sure that the ground-level floor was raised one meter. This could be done without cutting off the top floor, which would have decreased the profitability of the project. Examples of other technical solutions were to build waterproof basements and increase the dimensions of pipelines for runoff water. There was, however, some thinking about more innovative solutions. Some informants talked about the possibility for open solutions for surface water management. In none of the municipalities were areas deregulated or withdrawn from development because of expected climate change.

Articulation

In a formal hierarchy a measure is undertaken when formal routines and guidelines are articulated and written down and are applied to all equal cases. In only one of our case municipalities, Hammerfest, were such general guidelines formulated. A collective understanding about what needed to be done seemed absent in most of the municipalities. In Ålesund, for example, the environmental executive recognized the necessity of climate change adaptation. He could not, however, successfully pass this awareness down through the municipal organization to articulate adaptation strategies. Articulated routines addressing climate change existed to a very varying extent. In most municipalities, developers were given instructions pertaining to individual projects. Only in Hammerfest were these prescriptions incorporated into their upcoming general municipal plan.

Feedback

A hierarchical governance mode seems to be unable to cope with a problem when no authoritative signals are sent from the higher levels. In our case, study municipalities, the hierarchical governance mode very much runs aground even before the first stage in the learning cycle: no clear and authoritative signals are sent from above leading to a variety of responses: bewilderment, inaction, but also in some cases local actions despite a lack of clear signals. An environmental officer remarked that he had "stopped missing the state," recognizing the need for the municipality to act on its own initiative.

Markets

While embedded in a highly formalized, hierarchical planning system, the profit motive is clearly a key engine of urban development. The demand for homes and commercial property as well as for public office space is met through market-based coordination. In this context, as noted, the price mechanism is the predominant signal bearer. Our observations indeed indicate that developers are highly attentive to changes in demand and prepared to alter their supply accordingly. Yet this mechanism seems to have come into play only to a very limited extent in the five towns included in this study.

Interpretation of Signals

As far as we observe, the five local markets have sent no signals at all to developers about willingness to pay for climate-proofed buildings and environments due to expected sea-level rise and flooding. On the contrary, exposed buildings along the rivers and coastlines are more popular than ever, and prices are still rising. The situation is the same regarding vulnerability for more extreme weather conditions like heavy wind and precipitation. The only adaptation measures taken are purely technical solutions, and some exposed municipalities exceed national standards. We even observed an example of a developer trying to sell exposure to harsh climate as a competitive advantage!

However, there was a certain degree of anticipation among developers to the effect that the customers may be increasingly preoccupied with security in the future. Moreover, the developers themselves were afraid of possible future costs caused by climate damage, and they knew the insurance companies had started considering this subject as well. Still, they did little, partly because they were thinking mostly in the short term. This can be illustrated by the case of one large housing developer in northern Norway. The company was split into two separate economic entities: one concerned with developing new housing, the other with long-term real estate management. When they considered buying into a waterfront project in a highly exposed location, the management part of the company was hesitant, but as most of the flats were already sold, this hesitation was not decisive. As they jokingly argued, "If we get trouble, the whole central area in the town will also be in trouble."

Some developers did receive signals about climate change adaptation. However, these signals were not communicated through the price mechanism but from elsewhere – notably, from professional networks, from other towns with tighter regulations, or from the media. One developer noted Al Gore's movie *An Inconvenient Truth* as well as the National Geographic TV channel as key sources of such signals.

Search

As expected, the developers were negative to adaptation measures that made their projects more expensive. One developer stated that it was fair enough if such measures were brought to their attention beforehand, providing all developers were treated equally. In some of the waterfront projects we observed, where adaptation measures were considered in an early stage in the process, increasing costs were not incurred. In one case, the costs were indeed reduced: in Hammerfest, elevation of the ground floor helped the developer avoid stirring up polluted substances in the ground, the disposal of which would otherwise have been costly. In most cases, the task of searching for solutions was given to professional consultants, and it would have been up to them to propose climate-adapted solutions.

Articulation

Climate change–related adjustments were observed in certain projects. These came about, as far as we understand, either as a result of municipal requirements or the consultant's professional norms and knowledge. However, such adjustments are constricted by economic considerations. If, for example, a ground floor has to be elevated to avoid flooding, while the maximum height of the building is unchanged, a whole floor of the building may be lost. This could jeopardize the profit margin of the project. The absolute limit for carrying out such climate change–related adjustments is that the project would break even.

However, we also found one example of a developer who wanted to turn adaptation and innovative solutions into a competitive advantage. One reason for this was that the developer intended to be a long-term investor. The company was founded with the purpose of developing a former shipyard, and the project would make up a large proportion of future urban development in Fredrikstad. Furthermore, the market clearly gave incentives to include adaptation measures. Households in the town have historically experienced flooding on a regular basis. Runoff water and fluvial flooding has for a long time created problems in Fredrikstad, and the municipality had already required private households to take certain steps to prevent flooding via the toilets. The municipality also had been made subject to liabilities by insurance companies seeking regress for their payments for damage due to flooding. The insurance companies claimed that the municipality had acted with negligence in allowing construction in some low-lying areas. Thus risk of flooding had already become an economic issue for households. This situation allowed one developer in Fredrikstad to argue that raising the building and making the underground parking garage water tight were costly but necessary:

I am not concerned about climate change with regard to this area. I feel we have control, when thinking about the expert knowledge we have consulted and the reports made. It is incredibly costly, but it is necessary to handle this. We have to be in the forefront about adaptation. It is a question of credibility and reputation. It is simply a necessity.

Feedback

The scarce and apparently incidental occurrences of market-driven climate change adaptation observed in the cases does not indicate any great extent of collective learning going full circle (compare the model). Yet the time perspective should be kept in mind: because the case study projects are still under way, it is in all probability too early for developers to assess how particular adaptation solutions worked, to adjust future practice.

Networks

Several networks were observed in the case study towns. As noted previously, some of the towns participated in the Cities of the Future network or in NORADAPT (see note 2). Furthermore, horizontal cooperation among the leaders of different municipal departments was widespread, not least in the form of project organization, which is usual in urban planning. This was observed in Bodø, which had set up a special multilevel organization to coordinate the work on the plan for the town center. Moreover, municipal staff members participated in professional networks as a part of their jobs, such as seminars on climate adaptation arranged by the county governor. They also participated in their respective professional networks, together with the developers' consultants – architects, engineers, landscape architects. There was, finally, network-like contact of both a formal and an informal nature throughout the regulation processes, involving the municipal administration and the developers. Such "networking" is quite normal in urban development (Falleth et al. 2010).

Interpretation of Signals

There are different networks operating for interpretation of signals and for search for solutions, respectively. A distinction can be made between networks inside of the municipality, networks that involve municipal and nonmunicipal actors, and networks involving several municipalities. We also find that professional networks play a key role. To what extent does interpretation of climate change signals take place within these networks? In the interaction between municipal staff and the developers, we have found no evidence of interpretation of climate change signals. As far as we could see, climate change adaptation had not been discussed in these forums. However, a lot of signals on the necessity of adaptation are received through professional networks and networks between towns, particularly through the national project Cities of the Future.

Hammerfest, for example, participated actively in cooperation projects to further the town's ambitions to become a pilot town in climate change adaptation. Through this cooperation they had access to downscaled climate change scenarios as well as to climate change researchers from whom they could get additional information. There are also some examples of signal interpretation in horizontal networks between different municipal departments that are responsible for adaptation-relevant tasks. The question is how these signals are passed on to the organization. Very often the signals stop with the single person participating in the network. An example of this is the environment executive in Ålesund, who participated in regional and national seminars on climate change adaptation but who did not succeed in passing on the knowledge and the awareness to the rest of the municipality. There are, however, examples of signals from networks having materialized and been articulated. Sarpsborg's participation in the network Cities of the Future, for instance, gave rise to the inclusion of certain specific goals in the municipal master plan. Moreover, consultants have been observed to disseminate signals as well as solutions in their contacts with developers and municipalities.

Search

The networks are obvious arenas for search for viable solutions. Participants look at each others' solutions, learn from each other, and discuss forms of "best practice." The municipalities imitate each other. This in some cases takes place in a formally organized context, such as through the network Cities of the Future. The developers also search for solutions in their networks. As they often carry out projects in several municipalities, they can transfer knowledge from one place to the next.

In our cases, the search for solutions was very much driven by professions, like architects. Professional networks, accordingly, were found to be of some importance. For instance, the large urban development project "Værste" in Fredrikstad did make plans for open solutions for surface water management. This probably came about as a consequence of the architects' professional networks. The development project was localized close to the consulting firm Cowi, and the professional awareness of adaptation issues was communicated through frequent interaction between the project staff and consultants from Cowi.

There are also examples of networks inside of municipalities that are activated in the search for solutions. One example is the regulation of the central waterfront in Ålesund. It was decided to elevate the minimum building permission from +2.2 meters (which is the current contour line) to between +2.7 and 3.2 meters. This decision was made following an internal dialogue between the harbor master, the chief administrative officer, the chief of preparedness, and the planning architect.

Articulation and Feedback

The networks analyzed in this study are not arenas for decision making. Policies and measures can, as a consequence, be articulated only through markets or hierarchies. The network participants need to take the knowledge they have obtained back to their own organizations to get it articulated. However, networks may potentially serve as important channels for feedback.

Governance, Learning, and Adaptation: An Engine Running Idle?

Our initial research question was the following: how do different mechanisms of social coordination affect collective learning for urban climate change adaptation?

A summary conclusion could be that the case studies have provided ample evidence on different mechanisms of social coordination, a more limited amount of evidence on climate change adaptation, and very little evidence on collective learning.

A key observation is that adaptation in urban planning in the five towns of this study seems to be incidental. Specific measures seldom, if ever, appear to be grounded in established routines, policies, or strategies with any degree of permanence or comprehensiveness. Specific adaptation measures seem on the contrary to have been introduced somewhat arbitrarily, in the context of each specific development project and for reasons that appear circumstantial – notably, the involvement of one or more individuals with particular awareness and knowledge about climate change issues. If true, these contentions carry implications for the occurrence of collective learning as well as for the impact of the mechanisms of coordination.

If specific adaptation measures are indeed incidental and circumstantial, they can by definition not be seen as results of collective learning. We initially defined collective learning as changes in routines, and if measures are not guided by such routines (to the extent that these exist at all), there is no reason to believe that collective learning has occurred. This could for instance explain why one project contains adaptation measures while a similar project in the same town does not. Any learning that takes place seems to be individual rather than collective.

There seems to have been few instances of the learning cycle going a full round – hence the meager results on "feedback" in the preceding section. This can of course partially be attributed to the fact that adaptation was an emerging issue, and practical "experiences" have yet to be made. We have furthermore studied how this emerging issue has impacted on projects about to be implemented and it is clearly unreasonable to expect "feedback" before the buildings have even been inhabited. Yet the analysis of coordination through hierarchies, markets, and networks does not really suggest that this is solely a matter of time. The relative scarceness of adaptation measures cannot be attributed to an absence of institutions and arenas for social coordination – on the contrary, urban planning and development takes place in a context where hierarchies, markets, *and* networks are abundant, and where they coexist and interplay even more markedly than in many other sectors of society. Yet very little is happening.

Our conclusion is that quite elaborate systems for social coordination certainly are in place but are largely running idle as far as adaptation is concerned. Perhaps this can be explained by considering the "engines" that could drive learning and change in each of them. If there is no political awareness and will, public hierarchies will not act. Correspondingly, if there is little hope for profit, private markets will fail to deliver adaptation. Networks appear to alleviate some of these effects, but mostly by stimulating *individual* learning. If collective learning is to be a mechanism for comprehensive and systematic climate change adaptation, these "engines" will have to be turned on a lot more often and with greater throttle than what appears to be the case today. This finding corresponds with findings in other countries, for example, Denmark (Wejs et al. 2014).

This lack of collective learning and little evidence of adaptation poses severe limitations to urban planning as a tool for adaptation, whether planning made by developers at the detailed level or on the strategic level of municipal planning. This is problematic, as more people live in urbanized areas and urban planning is potentially a very important element in climate change adaptation. National authorities play a vital role in this respect. They are in the position to send clear signals to local government to act and thus initiate organizational learning also when there is no dedicated individual to bring it in.

Acknowledgments

The chapter is based on research funded by the Research Council of Norway as part of the project "The Potentials of and Limits to Adaptation in Norway" (PLAN).

References

Alexander, E. R. 1995. *How Organizations Act Together: Interorganizational Coordination in Theory and Practice*. Amsterdam: OSA.
Arrow, Kenneth J. 1974. *The Limits of Organization*. New York: Norton.
Berger, P. and T. Luckmann. 1966. *The Social Construction of Reality: A Treatise in the Sociology of Knowledge*. London: Penguin Books.
Berkhout, F., Hertin, J. and D. M. Gann. 2006. "Learning to adapt: Organisational adaptation to climate change impacts." *Climatic Change* 78 (1): 135–56.
Bouckaert, G., Peters, B. G. and K. Verhoest. 2010. *The Coordination of Public Sector Organizations: Shifting Patterns of Public Management*. Houndsmill, UK: Palgrave Macmillan.
Dahl, R. 1990. *After the Revolution? Authority in a Good Society*. New Haven, CT: Yale University Press.
Falleth, E., Hanssen, G. S. and I. L. Saglie. 2010. "Challenges to democracy in market-oriented urban planning in Norway." *European Planning Studies* 18 (5): 737–54.
Gulick, L. 1937. "Notes on the theory of organization." In *Papers on the Science of Administration*, edited by Gulick, L. and L. F. Urwick, 1–45. New York: Institute of Public Administration.
Hansen, T. and J. Guttu. 1998. *Fra storskalautslipp til frislepp: beretning om Oslo kommunes utbyggingspolitikk 1960–1989* [From large scale release to unbounded release: The story about development policy in Oslo 1960–1989]. Oslo: Norwegian Building Research Institute.
Hanssen-Bauer, I., ed. 2009. *Klima i Norge 2100. Bakgrunnsmateriale til NOU. Klimatilpasning* [Norwegian climate in 2100: Note on background material for public assessment]. Oslo: Norskklimasenter – Departementenes servicesenter.
Haus, M. and H. Heinelt. 2005. "How to achieve governability at the local level?" In *Urban Governance and Democracy: Leadership and Community Involvement*, edited by Haus, M., Heinelt, H. and M. Stewart, 12–39. London: Routledge.
Holsen, T. 2007. "Fysisk planlegging mellom offentlig styring og privat initiativ" [Physical planning between public governance and private initiative]. In *Areal og eiendomsrett*, edited by O. Ravna, 103–128. Oslo: Universitetsforlaget.
Hyland, J. L. 1995. *Democratic Theory: The Philosophical Foundations*. Manchester: Manchester University Press.

Jessop, B. 1998. "The rise of governance and the risks of failure: The case of economic development." *International Social Science Journal* 50 (155): 29–45.
Jessop, B. 2002. *The Future of the Capitalist State*. London: Polity Press.
Kooiman, J., ed. 1993. *Modern Governance: New Government–Society Interactions*. London: Sage.
Levitt, B., and J. G. March. 1988. "Organizational learning." *Annual Review of Sociology* 14: 319–40.
Mäntysalo, R. and I.-L. Saglie. 2010. "Private influence preceding public involvement: Strategies for legitimizing preliminary partnership arrangements in urban housing planning in Norway and Finland." *Planning Theory and Practice* 11 (3): 317–38.
March, J. G. and J. P. Olsen. 1976. *Ambiguity and Choice in Organizations*. Bergen: Universitetsforlaget.
March, J. G. and H. Simon. 1958. *Organizations*. Cambridge: Blackwell.
Næss, P., Næss, T. and A. Strand. 2009. *The Challenge of Sustainable Mobility in Urban Planning and Development in Oslo Metropolitan Area*. TØI-rapport 1024/2009. Oslo: Institute of Transport Economics. [The] Norwegian Planning and Building Act of 14 June 1985, No 77.
Olsen, J. P. 1978. *Politisk organisering* [Political organization]. Oslo: Universitetsforlaget.
Osborne, D. and T. Gaebler. 1992. *Reinventing Government: How the Entrepreneurial Spirit Is Transforming the Public Sector*. Reading, MA: Addison-Wesley.
Ouchi, W. G. 1979. "A conceptual framework for the design of organizational control mechanisms." *Management Science* 25 (9): 833–48.
Ouchi, W. G. 1980. "Markets, bureaucracies, and clans." *Administrative Science Quarterly* 25 (1): 129–41.
Peters, B. G. 2013. "Toward policy coordination: Alternatives to hierarchy." *Policy and Politics* 41 (4): 569–84.
Pierre, J., and B. Guy Peters. 2000. *Governance, Politics and the State*. New York: St. Martin's Press.
Rhodes, R. A. W. 1997. *Understanding Governance: Policy Networks, Governance, Reflexivity and Accountability*. Maidenhead, UK: Open University Press.
Samuelson, P. A. and W. D. Nordhaus. 1989. *Economics*. New York: McGraw-Hill.
Schmitter, P. C. 2002. "Participation in governance arrangements: Is there any reason to expect it will achieve 'sustainable and innovative policies in a multilevel context'?" In *Participatory Governance: Political and Societal Implications*, edited by Grote, J. R. and B. Gbikpi, 51–69. Opladen: Leske + Budrich.
Scott, W. R. 1987. *Organizations: Rational, Natural and Open Systems*. Englewood-Cliffs, NJ: Prentice-Hall.
Sehested, K., ed. 2005. *Bypolitik: mellom hierarki og netværk* [Urban politics: Between hierarchy and network]. København: Akademisk Forlag.
Simon, H. A. 1957. *Models of Man*. New York: Wiley.
Smith, A. (1776) 1957. *Selections from the Wealth of Nations*. Edited by G. J. Stigler. New York: Appleton Century Crofts.
St.meld. nr. 31. (1992–1993). *Den regionale planleggingen og arealplanleggingen* [Regional and spatial planning]. Oslo: Miljøverndepartementet.
Torfing, J. 2007. "Introduction: Democratic network governance." In *Democratic Network Governance in Europe*, edited by Marcussen, M. and J. Torfing, 1–20. Basingstoke, UK: Palgrave Macmillan.
Van Kersbergen, K. and F. Van Waarden. 2004. "'Governance' as a bridge between disciplines: Cross-disciplinary inspiration regarding shifts in governance and problems of governability, accountability and legitimacy." *European Journal of Political Research* 43 (2): 143–71.

Weber, M. (1924) 1947. *The Theory of Social and Economic Organization*. Edited by Henderson, A. H. and T. Parsons. Glencoe, IL: Free Press.

Wejs, A., Harvold, K., Larsen, S. V. and I.-L. Saglie. 2014. "Legitimacy building under weak institutional settings: Climate change adaptation at the local level in Denmark and Norway." *Environmental Politics* 23 (3): 490–508.

Williamson, O. E. 1975. *Markets and Hierarchies, Analysis and Antitrust Implications: A Study in the Economics of Internal Organization*. New York: Free Press

Winsvold, M., Stokke, K. B., Klausen, J. E. and I.-L. Saglie. 2009. "Organizational learning and governance in adaptation in urban development." In *Adapting to Climate Change: Thresholds, Values, Governance*, edited by Adger, W. N., Lorenzoni, I. and K. O'Brien, 476–90. Cambridge: Cambridge University Press.

5

The Challenge of Governing Adaptation in Australia

STEVE WALLER AND JON BARNETT

Adaptation to climate change is work in progress. The focus of much adaptation research, policy, and practice is on reducing the "adaptation deficit" of the highly vulnerable in the developing world (Moser and Ekstrom 2010). Nevertheless, developed countries, although assumed to have a high adaptive capacity, are still evolving the institutions to govern and guide adaptation. The relative youth of these institutions, along with the novelty of adaptation as an object of governance, undoubtedly influences the lack of coordination and purpose noted in many developed country adaptation strategies. Indeed, if strategy is defined as the "pattern in a stream of decisions," then current adaptation activities in many developed countries cannot yet be regarded as strategic (Mintzberg 1978; Narayanan et al. 2011). This lack of strategic adaptation, even in developed countries, reinforces the view that having a high adaptive capacity is a necessary, but not sufficient determinant of action (Adger and Barnett 2009; Burch 2009; Tompkins et al. 2010).

This chapter describes and analyzes adaptation policy in Australia and shows how insights from the organizational change literature can serve as a useful bridge between current technical-rational approaches to decision making that assume policies develop in a linear manner, and alternative adaptive models for addressing social change processes. The following section argues that adaptation is best thought of as a process of change – as opposed to a set of definitive outcomes – and it describes four key elements of such a change process: scoping, readiness, implementation, and monitoring. These four elements provide a framework for assessing the nature of adaptation policy in Australia. The second section describes the content and direction of adaptation policy in Australian jurisdictions. The third section assesses the extent to which those policies that specifically sought to affect adaptation have indeed achieved meaningful change. The chapter concludes by considering the importance of creating shared meaning and purpose for adaptation as well as the dangers of force fitting adaptation processes into prevailing models of decision making and change management.

Adaptation: From Outcomes to Processes

For the purposes of this chapter, adaptation is understood as a higher-order concept that describes the processes of change that improve the anticipated fit of a socioecological system to its most probable climate futures. Its purpose is to reduce the system's vulnerability or enhance its resilience to the risks of climate change, whether natural or anthropogenic.

Because climate change is expected to continue for centuries, adaptation "success" has meaning only within specific temporal contexts: policies or actions previously considered successful may prove to be maladaptive in the light of later review (Barnett and O'Neill 2010; Moser and Ekstrom 2010). The paucity of metrics for adaptation adds to the complexity of measuring its success (Moser and Boykoff 2013). Indeed, the very logic of evaluating adaptation outcomes is difficult. Because uncertainty and complexity are inherent features of the adaptation process, ex ante success criteria are unlikely to be relevant. So, if success factors cannot meaningfully be set at the start of the process, then outcomes must be evaluated at some future point against counterfactual, alternative courses of action – a logically unsound and unhelpful practice. Clearly, then, in many cases, arguments for choosing one adaptation decision over another based on probable outcomes are flawed and add to dissatisfaction with an understanding of adaptation as merely a set of outcomes or outputs (Barnett and O'Neill 2010).

However, considering adaptation as a process of change reframes how success is evaluated, shifting emphasis to the nature and performance of the change process rather than to the more traditional foci of policy inputs, established process compliance, outputs, or outcomes (Head and Alford 2008). For example, embedding small-scale, "learning-by-doing" experiments in the adaptation process may inform more meaningful evaluation criteria. In other words, function takes precedence over form (Dovers and Hezri 2010) and successful adaptation becomes "adapting well" (Tompkins et al. 2010). This "adapting well" model considers the adaptation process to be nested inside a wider process of change. Adaptation, by definition, is a metaphor for change, and adaptation interventions planned without a simultaneous change process are unlikely to gain traction. Understanding adaptation as a process therefore entails thinking about decision-making processes.

It is important, first, to distinguish between decisions and policies. A decision can be defined as "a commitment to a certain course of action." Policies are subclasses of decisions and, following the definition of Pielke (2007, 27), are defined through their collective nature, that is, "a group's commitment to a certain course of action." The circular "Define-Assess-Options-Implement-Monitor-Feedback" sequence has typically been offered as a guide for adaptation decision making (Australian Government 2006; Füssel and Klein 2006; Willows and Connell 2003). Such models have been referred to as "technical-rational" models (Beck 2010) due to the nature of the assumptions that underlie them. Notably, they assume that the problem to be solved can be clearly defined; there are set objectives for policy; adequate information exists on cause and effect; and optimal problem-solving options can be identified and implemented cleanly (Head and Alford 2008; Lempert and Collins 2007).

However, adaptation policies most often occur in a complex decision-making environment, where these assumptions may not universally hold true (Rittel and Webber 1973). Given this, new models of decision making seem necessary (Funtowicz and Ravetz 1993), and there is considerable agreement in the adaptation literature about the characteristics of these new adaptive decision-making models: polycentric governance, broad-based participative decision making, and a social environment that permits experimentation and learning (Brunner and Steelman 2005; Head and Alford 2008; Park et al. 2012; Stafford-Smith et al. 2011). Nevertheless, it remains to be seen if these types of decision models are ultimately implementable, because even the most conservative of these constitutes a radical departure from the prevailing technical-rational approach to decision making (Dovers and Hezri 2010; Head and Alford 2008).

It is surprising that social change research, which strongly informs mitigation policies like energy efficiency, has not been much applied to the issue of adaptation. Instead, adaptation research (and policy) has tended to focus on discrete changes in sectors, policies, and to some degree, institutions. Of the various approaches to social change, we draw here on research on organizational change, as this appears to be an appropriate lens through which to examine adaptation decision making. Indeed, as Berkhout et al. (2006) suggest, "organisations, such as business firms, are the primary social units within which processes of adaptation will take place." In the sense that organizations are reflective of prevailing societal values and beliefs (Hofstede 2001), organizational and social changes can be complementary. In other words, organizational change is to a large degree about adaptation, and to the extent that it is not, the lessons from research on organizational change are germane to adaptation processes.

The organizational change literature draws on diverse theories of change that can be categorized according to their underlying ontology of change, for example, there are theories that see change as matters of evolution, teleology, dialectics, life cycles, cognition, culture, and complex systems theory (Hudson 2000; Kezar 2001). This eclecticism of approach gives greater scope in researching and explaining organizational change (van der Ven and Poole 1995). Nevertheless, organizational change research has identified certain activities that increase the probability of successful organizational change. These activities can be grouped according to which part of the change process they support. In what follows, we group change activities into four nonsequential categories: scoping, readiness, implementation, and monitoring. Though their relative importance and sequencing is contested, general agreement exists on each activity's characteristics (Kotter 1995; Whelan-Berry and Somerville 2010).

Scoping

Scoping entails creating a shared understanding of the problem and its main characteristics and has three main components. First, *diagnosis* involves the comparison of the "actual" outcomes (e.g., modeled projections of climate change) with the expected outcomes (typically business as usual). The latter include current social goals, objectives, and expectations. Second, *urgency* is needed to sustain long-term change (Kotter 1995). It is typically generated in a top-down manner and requires a consistent problem focus. Finally, *vision'* focuses on outlining the desired state after change, providing a paradigm-breaking decision environment that facilitates the reframing of resource allocation patterns such that implementation follows.

Readiness

Readiness comes through building the right leadership coalition and assessment/preparation of the change subjects (organizational or social) for the change process. In particular, *leadership* should possess the requisite authority, stature, and connections to formal and informal networks necessary to enable change. The prevailing *culture*, the "fundamental assumptions; values; behavioural norms; expectations and larger patterns of behaviour," need to align with the proposed change (O'Reilly et al. 1991, 491). Similarly, congruence of the change approach with prevailing political cultures or governmentalities is critical both to gain leadership commitment and to minimize resistance to change (Gerkhardt et al. 2008; Sminia and Van Nistelrooij 2006).

Implementation

Implementation can be subdivided into three activities. First, *planning* involves managing potential barriers to change, for example, by coopting existing institutions and policies to reduce resistance (Dovers 2009; Whelan-Berry and Somerville 2010). Second, *implementation* focuses on sustaining motivation and commitment, building momentum, and remaining flexible to changing circumstances. Examples may be acquiring small wins through pilot projects. The third activity, governance, focuses on maintaining the flexibility and legitimacy of the change process and ensuring that its goals and resource allocations remain transparent.

Monitoring

Monitoring captures, converts, and disseminates the tacit, explicit, and cultural knowledge acquired during the change process (Choo 2002). In the case of adaptation, monitoring is difficult, given the absence of suitable indicators (Barnett et al. 2008).

We propose that these four groups of activities compose a framework through which to assess the effectiveness of adaptation processes (i.e., do they effectively scope the problem, build readiness, implement change, and monitor progress?). Before applying this framework to assess Australian adaptation policies and analyzing their effectiveness, a description of these policies is necessary.

Adaptation in the Australian System of Government

This section describes the Australian system of government and the evolution of adaptation policy within it, focusing in particular on developments at the national and state levels. Australia is bound together under a Constitution defining the relationship between the federal or commonwealth government (the "Australian government") and the six states (New South Wales, Queensland, South Australia, Tasmania, Victoria, and Western Australia) and two territories (the Northern and Australian Capital Territories). The latter are autonomous in matters including education, primary industries (including agriculture and natural resources), planning, health, justice, and environment. In all states and territories, the Australian government has direct powers over matters including defense, taxation and immigration, corporations, and external affairs. The third tier in the governmental framework of Australia includes around 600 local governments, deriving their legal foundation from various state and territory acts of parliament (Pillora 2010).

The principal mechanism of cooperation between the Australian and state levels of government is the irregular convocation of the first ministers, known as the Council of Australian Governments (COAG), meeting to "initiate, develop and monitor the implementation of policy reforms of national significance which require cooperative action by Australian governments" (COAG 2011). In specific policy areas, COAG is assisted by Ministerial Councils. In 2011, these were reformed to promote greater effectiveness. The new system subdivides COAG councils into Standing Councils (for ongoing issues of national significance), Select Councils (for time-limited issues of reform), and Legislative and Governance fora.

While Australia has a history of climate policy engagement from the late 1980s, the first national climate change strategies to canvas adaptation were the post–Rio Summit, the National Greenhouse Response Strategy (NGRS), and the National Greenhouse Strategy (NGS) (Australian Government 1992, 1998). In identifying adaptation as a discrete activity, facilitated by discrete policies, the latter created a precedent that persists to this day; adaptation, like mitigation, is something apart from other policies and arms of government. The NGS set other precedents too: first, the "need for action to be informed by research" is still a dominant focus of policy, no doubt because of the ongoing perception that mitigation and adaptation policies carry with them very significant economic and political risks (Australian Government 1998, viii). Second, the NGS pushed a "partnership" approach with states and territories, for instance, "[to] recognise the important role which partnerships between governments, industry and the community play in an effective national response to climate change" (3). Finally, it singled out priority sectors and regions for research, including efficient transport, sustainable urban planning, industrial wastes, land use, development of sinks and adaptation.

In the May 2004 budget, the A$14 million, four-year National Climate Change Adaptation Program (NCCAP) became the first coordinated national adaptation policy (Australian Government 2005). This was part of the Australian government's larger A$1.8 billion climate change response effort. The NCCAP duly provided seed funding for a preliminary national climate change risk and vulnerability assessment, which has become known as the Allen Report (Allen Consulting Group 2005; Westerhoff et al. 2010). The risk-based, sectoral and econocentric approach to adaptation adopted by the Allen Report has become a common frame for Australian adaptation policy, although it has since been criticized for adopting a simplistic and confusing conceptual framework for vulnerability (Nelson et al. 2010). In 2006, this risk management framing of adaptation was reinforced by the release of the influential *Climate Change Impacts and Risk Management: A Guide for Business and Government* (Australian Government 2006).

A key narrative in the Allen Report was the barrier to risk-based decision making created by a perceived information deficit about adaptation. The report made three important recommendations to address this: the development of a national adaptation framework, the creation of a national adaptation research institution, and tighter coordination of the Commonwealth Scientific and Industrial Research Organisation (CSIRO) and BOM (Bureau of Meteorology), which provide the bulk of climate change science in Australia.

The first recommendation, for a National Climate Change Adaptation Framework (NCCAF), came to fruition when finally endorsed by COAG in 2007 (COAG 2007). It was a five- to seven-year plan for increased adaptation research to reduce climate vulnerability and build adaptive capacity through the promotion of research, impact assessment, and strategic actions in priority sectors and regions. At the end of 2014, NCCAF still awaits the development and agreement of a separate action plan by COAG before it can be implemented, a task that now seems to have been largely abandoned after seven years of intermittent activity.

The second recommendation, to establish the National Climate Change Adaptation Research Facility (NCCARF), received an initial A$20 million funding under the Climate Change Adaptation Program (CCAP) announced in the 2007–8 budget (Australian Government 2007), after strong recommendation in the NCCAF. The preparation of National Adaptation Research Plans (NARPs) for priority sectors was a clear goal of the NCCARF, aiming to "identify critical gaps in the information needed by [vulnerable] sectoral decision-makers, and set national research priorities." This would be facilitated by cross-disciplinary collaboration through moderated Adaptation Research Networks. At the end of 2011, NARPs had been agreed for all eight sectors with an additional NARP added for indigenous adaptation issues (NCCARF 2011). In 2014, NCCARF has completed the research programs developed in the nine NARPs and even updated some. It is well into the extension and communication phase of its program, albeit with a much reduced funding allocation from the Australian government (NCCARF 2014).

The third recommendation, for better institutional coordination, was addressed through CCAP's allocation of A$43.6 million to establish a Climate Adaptation Flagship located within CSIRO. Much of the research effort of the Flagship has been in the agricultural sector where it has a significant presence, but more generally it has also contributed significantly to increasing adaptation awareness throughout government. Its emphasis on improved climate projections, impact assessments, and adaptation technologies differentiates it from NCCARF, which aims to build adaptive capacity through the coordination of targeted, multidisciplinary work programs (PMSEIC 2007). The addition of these two new adaptation research institutions to a burgeoning climate science and modelling establishment inevitably led to coordination and prioritization challenges. In 2009, the Australian government created an institutional framework, chaired by its chief scientist, to coordinate its adaptation research (Australian Government 2009).

Also in 2009, the Australian government released a position paper called *Adapting to Climate Change* (Australian Government 2010). More a statement of direction than actual policy, this paper draws on climate science to illustrate that the risks of climate change are "serious and pervasive" and, despite residual uncertainty, "the time to act is now." It reiterates the recurring theme of "partnership" between governments (through COAG), business, and society and clarifies its role in providing a "strong flexible economy," promoting national policy consistency, managing its own assets and programs, and providing public good adaptation information.

The "partnership" model has been the dominant approach to adaptation policy since the NGRS. It emphasizes COAG as the main platform on which partnerships between Australian governments happen and confirms this should also be the case for adaptation (Australian Government 2010). However, COAG has proved to be a somewhat ineffective platform for adaptation. After seven years, the action plan for the NCCAF is still to be agreed by COAG jurisdictions after years of work by COAG Climate Change Working Groups.

Nevertheless, the Australian government has funded several of its recommendations, for example, a digital elevation model of the Australian coastline and a revised *Rainfall and Run-off Handbook*. Furthermore, many sectoral/regional adaptation plans developed by Ministerial Councils have been endorsed by COAG, including for coastal zone management (2006), forestry (2009), agriculture (2006), biodiversity

(2004), emergency management (2009), and tourism (2008). However, agreement does not guarantee action, with the extent of plan implementation remaining highly conditional on the budgetary circumstances of each jurisdiction (COAG 2007, 2011).

At regional and subregional scales, somewhat distinct climate and adaptation research programs have emerged, driven by observed changes in regional climate and the slow progress of national climate science in providing climate knowledge relevant to policy making. The earliest of these, the Indian Ocean Climate Initiative (IOCI) collaboration between the WA government, CSIRO, and BoM, was formed to investigate the climatic drivers behind an observed step change in rainfall and runoff to water storages in the state's southwest that began in the mid-1970s (Government of Western Australia 2008, 2012).

Similar collaboration models have since been used for adaptation-related research across Australia, for example, in the Western Australian Marine Science Initiative (WAMSI) and the South Eastern Australia Climate Initiative (SEACI). A number of state-based research institutions have also been created to provide a local focus to adaptation. For example, the Queensland Climate Change Centre of Excellence (QCCCE) and the Victorian Climate Change Adaptation Research Centre (VCCAR) coordinate the adaptation research priorities of their respective governments through collaborative partnership models (Government of Queensland 2007; VCCAR 2012).

At a subregional scale, neighboring local governments have sometimes pooled their efforts to address their common adaptation concerns. For example, the Sydney Coastal Councils Group (a group of fifteen New South Wales coastal councils) has hosted the CCAP-funded Systems Approach to Regional Climate Change Adaptation Strategies in Metropolises collaboration that investigated vulnerability, impact assessment, and adaptive capacity (Smith et al. 2008), and the South East Councils Climate Change Alliance works collectively to develop a common approach to adaptation around Westernport Bay in Victoria.

Adaptation policies have also developed in a somewhat ad hoc fashion at the state level. Some states have high-level climate change strategies or legislation that have historically incorporated adaptation (Government of Victoria 2005; Government of Western Australia 2004; Government of Queensland 2007, 2009). In Victoria, for example, the Climate Change Act 2010 mandates that some statutory adaptation-related decisions consider climate change but curiously omits decisions under the Environment and Planning Act (1987). Similarly, the South Australian government, under its Climate Change and Emissions Reporting Act 2007, developed an innovative draft Climate Change Adaptation Framework in 2010 that emphasized regional vulnerability assessments (IVA) and a participative, local approach to adaptation decision making. Unfortunately, the adaptation strategy consultation papers in Victoria and Queensland have not survived various electoral changes, and there appears to be little appetite, or capacity, in governments for the resurrection of state- or national-scale adaptation plans.

At lower jurisdictional levels, policies have responded to the planning demands of rapidly populating coastal areas. Most jurisdictions have implemented coastal policies that require the consideration of sea-level rise (SLR), even though there is no national consistency in SLR benchmarks (Gurran et al. 2011). Sea-level rise has been the subject of court proceedings in the states of Victoria, South Australia, and New South Wales (NSW) (de Wit and Webb 2010; Gurran et al. 2011). In Victoria, challenges in the Victorian Civil and Administrative Tribunal have upheld a requirement for local councils to consider SLR in making development decisions (de Wit and Webb 2010), while in NSW the liability of local councils for SLR development approvals has been limited by courts, provided the decisions were made in good faith (Gurran et al. 2011).

Assessing the Effectiveness of Australian Adaptation Policies

If adaptation is viewed as a process of change, then Australian adaptation policy has, to date, concentrated on some scoping activities and largely ignored readiness building, implementation, and monitoring.[1] We now examine how this neglect of the entire change process has limited the extent and effectiveness of adaptation policy making in Australia.

Scoping the Challenge

Australian adaptation policy implemented thus far has almost exclusively sought to diagnose the climate change problem, in the form of substantial investments in climate science and modeling and risk and vulnerability assessment. Because the purpose of diagnosis is to provide a rationale for change, or to monitor it, the development of enhanced research and assessment capability is a legitimate undertaking, provided it informs the development of other policies that encourage adaptive change. However, diagnosis has, so far, tended to be the end point of adaptation policy development, which has implications for the future of adaptation policy in Australia.

Gaining insight into the reasons for this necessitates examining the nature of the adaptation science–policy relationship in Australia, which for the Australian government is seen in terms of what Beck (2010) calls a "linear chain of explanation" (LCE). For example, the Australian government considers that "climate change science is entering a new phase of complexity as decision makers and the public demand greater insight into likely impacts and the effort required for mitigation and adaptation" (Australian Government 2009). Thus the intention of the Australian government is to "establish the facts," considering this to be a necessary precondition for policy formulation.

---

[1] Having said this, some implementation of policies with multiple objectives, including adaptation, has occurred. Indeed, in the Murray-Darling Basin and the Great Barrier Reef, these policies, if successful, could result in transformative changes (Lane and Robinson 2009). These multiobjective policies are not further discussed because of the difficulty of attributing the degree of change to each driver.

This notion that policy develops in a linear manner after the facts are established rests on three assumptions. First, it assumes that the "policy relevance" of climate information is improved by more research that supposedly reduces uncertainty about climate impacts. Policy relevance is ambiguous but in the Australian context is code for risk assessments of high enough resolution to provide an unequivocal economic justification for action (COAG 2007, 3). This justification is necessary because of the foundational framing of Australian climate policy as, for example, "the pursuit of greenhouse action consistent with equity and cost-effectiveness and with multiple benefits" (Australian Government 1998, viii), so that adaptation becomes "a mechanism to manage risks, adjust economic activity to reduce vulnerability and to improve business certainty" (COAG 2007, 3).

Various types of cost-benefit analysis (CBA) are used to support decisions for adaptation in Australia, in common with most other liberal economies (Pidgeon and Butler 2009, 682). However, CBA has problems in dealing with uncertainty and surprises and in deriving and incorporating nonmonetary values and preferences (Hallegatte 2009). Relying on CBA can leave adaptation policy open to challenge and uncertainty. Thus policy makers default to economic no-regrets policies or delay policy making (Productivity Commission 2011).

The second assumption of the LCE is that "more and better science will solve political disagreements" (Beck 2010, 2). However, existing analytic tools do not adequately take into account the different values placed on adaptation outcomes by individuals and groups that underlie most disagreements about climate change (Hulme 2009; O'Brien and Wolf 2010). The different and at times incommensurate values of individuals and groups across society often create dispute over the prioritization of adaptation response options (Adger et al. 2009; Pidgeon and Butler 2009; Pielke et al. 2007). Reconciliation of plural views and uncertainty may cross electoral cycles and cost disproportionate amounts of political capital (Head and Alford 2008; O'Brien and Wolf 2010). Thus complexity and negative risk–reward outcomes for politicians certainly contribute to the observed lack of policy champions. Reducing uncertainty about impacts through greater investments in climate change science does nothing to help address this problem of value trade-offs and may help to justify avoiding them, meaning that responses that are currently politically tractable may become intractable later when the window for decision making narrows.

The final assumption behind the LCE, that separating science from politics makes for rational policy making, has probably never held true in Australia. Climate science was politicized in early adaptation policies, such as the NGS, through the selection and implementation of important economic sectors for adaptation research over regional or governance/institutional approaches (Australian Government 1998; Beck 2010; COAG 2007; Pielke 2007). Politics, conversely, has been "scientized" through the normative preconditioning to acquire scientific certainty before adaptation policy making (Australian Government 2009).

So, the expectation that adaptation policies will be economically justified and scientifically certain has resulted in a policy preference for investing a lot in in science, and not much else. The resultant policy inertia risks creating a science-centered path dependency where the risk taking and innovation required to address complex problems are inhibited (Auld et al. 2007; Burch 2009). Change requires more than diagnosis of the problem.

The organizational change literature also specifies vision as a "critical early step in the change process." It stresses the need for a clear and compelling picture of an ultimate future to specify how "particular aspects, characteristics, or outcomes of the organisation" will look postchange (Gerkhardt et al. 2008; Whelan Berry and Somerville 2010). There are two parts to visioning: top-down creation of desired end states and bottom-up acceptance of these (Whelan Berry and Somerville 2010, 187).

Australian adaptation policy provides few such vision statements. Typically, a picture of the "desired state" has to be inferred from statements of mission or purpose (e.g., Australian Government 1998, 3; Government of Queensland 2007, 9). The 2010 position paper by the Australian government proposed a vision for adaptation policy that substantially aligns it with neoliberal governmentality (Australian Government 2010). In this, adaptation policy making was cast as an economic rather than social activity, invoking the primacy of markets (e.g., water trading) as solutions. The role of government in these markets is seen as regulating for efficient operation and covering market failures. Indeed, the neoliberal goal of smaller government remains possible because adaptation is framed as a risk that is manageable by existing institutions informed by conventional science and devolved to individuals and groups, enabling "governing at a distance" (Lockwood and Davidson 2010). Nevertheless, consistent with Australia's hybridized neoliberalism (O'Neill and Argent 2005), it also acknowledges the need to retain a strong social safety net "to assist those who may otherwise have difficulty in adapting" (Australian Government 2010, 9).

In Australia, there is little beyond the vague description of end states as being systems of efficient responses through markets. There has been even less effort and building acceptance of this across society (or, indeed, any sector of it). Nor has there been much effort to build a sense of urgency about the need for adaptation. So, although the Queensland government seems convinced of "the overwhelming need to take urgent action against climate change" (Government of Queensland 2007, 2) and the Australian government (2010, 6) claims "the time to start acting is now," this rhetoric of urgency has not been accompanied by policies that aim to spread this message, nor has it been reinforced by political or community leaders, which the organizational change literature suggests are critical to sustaining long-term change processes (Kotter 1995).

Leaders can promote urgency by the consistent presentation of a troubling message (Kotter 1995). This does not mean creating fear but consistently highlighting the reasons for change (Grothmann and Patt 2005). The high vulnerability of Australia to climate change is well described (Australian Government 2010; Garnaut 2008), but little used by leaders in the media to arouse urgency for adaptation. In contrast, the ferocious domestic debate about mitigation policy is an example of urgency being derived from the constant attention of political and financial elites engaged in a high-stakes economic contest over climate policies (Stevenson 2009).

Sustaining urgency for long-term change is best done through the exploitation of policy windows, such as extreme events (Howlett 1998). These provide opportunities to reinforce climate risks and promote adaptive policies. Extreme events, such as the 2011 Queensland floods, open windows to address policy deficiencies and provide social permission for experiments. In the preceding example, the Queensland government was able to discuss alternate climate risk mitigation strategies, even including the role of the state as insurer of last resort (Government of Queensland 2011).

In short, the majority of Australian adaptation policy pays little attention to developing a vision and sense of urgency. Where present, vision has been found to be economic in nature, with little attention to the cultural, social, or cognitive aspects that promote a shared understanding of adaptation, while a profound disengagement of leadership from adaptation is a major contributor to lack of urgency.

Building Readiness

There is not a great deal that can be said about attempts to build readiness for adaptation in Australian policy, because there have been few if any such attempts. The previous section drew attention to the neoliberal framing of Australian adaptation policy. This frame is congruent with what Mercer et al. (2007, 276) suggests is the "prevailing policy regime" in Australia – that of "statist developmentalism" or government-led promotion of the extensive use and development of natural capital. This regime, common to many settler-capitalist societies, resists the development of policies that challenge traditional patterns of natural resource exploitation. Resistance by state policy makers to the recent Murray Darling Basin Plan proposals to increase environmental flows may, in part, be expressive of this aspect of Australian culture.

So, there is a sort of cultural path dependency that limits adaptive capacity in Australia, dictating what adaptation responses are likely to be seriously considered (Burch and Robinson 2007, 314). In Australian culture, what appears capable of serious consideration are adaptation policies that do not compromise existing patterns of resource use or development and that are biased toward economic, rather than social, activities (e.g., giving primacy to sectors rather than regions) and, as previously discussed, align with the prevailing neoliberal rationality.

An important question, then, is how adaptation policy might better fit with the prevailing culture. Talking about mitigation policy, Stevenson (2009, 171) is pessimistic about political or cultural change, noting that "political economic norms . . . are inextricably linked to the national episteme [statist developmentalism]." Mercer et al. (2007, 284) suggests, for water policy, a "return to governing and end the non-interventionist neo-liberal model of governance... radical extension of the time frames addressed by decisions, the adoption of long term thinking, and an end to the tyranny of small decisions." Thus developing effective adaptation policy seems to be at odds with the prevailing neoliberal and economically rationalist policy regimes in Australia. It follows, then, that building readiness will require leadership, time, and concerted changes to both social and political cultures.

Nevertheless, to the extent that neoliberal approaches seem to devolve responsibility for risk management to lower levels of government, and to civil society, they do offer opportunities for local and nonstate actors to increase agency in adaptation and build readiness for change. However, this potential remains unrealized. At present, this devolution of responsibility seems merely to carry with it an increased risk of liability (McDonald 2007), and in most cases the strongest policy imperative remains promoting flexibility in markets.

Conclusions

The analysis of Australian adaptation policy illustrates that the combination of relative wealth, well-established institutions, and high levels of skills and capabilities found in developed nations does not necessarily equate to adaptation action. Despite these advantages, and the accompanying projections of high climate vulnerability, it is the lack of attention by Australian governments at all levels to the cultural and processual aspects of adaptive change that appears to be primarily responsible for the observed slow and ad hoc adaptation response.

Australian adaptation policies are abstract and focused on acquiring sufficiently detailed climate and impact knowledge to economically justify controlling risks. In this sense, current policies attempt to force fit the adaptation process to prevailing cultural and political rationalities. However, adaptation is, at heart, a social activity, and neglecting opportunities to create a shared meaning and purpose for adaptation in organizations and societies will inevitably lead to policy failure. Increasing the efficacy of the adaptation process is possible through attending to activities that favor the social acceptance of change, including the creation and acceptance of vision, sustaining urgency and fitting policy to prevailing culture. It is through greater attention to the process of adaptive change that Australia's high adaptive capacity will be realized in meaningful outcomes.

References

Adger, W. N. and J. Barnett. 2009. "Four reasons for concern about adaptation to climate change." *Environment and Planning A* 41 (12): 2800–2805.

Adger, N. W., Lorenzoni, I. and K. O'Brien, eds. 2009. *Adapting to Climate Change: Thresholds, Values, Governance.* Cambridge: Cambridge University Press.

Allen Consulting Group. 2005. Climate Change Risk and Vulnerability: Promoting an Efficient Adaptation Response in Australia. Report for the Australian Greenhouse Office, Department of Environment and Heritage.

Auld, G., Bernstein, S., Cashore, B. and K. Levin. 2007. "Playing it forward: Path dependency, progressive incrementalism, and the 'super wicked' problem of global climate change." *Conference Papers – International Studies Association*, 1–25.

Australian Government. 1992. *National Greenhouse Response Strategy*. Canberra: Australian Govt. Pub. Service.

Australian Government. 1998. *National Greenhouse Strategy: Strategic Framework for Advancing Australia's Greenhouse Response*. Canberra: Australian Greenhouse Office.

Australian Government. 2005. "The National Climate Change Adaptation Program." In Department of Environment and Heritage (ed.). Canberra: Australian Greenhouse Office.

Australian Government. 2006. "Climate change impacts and risk management: A guide for business and government." In *Australian Greenhouse Office*, Department of Environment and Heritage, Canberra: Australian Greenhouse Office.

Australian Government. 2007. *Climate Change Adaptation Program*. Canberra: Department of Climate Change and Energy Efficiency.

Australian Government. 2009. *Australian Climate Change Science: A National Framework*. Canberra: Department of Climate Change and Energy Efficiency.

Australian Government. 2010. *Adapting to Climate: Change in Australia An Australian Government Position Paper*. Canberra: Commonwealth of Australia.

Barnett, J. and S. O'Neill. 2010. "Maladaptation." *Global Environmental Change* 20 (2): 211–13.

Barnett, J., Lambert, S. and I. Fry. 2008. "The hazards of indicators: Insights from the Environmental Vulnerability Index." *Annals of the Association of American Geographers* 98 (1): 102–19.

Beck, S. 2010. "Moving beyond the linear model of expertise? IPCC and the test of adaptation." *Regional Environmental Change* 11 (2): 1–10.

Berkhout, F., Hertin, J. and D. Gann. 2006. "Learning to adapt: Organisational adaptation to climate change impacts." *Climatic Change* 78 (1): 135–56.

Brunner, R. and T. Steelman. 2005. *Toward Adaptive Governance*. New York: Columbia University Press.

Burch, S. 2009. "Transforming barriers into enablers of action on climate change: Insights from three municipal case studies in British Columbia, Canada." *Global Environmental Change* 20 (2): 287–97.

Burch, S. and J. Robinson. 2007. "A framework for explaining the links between capacity and action in response to global climate change." *Climate Policy* 7 (4): 304–16.

Choo, C. W. 2002. "Sensemaking, knowledge creation, and decision making." In *The Strategic Management of Intellectual Capital and Organizational Knowledge*, edited by Choo, C. W. and N. Bontis, 79–88. Oxford: Oxford University Press.

COAG. 2007. *National Climate Change Adaptation Framework*. Canberra: COAG.

COAG. 2011. *Handbook for COAG Councils: A Guide for Best-Practice Operations for COAG Council Secretariats*. Canberra: COAG.

de Wit, E. and R. Webb. 2010. "Planning for coastal climate change in Victoria." *Environmental Planning Law Journal* 27 (1): 23–35.

Dovers, S. 2009. "Normalising adaptation." *Global Environmental Change* 19: 4–6.

Dovers, S. R. and A. A. Hezri. 2010. "Institutions and policy processes: The means to the ends of adaptation." *Wiley Interdisciplinary Reviews: Climate Change* 1 (2): 212–31.

Funtowicz, S. and J. Ravetz. 1993. "Science for the post-normal age." *Futures* 25 (7): 739–55.

Füssel, H.-M. and R. Klein. 2006. "Climate change vulnerability assessments: An evolution of conceptual thinking." *Climatic Change* 75 (3): 301–29.

Garnaut, R. 2008. *The Garnaut Climate Change Review: The Final Report*. Port Melbourne, Australia: Cambridge University Press.

Gerkhardt, M., Frey, D. and P. Fischer. 2008. "The human factor in change processes: Success factors from a socio-psychological point of view." In *Change 2.0*, edited by Klewes, J. and R. Langen, 11–25. Berlin: Springer.

Government of Queensland. 2007. *ClimateSmart Adaptation 2007–12: An Action Plan for Managing the Impacts of Climate Change*. Brisbane: Department of Natural Resources and Water.

Government of Queensland. 2009. *ClimateQ: Toward a Greener Queensland*. Brisbane: Department of Environment and Natural Resources.

Government of Queensland. 2011. *Climate Change: Adaptation for Queensland: Issues Paper*. Brisbane: Department of Environment and Resource Management.

Government of Victoria. 2005. *Victorian Greenhouse Strategy Action Plan Update*. Melbourne: Government of Victoria.

Government of Western Australia. 2004. *Western Australian Greenhouse Strategy*. Perth: Government of Western Australia.

Government of Western Australia. 2008. *Climate Change, Vulnerability and Adaptation for South West Western Australia 1970 to 2006: Phase One of Action 5.5, Western Australian Greenhouse Strategy*. Perth: Department of Agriculture and Food.

Government of Western Australia. 2012. *Indian Ocean Climate Initiative – Western Australia*. Perth: Government of Western Australia.

Grothmann, T. and A. Patt. 2005. "Adaptive capacity and human cognition: The process of individual adaptation to climate change." *Global Environmental Change Part A* 15 (3): 199–213.

Gurran, N., Norman, B., Gilbert, C. and E. Hamin. 2011. *Planning for Climate Change Adaptation in Coastal Australia: State of Practice*. Sydney: National Sea Change Taskforce.

Hallegatte, S. 2009. "Strategies to adapt to an uncertain climate change." *Global Environmental Change* 19 (2): 240–47.

Head, B. and J. Alford. 2008. "Wicked problems: The implications for public management." Paper presented at the International Research Society for Public Management Twelfth Annual Conference, March 26–28, Brisbane.

Hofstede, G. 2001. *Culture's Consequences: Comparing Values, Behaviors, Institutions, and Organizations across Nations*. Thousand Oaks, CA: Sage.

Howlett, M. 1998. "Predictable and unpredictable policy windows: Institutional and exogenous correlates of Canadian federal agenda-setting." *Canadian Journal of Political Science/Revue canadienne de science politique* 31 (3): 495–524.

Hudson, C. G. 2000. "From social Darwinism to self-organization: Implications for social change theory." *The Social Service Review* 74 (4): 533–59.

Hulme, M. 2009. *Why We Disagree about Climate Change: Understanding Controversy, Inaction and Opportunity*. Cambridge: Cambridge University Press.

Kezar, A. 2001. "Understanding and facilitating organizational change in the 21st century." *ASHE-ERIC Higher Education Report* 28 (4): 1–147.

Kotter, J. P. 1995. "Leading change: Why transformation efforts fail." *Harvard Business Review* 73 (2): 59–67.

Lane, M. B. and C. J. Robinson. 2009. "Institutional complexity and environmental management: The challenge of integration and the promise of large-scale collaboration." *Australasian Journal of Environmental Management* 16 (1): 16–24.

Lempert, R. J. and M. T. Collins. 2007. "Managing the risk of uncertain threshold responses: Comparison of robust, optimum, and precautionary approaches." *Risk Analysis* 27 (4): 1009–26.

Lockwood, M. and J. Davidson. 2010. "Environmental governance and the hybrid regime of Australian natural resource management." *Geoforum* 41 (3): 388–98.

McDonald, J. 2007. "A risky climate for decision-making: The liability of development authorities for climate change impacts." *Environmental Planning law Journal* 24 (6): 405–17.

Mercer, D., Christesen, L. and M. Buxton. 2007. "Squandering the future – climate change, policy failure and the water crisis in Australia." *Futures* 39 (2): 272–87.

Mintzberg, H. 1978. "Patterns in strategy formation." *Management Science* 24 (9): 934–48.

Moser, S. C. and M. T. Boykoff. 2013. *Successful Adaptation to Climate Change: Linking Science and Policy in a Rapidly Changing World*. Oxon: Routledge.

Moser, S. C. and J. A. Ekstrom. 2010. "A framework to diagnose barriers to climate change adaptation." *Proceedings of the National Academy of Sciences of the United States of America* 107 (51): 22026–31.

Narayanan, V. K., Zane, L. J. and B. Kemmerer. 2011. "The cognitive perspective in strategy: An integrative review." *Journal of Management* 37 (1): 305–51.

NCCARF. 2011. *Summary: NCCARF Strategy 2010–2012*. Brisbane: National Climate Change Adaptation Research Facility. http://www.nccarf.edu.au/sites/default/files/11136%20NCCARF%20Corporate%20Strategy.pdf.

NCCARF. 2014. *A Short History of NCCARF*. Brisbane: National Climate Change Adaptation Research Facility. http://www.nccarf.edu.au/content/short-history-nccarf#s7.

Nelson, R., Kokic, P., Crimp, S., Meinke, H. and S. M. Howden. 2010. "The vulnerability of Australian rural communities to climate variability and change: Part I – Conceptualising and measuring vulnerability." *Environmental Science and Policy* 13 (1): 8–17.

O'Brien, K. L. and J. Wolf. 2010. "A values-based approach to vulnerability and adaptation to climate change." *Wiley Interdisciplinary Reviews – Climate Change* 1 (2): 232–42.

O'Neill, P. and N. Argent. 2005. "Neoliberalism in Antipodean spaces and times: An introduction to the special theme issue." *Geographical Research* 43 (1): 2–8.

O'Reilly III, C. A., Chatman, J. and D. F. Caldwell. 1991. "People and organizational culture: A profile comparison approach to assessing person–organization fit." *Academy of Management Journal* 34 (3): 487–516.

Park, S. E., Marshall, N. A., Jakku, E., Dowd, A. M., Howden, S. M., Mendham, E. and A. Fleming. 2012. "Informing adaptation responses to climate change through theories of transformation." *Global Environmental Change* 22 (1): 115–26.

Pidgeon, N. and C. Butler. 2009. "Risk analysis and climate change." *Environmental Politics* 18 (5): 670–88.

Pielke, J. R., Prins, G., Rayner, S. and D. Sarewitz. 2007. "Climate change 2007: Lifting the taboo on adaptation." *Nature* 445 (7128): 597–98.

Pillora, S. 2010. *Australian Local Government and Climate Change Working Paper 1*. Sydney: Australian Center of Excellence for Local Government, University of Technology.

PMSEIC. 2007. *Climate Change in Australia: Regional Impacts and Adaptation – Managing the Risk for Australia*. Canberra: Prime Minister's Science, Engineering and Innovation Council.

Productivity Commission. 2011. *Barriers to Effective Climate Change Adaptation: Productivity Commission Issues Paper*. Melbourne: Productivity Commission.

Rittel, H. and M. Webber. 1973. "Dilemmas in a general theory of planning." *Policy Sciences* 4 (2): 155–69. Reprinted in *Developments in Design Methodology*, edited by N. Cross, 134–44, Chichester: Wiley and Sons, 1984.

Sminia, H. and A. Van Nistelrooij. 2006. "Strategic management and organization development: Planned change in a public sector organization." *Journal of Change Management* 6 (1): 99–113.

Smith, T. F., Brooke, C., Measham, T. G., Preston, B. L., Goddard, R., Withycombe, G., Beveridge, B. and C. Morrison. 2008. *Case Studies of Adaptive Capacity: Systems Approach to Regional Climate Change Adaptation Strategies*. Sydney: Sydney Coastal Councils Group Inc.

Stafford-Smith, M., Horrocks, L., Hervey, A. and C. Hamilton. 2011. "Rethinking adaptation for a 4°C world." *Philosophical Transactions of the Royal Society A: Mathematical, Physical and Engineering Sciences* 369 (1934): 196–216.

Stevenson, H. 2009. "Cheating on climate change? Australia's challenge to global warming norms." *Australian Journal of International Affairs* 63 (2): 165–86.

Tompkins, E. L., Adger, N. W., Boyd, E., Nicholson-Cole, S., Weatherhead, K. and N. W. Arnell. 2010. "Observed adaptation to climate change: UK evidence of transition to a well-adapting society." *Global Environmental Change* 20 (4): 627–35.

Van de Ven, A. H. and M. S. Poole. 1995. "Explaining development and change in organizations." *Academy of Management Review* 20 (3): 510–40.

VCCAR. 2012. "Victorian climate change adaptation research." http://ww.vccar.org.au.

Westerhoff, L., Carina, E., Keskitalo, H., McKay, H., Wolf, J., Ellison, D., Botetzagias, I. and B. Reysset. 2010. "Planned adaptation measures in industrialised countries: A comparison of select countries within and outside the EU." In *Developing Adaptation Policy and Practice in Europe: Multi-level Governance of Climate Change*, edited by E. C. H. Keskitalo, 271–338. Rotterdam: Springer Science+Business Media B.V.

Whelan-Berry, K. S. and K. A. Somerville. 2010. "Linking change drivers and the organizational change process: A review and synthesis." *Journal of Change Management* 10 (2): 175–93.

Willows, R. I. and R. K. Connell, eds. 2003. *Climate Adaptation: Uncertainty and Decision-Making*. Oxford: UKCIP.

6

Emerging Equity and Justice Concerns for Climate Change Adaptation

A Case Study of New York State

PETER VANCURA AND ROBIN LEICHENKO

Questions of equity and justice are receiving growing attention within the vulnerability, impacts, and adaptation (VIA) literature (Adger et al. 2006; Marino and Ribot 2012; Thomas and Twyman 2005; Wolf 2011; O'Brien and Leichenko 2009). Whereas the early literature on climate change and equity has primarily emphasized differences between wealthy and poor nations surrounding responsibility for historical emissions of greenhouse gases, recent attention has also been directed at questions of equity related to the uneven impacts of and vulnerability to climate change (Mearns and Norton 2010; O'Brien and Leichenko 2006). These differential impacts and vulnerabilities, which manifest across countries, regions, firms, households, and social groups, emerge not only as the result of physical exposure to extreme events or shifts in temperature and precipitation patterns but also as the result of differences in capacity to adapt. Moreover, research on equity and justice issues in vulnerability and adaptation to climate change is increasingly multifaceted and incorporating dimensions such as capacity to engage in adaptation planning, the unintended impacts of adaptation plans and policies, and the inclusiveness of adaptation planning efforts (Marino and Ribot 2012).

This chapter investigates these emerging equity and justice concerns through a case study of New York State. Although equity issues surrounding vulnerability and adaptation are present in all types of countries and regions, rich and poor alike, the themes and questions raised in our study of New York are likely to be most reflective of conditions within highly urbanized, developed countries, such as the United States, Canada, and many of the countries of Western Europe. Our New York State study was conducted as part of a statewide project on climate change impacts, vulnerability, and adaptation (Rosenzweig et al. 2011). The aims of the larger study, which is referred to as ClimAID, were to examine VIA issues for eight economic sectors in New York State, including transportation, water resources, agriculture, communication, ecosystem services, energy, coastal zones, and public health. Our contribution to the ClimAID project, as described in Leichenko et al. (2011), entailed identification of crosscutting and sector-specific equity and justice issues associated with both vulnerability and adaptation within each sector and to investigate these issues via case examples (Leichenko et al. 2011). We draw on that work in this chapter to show that adaptation is not a neutral process but instead has equity dimensions that are part of the larger adaptive challenge of climate change.

Equity and Justice in Climate Change VIA

Climate justice has become a widely adopted concept that applies to many types of equity and justice concerns that arise in the context of climate change. Climate justice issues range from differences in levels of emissions and energy usage to differences in ability to pay for adaptation technologies. Notions of climate justice draw in part from work in the area of environmental justice. Although environmental justice efforts have historically focused on inequitable distribution and burdens associated with exposure to hazardous waste in urban areas, these efforts have broadened in recent years to include consideration of equity and justice in the context of global environmental and resources issues, particularly impacts and responses to climate change (Agyeman et al. 2003; Aylett 2010; Ikeme 2003; Lindley et al. 2011; McDermott et al. 2011; Paavola 2008; Werrity et al. 2007; Zsamboky et al. 2011; Brisley et al. 2012). Whereas the bulk of this research has called attention to vulnerable communities and inequities arising from the uneven distribution of climate impacts, increasingly equity and justice are central components to urban and regional case studies of climate adaptation. Among other work, some recent projects analyze adaptation in resource-poor communities of color (Douglas et al. 2012; Paolisso et al. 2012), assess how equity dimensions are incorporated in emerging adaptation policies (Brisley et al. 2012), and highlight local efforts to mitigate climate change while also closing the "climate gap" that puts special pressure on low-income communities and communities of color (Morello-Frosch et al. 2009).

In defining equity and justice for this study, we drew insights from both distributional and procedural elements identified within the environmental justice literature and related equity literatures (Reed and George 2011; Walker 2010; Walker and Burningham 2011). In terms of distributional questions, the study emphasized situations where particular groups may be systematically disadvantaged either in terms of differences in vulnerability or capacity to adapt to climate change. In terms of procedural elements, we emphasized the inclusion of equity considerations in adaptation discussions, mechanisms for broad participation in future adaptation planning and policy efforts, and incorporation of a wide range of stakeholders and participants in the larger ClimAID project effort (see Rosenzweig et al. 2011).

Operationalizing Equity and Justice

Our broad aims in the New York State study included consideration of potential inequalities associated with climate change along both traditional lines that have been identified within the environmental justice literature (e.g., underprivileged, minority groups), as well as along new lines that may emerge under an altered climatic regime (e.g., different-sized firms) or may result from the implementation of adaptation policies and plans (e.g., different types of property owners or land users). Although the research effort did not entail investigation of procedural equity issues, procedural elements of environmental justice were put in place as part of the execution of the larger ClimAID study via encouragement of broad stakeholder participation in focus groups, including members of environmental justice groups and others who may be disadvantaged as a result of climate change or negatively affected by adaptation policies.

A key task of the project was to identify key equity and environmental justice issues surrounding both vulnerability and adaptation actions. For each sector, we considered the following questions:

1. Are there preexisting socioeconomic or spatial inequalities that make certain regions, communities, or groups of individuals systematically more vulnerable to the impacts of climate change on the sector? What groups or areas are likely to shoulder a disproportionate share of the burden from these impacts? Potential differentiations by group include socioeconomic status, education, health/disability, race, age, gender, culture, or citizenship. Community differentiations to consider include relative segregation, access to health care, unemployment, poverty/wealth/assets.
2. Are there groups, communities, or regions that are less able to adapt to the impacts of climate change on the sector and therefore merit special attention during adaptation planning?
3. Within the range of adaptation strategies in the sector, which strategies are more likely than others to exacerbate underlying socioeconomic disparities? Could some strategies change social and ecological dynamics so as to create emergent or unintended disparities? Are there situations in which strengthening adaptive capacity in one area or for one group may in turn somewhere else create, reinforce, or exacerbate maladaptation or vulnerability (either in an absolute or relative sense)?
4. Are there certain groups, communities, or regions that may be systematically underrepresented during adaptation planning, unable to access or influence the process and procedures of decision making, or otherwise disempowered, unable, or disinclined to consider adaptation when it likely is in their interest to do so?
5. When designing adaptation strategies, are there ways to insert mechanisms that encourage or ensure fair outcomes, whether preventive (such as avoiding and adjudicating disputes), corrective and compensatory, or retributive (sanctions and penalties)?

These questions were addressed through a combination of descriptive and mapping analysis of existing data to explore potential inequalities in vulnerability or adaptive capacity and qualitative evaluation of adaptation plans to identify potential equity and justice issues. We also developed case studies employing a mix of historical proxies and social and demographic analysis.

The iterative process of incorporating equity questions into a multidisciplinary and stakeholder-engaged research approach meant that certain questions and analytical techniques had more salience in some sectors than others. In the next sections we present results for two representative sectors in New York State, agricultural and coastal zones. Reflective of the flexible research approach designed to cross sectors, the agriculture section focuses on inequities in adaptive capacity, narrowing in on the situation of small farmers. The coastal zones section analyzes broader population vulnerabilities as a baseline analysis for adaptation visioning and planning. A detailed review of climate change's effects on agriculture and coastal zones, while beyond the scope of this article, can be found in the full results for the analysis of these two sectors in Wolfe et al. (2011) (agriculture) and Buonaiuto et al. (2011) (coastal zones).

Agriculture in New York State: Emerging Equity and Justice Concerns

Although agriculture makes up only a small share of total economic output within economically advanced countries such as the United States, where the sector accounts for less than 2 percent of total employment, it nonetheless plays an important role in many states and subnational regions. Agriculture adds about $4.5 billion to New York State's economy, making the state a dominant agricultural producer in the northeastern United States (Wolfe et al. 2011). The economic base of a number of rural counties in northern and western New York is tied to agriculture and natural resource industries, reflected in higher resource-based employment. For example, more than 10 percent of employment in Yates County is based in agriculture and greater than 5 percent in a handful of other counties such as Wyoming, Lewis, and Orleans (Bureau of Economic Analysis 2006). These counties provide local, fresh produce across the state and support recreation and tourism. The geographic clustering of these local resource economies underscores how certain regions may face unique and disproportionate need for adaptation planning, raising critical questions about who is responsible for adaptation and how to equitably distribute adaptation-related resources.

The state's agricultural sector is characterized by a substantial dairy industry and a mixture of larger wholesale operations and smaller farms producing vegetables such as apples, grapes, corn, and cabbage (Wolfe et al. 2011). Climate change is likely to impact the sector through a variety of pathways, including increased summer heat stress on cool-season crops and livestock, more intense weed and pest pressure, and more frequent heavy rainfall and summer water deficits (Wolfe et al. 2011). The need to adapt to these changing conditions brings into focus a range of equity and environmental justice issues. In this section, we highlight key vulnerabilities and equity dimensions of agricultural adaptation strategies. Because many of the climate impacts may be felt as price increases on farming inputs, we focus on small farmers and develop a case study detailing differences in the adaptive capacity of dairy farmers.

Agricultural Vulnerabilities

In New York State, those most vulnerable to climate change include small family farms with little capital to invest in on-farm adaptation strategies, such as new infrastructure, stress-tolerant plant varieties, new crop species, or increased chemical and water inputs. Small family farms also are less able to take advantage of cost-related scale economies associated with such measures. Small farmers, particularly those in the dairy sector, already face severe competitive pressures due to rising production costs and flat or declining commodity prices. Indeed, within New York State, current trends suggest that the total number of dairy farms will decline from approximately 7,900 in 2000 to 1,800 in 2020, with most of this decline resulting from closure or consolidation of smaller farms (LaDue et al. 2003; U.S. Department of Agriculture 2007). Climate change is likely to exacerbate cost pressures on small farmers, particularly if adaptation requires significant capital investments, thus accelerating trends toward consolidation within the industry. Survival for many smaller farms will hinge, in part, on making good decisions regarding not only the type of adaptation measures to take but also the timing of the measures. The most vulnerable farmers will be those without access to training about the full range of strategies or those who lack adequate information to assess risk and uncertainty (Wolfe et al. 2011).

In addition to directly impacting crops and farming, the effects of climate change also may transform agricultural demand (Wolfe et al. 2011). These effects can be associated with both long-term regional disinvestment such as out of high-risk areas (floodplains) or one-time extreme events in areas with high demand for New York State produce (such as a hurricane in the New York metropolitan region). These conditions may disrupt supply chains, close retail centers, or otherwise cut consumer access to markets, with especially detrimental effects on low-income or mobility-constrained residents. Low-income farmers with insufficient information and training or without access to credit or infrastructure are particularly at risk when conditions demand immediate flexibility and require quickly lining up alternative supply lines and retail locations (Wolfe et al. 2011). Under such conditions, farmers may feel pressure to supplement incomes or diversify their business beyond agriculture but may lack the training or capital necessary to engage such strategies. Decreasing yields and the high costs of adaptation may translate into significant downstream job losses and cascading economic effects across rural communities. Low-wage, temporary, seasonal, and/or migrant workers are particularly exposed to these shifts.

Agricultural Adaptations

Examining equity in adaptation involves evaluating existing vulnerabilities and adaptive capacities, but it also requires evaluating the unintended outcomes, externalities (secondary consequences), and emergent processes of specific adaptation strategies. Successful adaptation by individual farmers or regions may create downstream inequities. As some farmers successfully adapt, other farmers may experience relative increases in inequality related to rural income and agricultural productivity. Certain industries (such as the grape and wine industries) also may consolidate in such ways that it becomes difficult for smaller businesses to enter the market. Increasing chemical inputs, such as fertilizers and pesticides, may create or exacerbate inequitable distributions of human health burdens or negatively affect waterways, disproportionately impacting low-income or natural resource–dependent communities involved in hunting- and fishing-related revenue. Furthermore, degrading land and community health could drive down property values, exacerbating geographic inequities. Finally, increasing natural resource use, whether it is water for irrigation or energy for cooling, is likely to raise utility prices. These increases are felt the most by low-income families who proportionally spend more on these basic goods than middle- and upper-income families (Wolfe et al. 2011).

Addressing and avoiding spillover effects in the implementation of adaptation measures requires engaging local communities and agricultural managers in each stage of the planning process. This includes mechanisms for expressing and addressing property disputes and conflicting claims to resources, collaborative regional planning across sectors and communities, and training or retraining to provide information regarding strategies and best practices. In particular, adaptation strategies focused at regional or state scales have the capacity to marginalize local actors who are unable to capitalize on social or economic networks or access policy-making procedures.

Because agriculture provides fresh produce across the state, equity issues exist along every part and process of the food-supply chain (Wolfe et al. 2011). For low-income communities throughout the state, the connection between climate change and issues of food justice is an area of growing concern. For example, climate stress on agriculture could impact the quality, accessibility, and affordability of local produce. This has implications for food security among low-income groups, those communities with fragile connections to markets offering nutritional options, or those otherwise burdened by preexisting poor nutrition. Increased incidence of extreme heat or prolonged droughts may also affect the cost structures and productivity of community gardens and other local food production systems that serve lower-income areas.

Equity in New York State's Dairy Regions

Vulnerability and capacity to adapt to climate change may vary substantially across different dairy regions in New York State due to differences in climate change exposure, regional cost structures, farm sizes, and overall productivity (Wolfe et al. 2011). Should climate change have a highly detrimental effect on dairy farming in the state overall, those regions with higher concentrations of dairy farms are likely to

Table 6.1. *Characteristics and production costs and revenues of dairy regions*

	West and Central Plateau	West and Central Plains	Northern New York	Central Valley	North Hudson and Southeastern New York
Average number of cows per establishment	168	673	372	289	210
Tillable acres	417	1241	848	707	495
Total cost of producing milk ($ per hundredweight)	$18.03	$17.16	$17.04	$17.33	$18.63
Average price received ($ per hundredweight)	$20.62	$20.09	$20.06	$20.77	$20.95
Difference between cost and price received	$2.59	$2.93	$3.02	$3.44	$2.32

Source: Cornell University, Dairy Business Summary and Analysis Project (Knoblauch et al. 2008).

experience a more substantial economic disruption. Conversely, farmers in regions with higher concentrations of farms may also have some advantages associated with external economies of scale that facilitate adaptation to climate change, such as ability to learn from other farmers in the area regarding best adaptation practices or pooling of resources for different types of services that are needed to foster adaptation.

In New York State, dairy farms are particularly abundant in Northern New York, Central Valleys, and the Western and Central Plateau region (Table 6.1). Annual sales across the different regions show wide variations. While total sales in most of the counties located in the Northern Hudson and Southeastern New York regions were less than $15 million in 2007, those with the highest sales were an order of magnitude greater. Sales in four counties in Northern New York and the Western and Central Plateau region each topped $100 million (U.S. Department of Agriculture 2007).

In addition to differences in numbers of farms and total sales, the major dairy regions within the state also exhibit different characteristics in terms of size and profitability. Examination of Table 6.1 suggests that larger farms are concentrated in the Western and Central Plain regions, where the average farm size has 673 cows and more than 1,200 tillable acres (Knoblauch et al. 2008). The smallest average farm size is in the Western and Central Plateau region, where the average farm within the sample has 168 cows on 417 tillable acres. Costs of milk production range from $17.04 per hundredweight of milk in Northern New York to $18.63 per hundredweight in the Northern Hudson and Southeastern New York regions. Some factors that account for these regional differences include land and labor costs, which are likely to be higher in the areas closer to metropolitan New York. The Western and Central Plateau region

Table 6.2. *Small versus large dairy farms in 2007 and 2008*

	Small farms (39 farms)			Large farms (83 farms)		
	2007	2008	% chg	2007	2008	% chg
Farm size						
Average number of cows	52	52	0.0	773	797	3.1
Total tillable acres	40	41	2.5	1,482	1,595	7.6
Milk sold, lbs.	975,626	976,710	0.1	1,8500,129	19,671,976	6.3
Costs						
Grain and concentrate purchases as a percent of milk sales	24%	31%	29.2	24%	30%	25.0
Total operating expenses per cwt. sold	$16.29	$17.51	7.5	$16.32	$17.87	9.5
Capital efficiency						
Farm capital per cow	$11,880	$12,576	5.9	$7,981	$8,772	9.9
Income and profitability						
Gross milk sales per cow	$3,817	$3,678	−4.4	$4,870	$4,753	−2.4
Net farm income (w/o appreciation)	$54,680	$28,117	−48.6	$939,605	$483,799	−48.5
Income per oper/manager	$20,267	−$5,257	−126	$388,494	$128,755	−67
Farm net worth	$498,120	$502,664	0.9	$4421159	$4658105	5.4

Source: Cornell University, Dairy Business Summary and Analysis Project (Karszes et al. 2009; Knoblauch et al. 2009).

(which also has the smallest farms) is another region with relatively high costs of $18.03 per hundredweight of milk sold. These differences in milk production costs across regions may influence capacity to adapt to climate change, particularly in cases where adaptation requires additional expenditures for energy and pest control due to higher summer temperatures.

Differences in farm and herd size are also potentially significant factors in determining vulnerability and capacity to adapt to climate change. Although it is difficult to know precisely how different-sized dairy farms will be affected by climate change, the effects of other types of shocks can help to illustrate which types of farmers might be more or less vulnerable to climate change. Comparison of the performance of small versus large farms in 2007 (a relatively profitable year) versus 2008 (a more challenging year due to spikes in input prices including feed, energy, and fertilizer) throughout the state provides a glimpse into how shocks affect farms of different sizes (Table 6.2). The study reveals significant differences in costs, milk sales per cow, capital efficiency, income, and profitability. All of these differences may affect the overall capacity of smaller farms to adapt to climate change, particularly if such adaptation requires significant new outlays of capital for purchase and installation of ventilation systems in dairy barns as well as additional costs associated with energy for operating this equipment.

Data from the Cornell dairy survey (Karszes et al. 2009; Knoblauch et al. 2009) reveal that both small and large farms experienced significant challenges in coping with these conditions in 2008, but small farms appear to have fared worse (Table 6.2). Small farms experienced no increase in milk sold and had a 4.4 percent decline in gross milk sales per cow from 2007 to 2008. By contrast, large farms increased sales by 6.3 percent and experienced a 2.4 percent decline in gross milk sales per cow. Small farms also experienced relatively larger increases in purchased input costs (29.2 percent for small farms compared to 25 percent for large farms) but smaller increases in total production costs (7.5 percent for small farms compared to 9.5 percent for large farms). While net farm income for both small and large farms declined by approximately 48 percent between 2007 and 2008, income per farm operator or manager declined much more precipitously for small farms. Overall, the typical small farm operator experienced an income loss of 126 percent, while large farm operators experienced losses of 67 percent.

Similar types of input price shocks may also occur under climate change, because more frequent extreme weather conditions could lead to higher feed and energy prices. Policies intended to reduce emissions may also contribute to higher energy prices, though such effects are likely to be more gradual as taxes or other mechanisms to mitigate climate change are put into place. In sum, the dairy sector varies across New York in regional character, production costs, and farm operation size. The case of a demand shock in 2008 demonstrates that some of these variations affect the capacity of certain operations to cope with challenging conditions. Using this event as a historical proxy for climate effects, the analysis in this case study suggests that small farms may be less able to withstand shocks related to climate change without some type of adaptation assistance. By extension, resource-dependent regions and low-income communities also face disadvantages that may benefit from targeted efforts to build resilience.

Coastal Zones: Equity and Justice in Vulnerability and Adaptation

New York State's coastal zone comprises parts of all five counties of New York City, extending along the north and south shore of Long Island and stretching up the Hudson River for more than one hundred miles to the Troy Dam. Coastal zone populations in New York State are increasing and becoming denser relative to other regions in the state. Coastal populations living within floodplains in New York City and Long Island tend to be more affluent than populations living outside the floodplains. These differences are largely due to amenity appeal, which makes property near the water more desirable. However, there are pockets of poverty across the New York State coastal zone, and many of these areas are home to concentrations of racial and ethnic minority populations.

Coastal Vulnerabilities

Key climate change–related risks from coastal zones include sea level and storm surge and increased likelihood of coastal storms (Buonaiuto et al. 2011). Flooding and natural hazards can disproportionately impact certain socioeconomic groups, especially racial and ethnic minorities and low-income communities (Wu et al. 2002; Fothergill et al. 1999). Often this is an expression of physical vulnerability, such as in pre-Katrina New Orleans, where low-lying areas at risk of inundation were home largely to African Americans. Frequently, physical vulnerability is compounded by intrinsic individual vulnerabilities – related to age and physical immobility, for example – as well as a host of community-level, contextual vulnerabilities that can surface in every phase from prevention through to relief, recovery, and reconstruction (Morrow 1999). For example, low-income communities are less likely to have access to a full range of preventive strategies, such as resources to fortify property, prepare emergency provisions, and acquire insurance (Morrow 1999; Yarnal 2007). Similarly, both real and perceived inequities have plagued rebuilding following past hurricanes and coastal storms, particularly when redevelopment planning takes a top-down form (Wright and Bullard 2007). Other frequent concerns include unequal access to emergency and recovery loan assistance and inadequate resources for compensation of health and property losses.

An important dimension of vulnerability in the context of coastal hazards and the emphasis of the following case study presented is access to transportation. Interruptions in public transportation or reductions in affordable options can concentrate the effects of infrastructure failures on populations dependent on public transport. Lower-income populations are disproportionately transport dependent (Pucher et al. 2003). At the same time, they tend to live farther from their places of work, so they are the most likely to be affected by lost wages during a protracted recovery (Chen 2007). Transport interruptions take a particular toll on working women, who tend to have less spare time because of child and family care and on average earn less than men (Root et al. 2000; Morrow 1999).

Some of these patterns are readily apparent in New York City. Of the three-quarters of a million New York City workers who commute more than an hour, two-thirds of them earn less than $35,000 per year (Byron 2008). Many low-income individuals living at the periphery of the outer boroughs, especially parts of Queens and Staten Island, have access only to unreliable, inefficient, or inconvenient public transit (New York Metropolitan Transit Council 2009). Even minor service disruptions may create hardship for them relative to individuals in areas with transit redundancies.

Coastal Zone Adaptations

Vulnerable communities are at risk of material injury associated with increasing magnitudes of flood events and throughout the relief and recovery process. The previous section reviewed some of the underlying conditions associated with this vulnerability. These underlying dimensions are also present in the context of adaptation efforts and interventions (Marino and Ribot 2012). For illustrative purposes, two widely considered adaptation strategies for coastal regions are discussed in this subsection – infrastructure adaptation and managed retreat – along with a review of key equity issues. These two strategies are frequently referenced as potential options for different regions of New York State (Buonaiuto et al. 2011; Jacob et al. 2011).

Coastal Zones: Infrastructure Adaptations

Building climate-secure hard infrastructure offers an amenity that may create new patterns of winners and losers. Key questions surround which communities will be protected, in what ways, and who bears the costs of construction. For example, when building a seawall or levee, where is it placed and whose property does it protect? What areas of a city or town are treated as critical while others are deemed nonpriorities? These equity issues extend into strategies that include "softer" design. Choosing which wetlands to restore, beaches to fortify with additional sand, or structures and lands to elevate is not a simple issue of exposure to risk. It involves making difficult decisions about distributing benefits and costs among communities and prioritizing some areas potentially at the expense of others.

Adapting coastal transportation infrastructure poses a particular challenge. Because transportation design is generally locked into the landscape, an array of equity dilemmas extends throughout the life cycle of the infrastructure. Immediate questions include, Where is a new or upgraded route going to be sited? Who will be displaced by the construction? In the medium term, what demographic groups or regions will benefit from the adaptation? Decommissioning highways, roads, and other infrastructure for climate protection could involuntarily leave communities isolated from job centers or otherwise stranded. Some communities may have increased traffic and demand for services, while others experience shrinkage. And in the long term, new transportation flows could induce new patterns of land use and mobility and new patterns of migration into and out of an area.

Questions of whether to retrofit and renovate old infrastructure versus building new climate-adaptive infrastructure may also raise equity issues. For example, an adaptation policy concentrating on designing new road construction outside of floodplains could be biased toward exurban, high-income fringe suburbs of the various cities at the expense of inner-ring suburbs and central city areas that are more set in place and would most benefit from other types of measures such as increasing bus capacity.

Coastal Zones: Managed Relocation

Managed relocation from floodplains is another adaptation strategy that is accompanied by a portfolio of equity concerns related to the specific measures employed in the policies, from the relocation incentives to the environmental restoration of reclaimed lands. In some cases, managed retreat may materialize less as proactive planning and more as reactive incrementalism or planned obsolescence, such as service cutbacks, squeezing areas into shrinkage, or "choking" growth. Under such conditions, those individuals and groups with the widest range of job and residence options and the ability to forecast policy changes would be first to migrate. Lower-income populations could find themselves at an adaptive disadvantage, because they lack either the capital to invest in new housing or the socioeconomic flexibility to allow them to transfer jobs and livelihoods locations. Relocation can be difficult for any business, but minority-owned businesses may be especially vulnerable. They tend to be smaller, less well capitalized, less able to get loans, and subject to discrimination.

During a managed retreat, upland areas could be transformed by migration and localized population pressures. These communities may experience gentrification, increased cost of services from in-migration, and burdens of displacement from lowland areas. The viability and cohesion of low-income communities tend to be vulnerable under these conditions. Retreat from the southern coast of Long Island, for example, where housing issues are already a critical concern, may displace households, increase housing demand, and push up property values, a process that may indirectly burden the low-income population.

Case Study: Vulnerability to Coastal Flooding

Vulnerability and adaptation to flooding is one of the key challenges facing local and regional governments in New York State and cuts across a number of the sectors in the ClimAID study (Buonaiuto et al. 2011; Jacob et al. 2011; Shaw et al. 2011). How different groups and regions fare reflects existing resource inequalities and their varying capacities to put in place precautionary strategies and manage the effects of flood events. To capture the contextual nature of these equity considerations, we examined the general impact of a one-in-one-hundred-year flood along urbanized coastal locations in New York City and Long Island. We also narrowed in on the effects of climate change on transport-dependent populations. In both case studies, we mapped vulnerabilities and described key equity concerns in the major adaptation strategies under consideration. In addition, we raised critical equity questions that could frame site-specific adaptation planning processes and looked for opportunities in existing management processes where equity concerns could be made more concrete.

In this section, we first highlight a number of the key findings from the vulnerability component of the analysis. We then focus on transportation disadvantages. Although

Table 6.3. *Characteristics of population in Census block groups in and out of one-hundred-year floodplain for the New York Coastal Zone*

	In flood zone	Out of flood zone
Median income	$56,132	$48,551
Median housing value, owner-occupied housing	$235,297	$229,149
Female head of households as percentage of total	20.2	26.5
% in poverty	12.8	17.1
% less than high school	18.7	25.5
% over 65	14.2	12.2
% African American	12.6	23.0
% Hispanic	15.4	22.1
% renter	41.5	54.1
% vacant housing	9.5	5.2
% foreign born	18.9	30.2

Source: U.S. Census 2000; authors' calculations.

the empirical research was conducted just prior to Superstorm Sandy in October 2012, the analysis presented here captures a snapshot of issues that subsequently came to prominence in the aftermath of the storm, including the vulnerability of particular residential populations and transportation disadvantages. This suggests that rapid, timely, and recurring equity assessments such as the one presented here can play a role in helping to identify those situations where negative outcomes may concentrate as a result of underlying inequities. It also draws attention to the need to design resilience and adaptation measures that take these conditions into account. We return to this linkage following the presentation of the analysis in the next paragraphs.

Many coastal communities on the south shore of Long Island are fairly affluent. Indeed, Table 6.3 indicates that residents within FEMA's one-hundred-year flood zone tend to have higher incomes, live in more expensive homes, and represent a lower minority population than those outside the floodplain. Examining the distribution of certain higher-risk subsets within this population can help locate potential environmental justice effects. Low-income households, for example, are confronted with constrained resource options for both long-term adaptation and immediate coping (Wu et al. 2002). In the coastal flood zone of New York City and Long Island, nearly 75,000 people live under the poverty line. More than 80 percent of this population resides in New York City, where wealthier and poorer neighborhoods often coexist in close proximity near the shore (e.g., Coney Island, Brighton Beach, the Rockaways) and are thus potentially equally exposed to the physical consequences of flooding from a major storm or hurricane. Equity issues may arise in the form of structural damage associated with variations in construction, ease of timely evacuation and availability of transportation, or the ability to recover after a storm.

Table 6.4. *Profile of the population residing in the one-hundred-year floodplain (ClimAID Region 4)*

	New York City	Nassau County	Suffolk County	Total
Population				
Total population	286,374	159,644	70,523	516,541
% over 65	14	15	14	15
% below poverty line	21	6	6	14
% African American	25	5	3	16
% Latino	23	9	7	16
% foreign born	27	13	9	20
Housing				
Occupied housing units	110,194	58,206	27,103	195,503
% renter occupied housing units	70	24	20	49
Aggregate value of owner-occupied housing ($)	8,255,214,696	13,342,438,484	6,171,271,384	27,768,924,564

Source: U.S. Census 2000; authors' calculations.

Previous research also has suggested that racial and ethnic minorities tend to be more exposed to hazard events and more vulnerable than nonminority populations (see, e.g., Fothergill et al. 1999). Hurricane Katrina provided a vivid reminder of this uneven burden in 2005 (Yarnal 2007). In coastal New York City and Long Island, just over 82,000 African Americans and nearly the same number of Latinos live in the one-hundred-year flood zone. Examination of Table 6.4 suggests that African Americans and Latinos are significantly overrepresented in New York City's flood zone relative to the distribution of the total population in coastal New York, which likely reflects a legacy of suburban settlement patterns on Long Island (i.e., fewer minorities the farther away from New York City). This population distribution, in combination with the disproportionately high concentration of poverty and the greater proportion of renters, suggests that New York City would face fundamentally different equity challenges than Nassau and Suffolk counties. In contrast, proportionally higher rates of homeownership and greater income may signal a measure of resilience across more wealthy regions of Long Island.

Based on estimates generated by Jacob et al. (2011) (and for current sea level), 90 percent recovery times for specific parts of the New York City metropolitan transport system would vary from a few days to almost a month. During a protracted recovery, differences in regional transport options and population specific transport disadvantages can exacerbate hardships. In general, populations and regions with diverse and redundant transport options would more easily cope and recover from transport systems failure. Further hardship through lost wages and commercial activity, among other pathways of stress, would confront transport-disadvantaged populations. These include communities constrained by geography to limited transport options, low-income households dependent on public transport, and individuals with limited mobility.

A number of key issues differentiate the transport advantages of populations on Long Island and in New York City. Areas in New York City have a significantly greater percentage of households with no car, and working residents in the city's floodplains are four times more likely than those on Long Island to use public transportation as their primary means of commuting.

Across census block groups, the percentage of people with access to a car ranges from less than 5 percent to more than 60 percent. Despite generally high rates of car ownership on Long Island, small pockets of low ownership are interspersed largely within Nassau County. A look at the demographic and socioeconomic makeup of a few of these census block groups underscores that car ownership is partly a function of underlying socioeconomic conditions. For example, a few such areas in Hempstead also have higher rates of poverty and lower average educational attainment compared to regional means. These conditions would act together as a group of stresses during a storm event, reinforcing the vulnerability of a person with no car.

Vulnerability and Adaptation in Post-Sandy New York

The coastal analysis was conducted before Superstorm Sandy, but preliminary observations in the storm's aftermath affirm the importance of such vulnerability analyses. For example, low-income renters were disproportionately impacted by the storm surge (Furman Center 2013). Communities and populations identified in our analysis as transport disadvantaged, such as low-income residents in Coney Island, indeed shouldered especially heavy transport-related hardship. In reports from Coney Island a week after the storm, the compound effect of subway flooding, power outages, and the closure of local businesses continued to create a public health crisis among elderly and disabled residents in public housing (Fractenberg and Upadhye 2012). Evidence of a wider crisis was acknowledged nine days after the storm, when Mayor Bloomberg announced renewed outreach to all public housing to reconnect stranded citizens to vital services (Lipton and Moss 2012). By way of comparison, most of the cross-river service between Brooklyn and Manhattan was restored within a few days through an impressive combination of subway tunnel restoration and a system of express shuttle buses.

The storm surge case study presented here reveals critical factors underlying social vulnerability to flooding. As such, it provides an underlying layer of information that can frame an analysis of the equity dimensions of different adaptation strategies. The two illustrative adaptation strategies discussed earlier have been prominent points of discussion and action in post-Sandy New York. There has been renewed attention to hard infrastructure projects, such as a proposed storm surge barrier in the Verrazano Narrows. A baseline equity analysis draws attention not simply to the physical spatial implication of such a barrier (Coney Island would be outside it, for example) but to the special pressures it might place on particular communities as a result of concentrated vulnerable populations. There also has been a surge in attention to transforming coastal land use, including ambitious and visionary options for flood-friendly urban spaces and strategies of managed retreat (Rebuild by Design 2014). The governor of New York has supported limited buyouts in residential areas of Staten Island (Kaplan 2013). Accompanying these new interventions are the potential for emerging or unintended inequities that should be incorporated into planning exercises.

Conclusions

As New York State and other regions address climate change, incorporating equity concerns and promoting broad and inclusive participation in adaptation decisions will be critical for ensuring that the outcomes of these decisions are both fair and economically feasible. Studies in the VIA tradition have begun to develop research frameworks and empirical work to further this discussion. In this chapter, we have advanced this work by adding a case study reporting on equity concerns in New York State. Our study framed equity and justice to include those communities that historically have been considered in the environmental justice literature, predominantly low-income and racial and ethnic minority communities, while at the same time exploring new lines of uneven climate impacts that might intersect other types of contextual inequities, such as disability and farm size. At times, our analytical approach drew on established traditions in vulnerability analysis and environmental justice. In the coastal zone sector, for example, a population-level vulnerability analysis established a data frame for considering issues of fairness and equity in adaptation choices. In the wake of Superstorm Sandy, the severity of storm impacts has drawn attention to a range of such choices, such as a storm barrier for lower Manhattan. A social vulnerability analysis offers a rapid way to frame and ground an equity-based impact analysis.

In the agricultural sector, conversely, we took an added step, extending analysis to adaptive capacity at a unit of analysis not generally considered, that is, small enterprises. A case study of dairy operations showed considerable variation in the size, profitability, and concentration of operations across that state's dairy regions. Using a historical proxy of a shock to the price of farming inputs, the case study underscores ways that smaller farmers may be disadvantaged when attempting to adapt to the types of shock that might be expected by climate change.

Together, these two analyses detail a wide range of underlying factors that contribute to social vulnerability across different regional and sectoral contexts. Adaptation responses are imagined, planned, and practiced against this uneven social backdrop. Those adaptation strategies that are specifically paired with efforts to address conditions and factors that contribute to vulnerability can generate social equity cobenefits. Disregarding these conditions risks exacerbating or creating new forms of inequity. Incorporating equity analyses into recurring regional adaptation assessments can play a role in helping plan appropriate strategies. Equity-driven assessments such as the one drawn on here can suggest ways to better understand, forecast, and monitor how proposed and implemented adaptation measures may either reduce or exacerbate those processes and conditions that create vulnerability. By better understanding fairness in adaptation and developing ways to incorporate equity dimensions in science-policy efforts, the benefits of adaptation interventions can be realized more fully.

Acknowledgments

This work is based in part on the study "Responding to Climate Change in New York State: The ClimAID Integrated Assessment for Effective Climate Change Adaptation" (Rosenzweig et al. 2011), and we gratefully acknowledge New York State Energy Research and Development Authority for financial support and the scientific support of ClimAID researchers who provided key scientific input to our equity analysis. We also thank two anonymous reviewers for their helpful comments and suggestions.

References

Adger, N., Paavola, J., Huq, S. and M. J. Mace, eds. 2006. *Fairness in Adaptation to Climate Change*. Cambridge, MA: MIT Press.

Agyeman, J., Bullard, R. and Evans, B., eds. 2003. *Just Sustainabilities: Development in an Unequal World*. Cambridge, MA: MIT Press.

Aylett, A. 2010. "Participatory planning, justice, and climate change in Durban, South Africa." *Environment and Planning A* 42 (1): 99–115.

Brisley, R., Welstead, J., Hindle R. and J. Paavola. 2012. *Just Adaptation Responses to Climate Change*. York: Joseph Rowntree Foundation. http://www.jrf.org.uk/publications/socially-just-adaptation-climate-change.

Buonaiuto, F., Patrick, L., Gornitz, V., Hartig, E., Leichenko, R. and P. Vancura. 2011. "Coastal zones." *Annals of the New York Academy of Sciences* 1244: 169–223.

Bureau of Economic Analysis. 2006. "Local area personal income." CA25N. http://www.bea.gov/regional/reis/.

Byron, J. 2008. "Bus rapid transit for New York?" *Gotham Gazette*, April 21. http://www.gothamgazette.com/article/sustain/20080421/ 210/2498.

Chen, D. 2007. "Linking transportation equity and environmental justice with smart growth." In *Growing Smarter: Achieving Livable Communities, Environmental Justice, and Regional Equity*, edited by R. D. Bullard, 299–322. Cambridge, MA: MIT Press.

Douglas, E., Kirshen, P. H., Paolisso, M., Watson, C., Wiggin, J., Enrici, A. and M. Ruth. 2012. "Coastal flooding, climate change and environmental justice: identifying obstacles and incentives for adaptation in two metropolitan Boston Massachusetts communities." *Mitigation and Adaptation Strategies for Global Change* 17 (5): 537–62.

Fothergill, A., Maestas, E. G. M. and J. D. Darlington. 1999. "Race, ethnicity and disasters in the United States: A review of the literature." *Disasters* 23 (2): 156–73.
Fractenberg, B. and J. Upadhye. 2012. "Disabled and elderly stuck in Coney Island building week after Sandy." DNAInfo.com, November 5.
Furman Center for Real Estate and Urban Policy. 2013. "Fact brief: Sandy's effects on housing in New York City." http://furmancenter.org/files/publications/SandysEffectsOnHousingInNYC.pdf.
Ikeme, J. 2003. "Equity, Environmental Justice and Sustainability: Incomplete Approaches in Climate Change Politics." *Global Environmental Change* 13 (3): 195–206.
Jacob, K., Deodatis, G., Atlas, J., Whitcomb, M., Lopeman, M., Markogiannaki, O., Kennett, Z., Morla, A., Leichenko, R. and P. Vancura. 2011. "Transportation." *Annals of the New York Academy of Sciences* 1244: 169–223.
Kaplan, T. 2013. "Cuomo seeking home buyouts in flood zones." *New York Times*, February 3.
Karszes, J., Knoblauch, W. and L. Putnam. 2009. "New York large herd farms, 300 cows or larger, 2008." Dairy Farm Business Survey, Department of Applied Economics and Management, College of Agricultural and Life Sciences, Cornell University.
Knoblauch, W., Putnam, L., Karszes, J., Murray, D. and R. Moag. 2008. "Business summary, New York State." Dairy Farm Business Summary and Analysis, Department of Applied Economics and Management, College of Agricultural and Life Sciences, Cornell University.
Knoblauch, W., Putnam, L., Kiraly, M. and J. Karszes. 2009. "New York small herd farms, 80 cows or fewer, 2008." Dairy Farm Business Survey, Department of Applied Economics and Management, College of Agricultural and Life Sciences, Cornell University.
LaDue, E., Gloy, C. and B. Cuykendall. 2003. "Future structure of the dairy industry: Historical trends, projections, and issues." Cornell Program on Agriculture and Small Business Finance, College of Agricultural and Life Sciences, Cornell University.
Leichenko, R., Klein, Y., Panero, M., Major, D. C. and P. Vancura. 2011. "Equity and economics." *Annals of the New York Academy of Sciences* 1244: 169–223.
Lindley, S., O'Neill, J., Kandeh, J., Lawson, N., Christian, R. and M. O'Neill. 2011. *Climate Change, Justice and Vulnerability*. York: Joseph Rowntree Foundation. http://www.jrf.org.uk/publications/socially-just-adaptation-climate-change.
Lipton, E. and M. Moss. 2012. "Housing agency's flaws revealed by storm." *New York Times*, December 9.
Marino, E. and J. Ribot. 2012. "Special issue introduction: Adding insult to injury: Climate change and the inequities of climate intervention." *Global Environmental Change* 22 (2): 323–28.
McDermott, M., Mahanty, S. and K. Schreckenberg. 2011. "Defining equity: A framework for evaluating equity in the context of ecosystem services." Working paper.
Mearns, R. and A. Norton. 2010. *Social Dimensions of Climate Change: Equity and Vulnerability in a Warming World*. Washington, DC: The World Bank.
Morello-Frosch, R., Pastor, M., Sadd, J. and S. Shonkoff. 2009. "The climate gap: Inequalities in how climate change hurts Americans and how to close the gap." Program for Environmental and Regional Equality (PERE), University of Southern California. http://college.usc.edu/geography/ESPE/perepub.html.
Morrow, B. H. 1999. "Identifying and mapping community vulnerability." *Disasters* 23 (1): 11–18.
New York Metropolitan Transportation Council. 2007. "Access to transportation on Long Island: A technical report." http://www.nymtc.org/project/LIS_access/documents/Final_TechRpt.pdf.

New York Metropolitan Transportation Council. 2009. "A coordinated public transit-human service transportation plan for the NYMTC region." http://www.nymtc.org/project/PTHSP/PTHSP_documents.html.

O'Brien, K. and R. Leichenko. 2006. "Climate change, equity, and human security." *Die Erde* 137 (3): 165–79.

O'Brien, K. and R. Leichenko. 2009. "Global environmental change, equity, and human security." In *Global Environmental Change and Human Security*, edited by Matthew, R. A., Barnett, J., McDonald, B. L. and K. L. O'Brien, 157–76. Cambridge, MA: MIT Press.

Paavola, J. 2008. "Science and social justice in the governance of adaptation to climate change." *Environmental Politics* 17 (4): 644–59.

Paolisso, M., Douglas, E., Enrici, A., Kirshen, P., Watson, C. and M. Ruth. 2012. "Climate change, justice, and adaptation among African American communities in the Chesapeake Bay region." *Weather, Climate, and Society* 4 (1): 34–47.

Pucher, J. and J. L. Renne. 2003. "Socioeconomics of urban travel: Evidence from the 2001 NHTS." *Transportation Quarterly* 57 (3): 49–77.

Rebuild by Design. 2014. Homepage. http://www.rebuildbydesign.org/.

Reed, M. G. and C. George. 2011. "Where in the world is environmental justice?" *Progress in Human Geography* 35 (6): 835–42.

Root, A., Schintler, L. and K. J. Button. 2000. "Women, travel and the idea of 'sustainable transport.'" *Transport Reviews* 20 (3): 369–83.

Rosenzweig, C., Solecki, W., DeGaetano, A., O'Grady, M., Hassol, S. and P. Grabhorn, eds. 2011. *Responding to Climate Change in New York State: The ClimAID Integrated Assessment for Effective Climate Change Adaptation. Annals of the New York Academy of Sciences* 1244.

Shaw, S., Schneider, R., McDonald, A., Riha, S., Tryhorn, L., Leichenko, R., Vancura, P., Frei, A. and B. Montz. 2011. "Water resources." In *Annals of the New York Academy of Sciences* 1244: 169–223.

Thomas, D. S. G. and C. Twyman. 2005. "Equity and justice in climate change adaptation amongst natural-resource-dependent societies." *Global Environmental Change* 15 (2): 115–24.

U.S. Department of Agriculture. 2007. "Census of Agriculture." http://ww.nass.usda.gov/ny.

Walker, G. 2010. "Beyond distribution and proximity: Exploring the multiple spatialities of environmental justice." *Antipode* 41 (4): 614–36.

Walker, G. and K. Burningham. 2011. "Flood Risk, vulnerability and environmental justice: Evidence and evaluation of inequality in a UK context." *Critical Social Policy* 31 (2): 216–40.

Wolf, J. 2011. "Climate change adaptation as a social process." In *Climate Change Adaptation in Developed Nations*, edited by Ford, J. D. and L. Berrang-Ford, 21–32. Dordrecht: Springer Netherlands.

Wolfe, D. W., Comstock, J., Menninger, H., Weinstein, D., Sullivan, K., Kraft, C., Chabot, B., Curtis, P., Leichenko, R. and P. Vancura. 2011. "Ecosystems." *Annals of the New York Academy of Sciences* 1244: 169–223.

Wright, B. and R. Bullard. 2007. "Washed away by Hurricane Katrina: Rebuilding a 'new' New Orleans." In *Growing Smarter: Achieving Livable Communities, Environmental Justice, and Regional Equity*, edited by R. D. Bullard, 189–211. Cambridge, MA: MIT Press.

Wu, S. Y., Yarnal, B. and A. Fisher. 2002. "Vulnerability of coastal communities to sea-level rise: A case study of Cape May County, New Jersey, USA." *Climate Research* 22 (4): 255–70.

Yarnal, B. 2007. "Vulnerability and all that jazz: Addressing vulnerability in New Orleans after Hurricane Katrina." *Technology in Society* 29 (2): 249–55.

Zsamboky, M., Fernández-Bilbao, A., Smith, D., Knight, J. and J. Allan. 2011. "Impacts of climate change on disadvantaged UK coastal communities." http://www.jrf.org.uk/work/workarea/climate-change-and-social-justice.

7

Transforming toward or away from Sustainability?

How Conflicting Interests and Aspirations Influence Local Adaptation

SIRI ERIKSEN AND ELIN SELBOE

The complexity involved in adaptation is becoming more and more apparent. There is increasing recognition that adaptation cannot be seen merely as a technomanagerial challenge involving incremental adjustment to technologies, regulations, policies, and practices to *live* with change. Instead fundamental shifts in societal systems are required (O'Brien 2012), "including value systems; regulatory, legislative, or bureaucratic regimes; financial institutions; and technological or biological systems" (IPCC 2012, 3). This implies treating climate change as an adaptive challenge and addressing the drivers of vulnerability, including the assumptions and values that underpin naturalized and hegemonic understandings of change and how they serve particular interests (see Chapter 1).

Transformative change thus involves contesting change and questioning the decisions and development pathways that determine the magnitude, extent, and type of change to the climate system (Pelling 2011). Hence climate change highlights the need for *deliberate* transformation, that is, consciously taking action to influence future change toward more sustainable pathways. Sustainability here implies both reducing emissions and human influence on the climate system and addressing the social, political, and cultural causes of vulnerability, including greater justice and equity (Eriksen et al. 2011; Handmer and Dovers 2009; IPCC 2012; O'Brien 2012; Pelling 2011). This inevitably means that adaptation involves normative judgments about what is a desirable, ethical, and sustainable future and a questioning of some strongly held values and beliefs (O'Brien 2012).

There is, of course, no one desirable future, nor one correct blueprint for how to get there. Interests and actions may be highly differentiated. First, the challenges that people face vary, and the characters and causes of vulnerability differ between groups. Vulnerability is shaped by multiple stressors in addition to climate change, such as processes related to economic globalization (Eakin 2006; Leichenko and O'Brien 2008; O'Brien and Leichenko 2000; O'Brien 2011; Reid and Vogel 2006). Because such processes vary both in space and time, along with the social, economic, cultural, technological, and political conditions, the vulnerability context is highly localized.

Second, aspirations and values, which reflect what is considered important and what constitutes quality of life, vary both between individuals, groups, and communities and over time (Adger et al. 2009; Graham et al. 2013; Næss 2011; O'Brien 2009; Wolf et al. 2013). Values can be defined as "trans-situational conceptions of the desirable that give meaning to behaviours and events, and influence perception and interpretation of situations and events" (Wolf et al. 2013, 548). Consequently, values underpin aspirations and interests, guide actions, and shape the perception and interpretation of different aspects of reality, such as the significance and meanings people give to their communities and ways of life. Values are critically important to the way people make sense of their social and environmental surroundings, events, and change and thus influence perceptions of and responses to climate change (O'Brien 2009).

Different aspirations and interpretations, conflicts of interests, competing values, and the manifold ways that people act to reach their goals form important parts of the adaptation process (Wolf and Moser 2011; Wolf et al. 2013). Competing and conflicting interests, aspirations, and values could create limits to adaptation, particularly if not acknowledged and addressed (Wolf et al. 2013). In Norwegian rural communities, like Øystre Slidre, people often share a particular cultural and place identity (Fresque-Baxter and Armitage 2012), local economic development pathways, and changing landscapes (Vittersø 2012), but motivations to run a farm, including the role of lifestyle versus income and production, values, and conceptions of sustainability vary (Vik and Blekesaune 2008).

Despite emerging literature on climate change, values, and adaptation (Adger et al. 2009, 2011, 2013; Graham et al. 2013; O'Brien and Wolf 2010), the role of values and value conflicts in adaptation to climate change is still underresearched (Denton et al. 2014; O'Brien 2009; Wolf et al. 2013). Values and practices are mediated by power relations (Heyd and Brooks 2009), and the power and influence to act and achieve desired goals vary between actors (Cote and Nightingale 2012). Adaptation is thus both a social and political process, making it essential to investigate whose values count in conflicts of interest and decision making (O'Brien 2009).

In the context of this study, adaptation involves social processes where people, individually and collectively, deal with multiple types of environmental and social change occurring simultaneously. Hence adaptation includes daily decision making and practices at individual and household levels to achieve aspirations and well-being in the face of multiple changes as well as planned adjustment to practices and frameworks at the level of municipal, regional, and national planning and international policies (Commission of the European Union 2009; Dannevig et al. 2012; Eriksen 2013; Juhola et al. 2012). While the focus of this study is on how local adaptation processes are related to aspirations and values, we acknowledge that these processes and related negotiations between different interests are nested within other social and political processes and take place in a context of social and political change (Agrawal and Perrin 2009; Eriksen and Selboe 2012).

As a starting point for understanding how values influence local adaptation, we are looking at what people's aspirations and definitions of a meaningful life are, and how they strive to achieve them. We address several questions: Given variations in vulnerability and aspirations, what does responding to climate change mean for different groups in a Norwegian context? How do aspirations and understandings of threats to these differ among people in a local community? Importantly, how do these different understandings shape action, and how do the outcomes of conflicts and negotiations between different interests either increase or reduce vulnerability and influence development pathways toward sustainability?

We investigate these issues through a study of local processes of adaptation in the Norwegian municipality of Øystre Slidre. Data collection involved eighty-nine interviews in 2009 and 2010. These included forty-seven semistructured interviews with key informants, including local politicians, bureaucrats, farmers, and actors engaged in the tourism industry or other local and regional institutions as well as forty-two structured household interviews with farmers.[1] The interviews focused on long-term changes in the community, climatic events and shifts, political processes and priorities, local cultural and social identity, on- and off-farm activities, changes in production and market conditions, economic activities, social networks, and strategies to manage social and environmental change. Aspirations, values, and motivations were recurrent themes, transcending all these topics, and became an increasing part of the study. Interview data were supplemented by information from informal conversations and observations during field trips as well as formal documents.

In this chapter, we first describe the vulnerability context of Øystre Slidre. Next, we present a typology of three archetypes of community members, portraying their differing aspirations and values. We enquire whether the three archetypes identified are related to particular understandings of the climate change problem and other perceived threats. We then analyze how these contrasting aspirations and understandings translate into action, both in terms of strategies by individuals and households and in terms of how converging or conflicting interests within the community are reflected in the priorities and politics of the municipality. The case of Øystre Slidre demonstrates that individuals and groups experience threats and opportunities differently, depending in part on their aspirations and understandings of the multiple stressors that they face. As a result, climate change means very different things to different individuals and groups within a relatively small community and inspires different actions and considerations. Dominant interests and actors are shaping particular development pathways in a process of intense strategizing and negotiation between conflicting interests. Household and community aspirations and actions interact with decisions

---

[1] The 42 farming households were randomly selected from a list of the 169 active farms in the municipality in 2009.

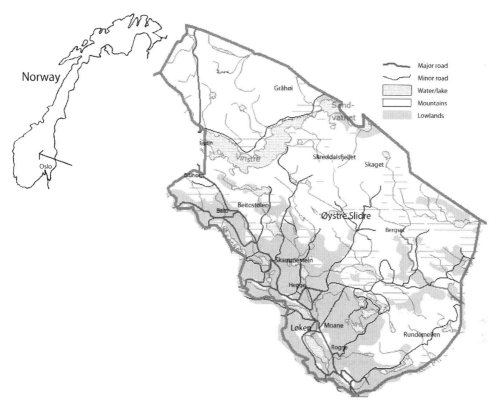

Figure 7.1. The location of Øystre Slidre. Developed from Norwegian Mapping Authority maps.

by municipal, regional, and national authorities (themselves shaped by particular values) to produce unpredictable, and sometimes unintended, adaptation outcomes.

Vulnerability and Transformation in Øystre Slidre

Øystre Slidre is a municipality in central southern Norway covering an area of 964 square kilometers (Øystre Slidre kommune 2014) and with a population of 3,204 (Statistics Norway 2014) (see Figure 7.1). It provides an illustrative example of local adaptation to climatic stressors, as the most important economic activities operate close to climatic thresholds, and climatic conditions have been shifting. For instance, the tourism industry is dependent on good skiing conditions in wintertime, such as early onset of snow and stable snow conditions throughout the winter. Also, the high altitudes at which people farm, 400 to 800 meters above sea level (masl), with areas up to 1,200 masl used for grazing, means there is a relatively short growing season, and the weather conditions prevalent from late April to mid-May (when the snow melts) to September are critical for agricultural production. Mountain ecosystems

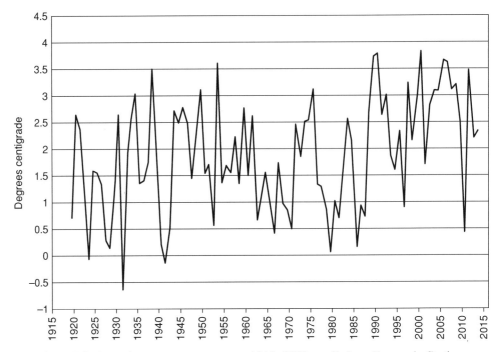

Figure 7.2. Annual mean temperatures, 1919–2013, at Løken Research Station, Øystre Slidre. The 1961–90 mean is 1.6°C. *Source*: Based on climate data from Løken Research Station and met.no.

are often particularly vulnerable to climate change because changes in temperatures lead to shifts in vegetation zones and growing seasons, winter tourism is sensitive to warming conditions, and government policies are often inappropriate for mountain systems (Dasgupta et al. 2014; NILF 1990; Porter et al. 2014).

Øystre Slidre has experienced a gradual warming over the last three to four decades. Data from Løken Research Station[2] show that there has been considerable temperature and intraseasonal variability historically. However, winters have become milder, and a general warming has accelerated since 1980 (Figure 7.2). This is consistent with an average warming of approximately 1°C in Norway over the past century (Norwegian Meteorological Institute 2014). Average annual temperatures in Norway are projected to rise another 2.5°C to 4.5°C over this century, with inland sites like Øystre Slidre projected to have temperature changes near the top end of this range (Hanssen-Bauer et al. 2009; Hanssen-Bauer 2010). The snowy season, which is currently about six months, is projected to become shorter, milder, and more unstable, while the growing season is expected to be extended (Benestad 2002; Hanssen-Bauer et al. 2003, 2009; Hanssen-Bauer 2010; Iversen et al. 2005).

---

[2] Located in the southern part of the municipality; see Figure 7.1.

Table 7.1. *Restructuring in the agricultural sector*

	Number of active farm units			Number of small farms (<5 ha)		Number of large farms (>50 ha)	
	Øystre slidre	Oppland region	Norway	Oppland region	Norway	Oppland region	Norway
1969	401	13,069	154,977	6,776	88,481	44	496
1999	247	7,054	70,740	1,248	14,517	128	1,577
2012	165	4,970	44,673	563	6,085	266	3,720

Source: Developed from Statistics Norway data tables 0846, 03313, 05971, http://www.ssb.no/statistikkbanken.

Among the local population, however, many societal and structural changes are seen as being at least as important as climate change in threatening economic activities, cultural identity, and quality of life. Some of the changes mentioned by farmer interviewees included already difficult and changing economic conditions such as increasing prices of agricultural inputs. The general development of the municipality and municipal (and regional) politics and decision making were perceived by many as disfavoring local agriculture and promoting the tourism industry and short-term economic gain, with insufficient concern for long-term effects. Other interviewees worried about the possibility of economic crisis, reduced employment opportunities, and population decline.

Farming, mainly animal husbandry, is the traditional mainstay in Øystre Slidre. However, an examination of agricultural data shows that the number of active farms has decreased significantly, in particular during the past decade (2000–2010) (see Table 7.1). This recent decline is part of a regional and national restructuring of the agricultural sector over the past decades, which has led to the number of farms being reduced by three quarters since 1959, increasing farm size and capital and production intensity and decreasing dairy farming[3] (Gjønnes 1998; Nersten 2001; Rognstad and Steinset 2010) (see Table 7.1). In Øystre Slidre, many have quit farming altogether, either selling their farms or renting out their land and milk production quota to remaining farmers. Others have turned their production from milk to meat or fodder production. For example, the number of dairy farms and goat farms in Øystre Slidre declined from 122 to 68 and from 15 to 8, respectively, between 1999 and 2012 (Statistics Norway 2013). Local agricultural production is now concentrated on fewer hands, but increases in production intensity and specialization means that milk production has remained relatively stable (Rognstad and Steinset 2010; Øystre Slidre kommune 2011b).

---

[3] The number of dairy farm units in Norway declined from 148,000 to 11,700 between 1959 and 2009 (Rognstad and Steinset 2010, 56).

The changes described have been accompanied by increased economic diversification in Øystre Slidre. The tourism sector has expanded rapidly over the last decades, in particular through the development of facilities in the Beitostølen area (see Figure 7.1), a well-known resort in Norway used by both domestic and European tourists. It has yielded opportunities for income for local landowners through the sale of plots for cabins and tourism development and has generated construction work, service jobs, and other tourism-related employment. Diversification options have not only been related to the growth of the tourism industry in the municipality; employment in the municipal administration, health, and education sectors has also been important. Now, almost all farming households have off-farm employment, securing additional income to sustain living standards and make investments. While 21 percent of employment in Øystre Slidre is provided by agriculture, four sectors that are directly and indirectly related to tourism – hotels and restaurants (16 percent), private services (10 percent), trade and shops (7 percent), and industry and construction (16 percent) – now constitute 49 percent of employment. Public services (such as municipal administration and social services) constitute the remaining 30 percent (Øystre Slidre kommune 2014).

Whereas public finances have remained stable, the local tourism industry is affected by international trends and shocks such as the 2008 financial crisis. Interviewees explained that although Norwegian tourists still visit the locality, there is a decline in the number of European visitors. In addition, the sales of cabins or rentals and sales of apartments has slowed down after a boom that lasted from the mid-1990s until 2001. Interviews with local building firms revealed that these firms have survived on a relatively steady supply of smaller jobs maintaining and renovating existing cabins. In local agriculture, the decrease in the number of active farms and the number of persons involved in agricultural production on these farms has led to changes in the social organization of adaptation (Eriksen and Selboe 2012). Although farmers continue to draw on local knowledge and social networks to plan and organize agricultural tasks that form part of local adaptation, there are now fewer people with whom to cooperate and with whom to build the required capacity to manage the short growing season and difficult weather events, such as wet summers. Adaptation strategies among farmers in Øystre Slidre are described in more detail in Eriksen and Selboe (2012); in the next sections we focus on how differing aspirations, values, and understandings of threats inspire particular actions and the implications for sustainability and deliberate transformation in the face of climate change.

A Typology of Aspirations, Understandings, and Actions

Interviews show that how people perceive climate change varies greatly within the local community. In particular, values and aspirations regarding what constitutes a desirable life for themselves and a good development for the community affect the way climate change is understood and how people act to manage climate variability in their daily lives. Three archetypes associated with sets of aspirations and values emerged from interview data.[4] The three archetypes represent a generalization of what was in fact a multitude of aspirations and values among informants, some intersecting or representing nuanced versions of the three main categories and a few representing more specific views, such as environmentalist. We have chosen to focus on the three most dominant understandings of local environment and development to discern contrasting values and interests and how conflicts and negotiations between them shape actions and decisions.

The first archetype comprises what we might call "lifestyle farmers," that is, those who saw farming as their main occupation and regarded farming traditions as an integral part of their identity. The farm had often been in their families for generations or, indeed, centuries. They emphasized that they had chosen farming as a lifestyle, often giving up holidays, higher incomes, and material standards of living for the opportunity to raise their children in the local environment, transferring knowledge, values, and, it was hoped, the farm to the next generation. Keeping animals and being able to seasonally live in the mountains while tending their animals were valued aspects of farming. While most had off-farm incomes, several expressed the wish to concentrate mainly on the farm and to reduce other economic activities. They saw agricultural production as contributing to local (community and family) traditions, social bonds and networks, and the cultural landscape and identity, hence sustaining what they saw as important community values. These attributes, they argued, were also important in making the area attractive to tourists.

As a contrast, the second archetype represents "business-oriented farmers," valuing individualistic success highly in addition to the common good of the community. They emphasized the modernizing aspect of farming, for example, being one step ahead of others in trying something new and being economically successful. Production was often more specialized, and entrepreneurship and gaining incomes comparable to other professions were important aspirations. In addition, the flexibility to take time off from farming and go on holidays, for example, was desirable. Informants often stressed the importance of their children having the freedom to choose other occupations rather than being locked into taking over the family farm.

The third archetype of community members represents "development-oriented actors." This includes municipal administrators, politicians, tourism entrepreneurs, and some farmers. These individuals tend to be concerned with economic growth and development, income generation, and stimulation of local businesses. The growth of the tourism industry was seen as instrumental in transforming Beitostølen into a modern tourism center, raising its profile as a resort and generating continued growth in employment in ways that neighboring municipalities had been unable to achieve.

---

[4] This typology of community members is developed inductively through qualitative analysis of interviews and conversations.

For local politicians, this was not only desirable in itself but part of a main strategy to maintain the current demographic stability in Øystre Slidre and avoid the declining number of inhabitants (particularly the young and resourceful) that has affected many small rural municipalities in Norway. Hence key aspirations and goals among informants in this group included ensuring continued tourism investment and developing new areas for construction of cabins as well as expanding settlement around Beitostølen and increasing its infrastructure and services. The image of Beitostølen as a snow-secure location was seen as an important part of local identity and the area's commercial viability.

There was a clear pattern in the way that climate change was understood and acted upon among community members of the three archetypes. In general, people in Øystre Slidre do not necessarily feel vulnerable to climate change. Most informants told us they feel certain that they will be able to manage potential or upcoming changes, "just like people have managed climatic changes in the past." For many, a lack of concern about climate change is due in part to a sense of confidence in their own adaptive capacity, and in part to uncertainty about the seriousness or magnitude of future climatic changes. Interviews and informal discussions indicate that although some people are aware of and "believe in" (anthropogenic) climate change, only a few feel that these changes do currently or will impact them directly. Consequently, only a few are engaged in mitigating and actively adapting to the situation. Such attitudes and nonresponses might be part of a socially organized denial of climate change (Norgaard 2006, 2011), where cultural norms of emotion, conversation, and attention reproduced through everyday practices decide what it is normal to think and talk about and naturalize silence and certain narratives about climate change and adaptation. However, the data also show that the perception of climate change and the need to adapt are closely related to values and conceptions of the local community and a desirable way of life.

The lifestyle farmers understood climate change as one of several threats to livelihoods and cultural identity. Some, though not all, in this group were concerned about climate change and saw it as representing or reinforcing a development pattern that is undermining small-scale farmer production, local culture, and sustainability. Their concern related mainly to local potential impacts. A sense of local (and global) solidarity, as well as local production and food security, were key values expressed. The threats that they recounted related also to the economic and political role of agriculture, such as declining agricultural producer prices, declining returns on agricultural investment, loss of grazing rights due to land conversion to tourism development, powerlessness in local decision making in the face of commercial tourism interests, environmental changes, and environmental policies such as the establishment of a conservation area limiting their activities in the mountains. They also mentioned the high dependence on government policies that vary with national political configurations. Policies are important in determining the economic viability of farming because they regulate producer prices and production quotas and provide financial support for production and some types of investments. Norwegian agriculture is market protected and relatively small scale and is characterized by multifunctionality beyond merely maximizing production. National policies are intended to give incentives to uphold a populated countryside, in addition to sustaining local agricultural production (Bjørkhaug and Richards 2008; Mittenzwei and Svennerud 2010). Furthermore, some farmers worried that future possibilities for food production and grazing are being destroyed, as many fertile areas of arable land were being used to build tourist accommodations and facilities, and agricultural land in the mountains was now taken out of use.

Most of these lifestyle farmers performed a range of activities that were adapted to climatic variability, even if not consciously directed at climate change mitigation or adaptation (Eriksen and Selboe 2012). Altering production systems to promote resilience and environmental stewardship was perceived as important in the face of climate change. Such actions are consistent with current strategies for managing climatic and environmental variability and maintaining the integrity of the environment (some farmers were also engaged in organic farming), a strong sense of place, and the cultural and social meaning of farming. Lifestyle farmers' responses to the combination of climatic and societal threats focused on limiting off-farm activities to focus efforts on running the farm and responding quickly to changes in weather conditions, active use of ecological diversity to respond to seasonal variability (moving animals between valley and mountain grazing areas), increasing the scale of production, innovation in production methods, creating local handicraft and niche production and businesses connected to their farm activities, building on local knowledge and social networks, and creating new forms of collaboration (Eriksen and Selboe 2012).

Among the more business-oriented farmers, concern was predominantly directed at threats to local economic activities from government regulations, economic and market fluctuations, and potential environmental and climate policies and less on climate change and other environmental changes. Their actions focused on rationalizing agricultural activities (investing in equipment and switching to labor-saving production forms), diversifying to off-farm activities with less reliance on farming and weather-dependent harvests, investing in their own equipment and hiring assistance to cope with seasonal and weather-dependent labor demands in agriculture, and engaging in new economic activities. As a contrast to the lifestyle farmers, who focused on investing and reorganizing their labor time according to the weather conditions, the predominant strategy for managing climatic variability among the business-oriented farmers involved using incomes from nonagricultural economic activities to invest in machinery, hire equipment or labor, or buy fodder (Eriksen and Selboe 2012). Like lifestyle farmers, they also moved animals between valley and mountain grazing areas, sought to increase the scale of production, and engaged in innovation in production methods as a way to manage environmental variability. However, this group was less preoccupied with adjusting farming activities to maximize production with seasonal and intraseasonal climatic variability, because farming was only one of several income strategies and only one of several sources of cultural identity and belonging.

The development-oriented actors generally saw climate change as a fairly insignificant factor to their desired goals for Øystre Slidre and emphasized the potential threat of demographic and economic decline in the face of structural economic changes. Developing tourism facilities to sustain economic growth and incomes was central to aspirations among this group. The idea and image of Øystre Slidre as a stable, cold, and snowy place in future fit best with these aspirations, rather than one of warming and unreliable winter conditions. Thus climate change was not perceived as a threat to local activities. Contrary to vehement insistence by some interviewees in this group that winters never fail in Øystre Slidre, the early winter of 2011–12 was so warm that an international Nordic skiing competition had to be canceled. Later in that season, much of the snow disappeared before the economically important Easter holidays. Interestingly, the importance of weather for the tourism sector is nevertheless reflected in the fact that the first item mentioned in the municipal annual report was an account of the warm wet autumn and late onset of snow, affecting the tourism sector and the skiing competition (Øystre Slidre kommune 2011a).

Although some actors in the tourism industry were very concerned about climate change, many were reluctant to admit that they worried about it. However, several informants mentioned the importance of snow for the tourism industry of the locality. For example, one interviewee explained that he knew that increasingly late onset of winter could bring problems that could not even be solved with the production of artificial snow, because such production depends on low temperatures. Still, he claimed that they would probably cope relatively better in Øystre Slidre than elsewhere: "It [temperature rise during winter/lack of snow] will be bad, but it will be 'less bad' here, as we are one of the most snow-secure localities!"

The development-oriented actors were particularly worried that regional and national policies on biodiversity and climate change might limit local construction of cabins and tourism development. The most widespread concern in this group was how the revegetation of mountain landscapes could affect tourism. Revegetation can be partly attributed to warming and partly attributed to changing grazing regimes. Reduced numbers of animals and species grazing and browsing in the mountains means that the vegetation is less effectively kept down than previously (Skarstad et al. 2008). Across all three archetypes, the revegetation of the mountain landscape was seen as a negative change to the cultural landscape, and there were some efforts to introduce strategic grazing and clearing of forest.

Adaptation to climate change was not seen as important among the development-oriented actors, whereas strong government measures to reduce energy consumption were mentioned as increasing the risk of losing tourism investments to other areas.

However, addressing increased energy use due to expansion of tourism facilities and reducing emissions from increased driving by tourists from urban areas to Øystre Slidre was considered outside the scope of local action (or responsibility). A plan for reducing greenhouse gas (GHG) emissions was passed in the municipal council in 2009 (Øystre Slidre kommune 2009). This was mainly implemented through plans for concentrating construction around Beitostølen (to reduce transport) as well as suggestions to expand ski lifts between different mountain areas (to limit driving between them). In interviews with the municipal administration, the 2009 plan came across as isolated and rarely actively invoked in other planning activities, in line with previous observations that such planning is often initiated top-down from the national level of government and political parties (Aall 2012).

Inherent Paradoxes in Local Development Pathways

Intense strategizing and negotiations between community members represented by the three archetypes shape particular development pathways for Øystre Slidre. Amid multiple aspirations among the local population, the climate issue becomes nested in conflicting interests regarding tourism development and land use, framed by national policies and international economic processes. Dominant in shaping local development directions are the strong alliances between people represented in archetypes 2 and 3, contributing to commercial development (agricultural and tourism) as near-hegemonic.

The development pathway of Øystre Slidre, as promoted by local government and economic actors, involves a strong focus on tourism-led short- and medium-term economic growth. Tourism development has taken the form of a steady increase in the number of cabins, hotels/apartments, and ski lifts over the past few decades. This is in line with national trends, where the number of cabins increased by on average 5,000 per year between 1973 and 2010 (Aall et al. 2011). Even after the end of the main boom, the number of cabins in Øystre Slidre increased from 2,735 in 2001 to 3,393 in 2014 (Statistics Norway 2014). This development is the result of local entrepreneurship and creativity within the tourism industry combined with decisions by municipal authorities facilitating such development, all within the context of a national trend of increasing demand for tourism facilities and cabins.

Clearly this local development path has many positive consequences, most notably demographic stability and economic growth as well as opportunities for economic diversification in a region where several municipalities have experienced outmigration and declining population numbers. Employment, as well as sales of land plots to tourism facilities, enables raising living standards and economic investment required to run and upgrade farms. The feeling of dependency on the tourism sector is pronounced, particularly among the development-oriented actors. Many local politicians and the municipal administrators share a common perception with the local tourism industry that this sector (and Beitostølen as a resort) is the "engine" of the municipality, justifying why efforts need to be concentrated on tourism to ensure continued economic growth and demographic stability. They recount how taxes from tourism-related activities enable the municipal administration to provide public services for the whole of Øystre Slidre, including schooling and health services. In the face of a farming sector that is perceived as declining (at least in number of farmers and employment), the dependence of Øystre Slidre on growth and development of local tourism becomes the mainstream discourse in the municipality.

Nevertheless, there are also paradoxes inherent in this pathway. Rising energy use and emissions contribute to climate change that may in the long term undermine the stable winter conditions on which tourism development depends and which forms part of the local cultural identity. Although the sector is increasingly facilitating summer and autumn tourism, the area is still intimately tied to winter tourism and is dependent on good snow cover for skiing, the conditions for which are likely to deteriorate under climate change. The tourism-led economic growth has driven a sharp increase in both energy use and GHG emissions. Total electricity use[5] in the municipality increased by 25 percent from 2000 to 2006, much of it due to electric heating of tourism accommodation. CO_2 emissions increased by 11 percent from 1991 to 2006, mostly due to increased traffic within the municipality (Øystre Slidre kommune 2009). This development is embedded in national-level development pathways. Consumption and energy use, which are closely linked to GHG emissions, have risen steadily in outdoor recreation and leisure activities in Norway over the past decades, driven by an increase in the number, size, and technical standard of cabins; increased driving to cabins in private cars; and a dramatic growth in consumption of equipment and clothing for such activities (Aall et al. 2011). These changes reflect a shift from people valuing simple (low-consumption) outdoor recreation to people preferring easier (low-effort) recreation such as high-standard cabins and easy accessibility by car. Despite the high increase in leisure transport and consumption, this issue has been largely overlooked in government policy (Aall et al. 2011).

In addition to leading to a pathway of increasing emissions, the high and increasing dependency on the tourism sector may also make Øystre Slidre vulnerable to global trends, such as the current financial crises leading to a decline in international tourism. In the case of Øystre Slidre, value changes – and climate change impacts on these – among groups outside Øystre Slidre, such as any loss of interest in skiing among domestic tourists due to loss of snow cover in more densely populated areas near the coast, could affect areas like Øystre Slidre profoundly. Although these issues are not examined in the current study, behavioral change among skiers in the face of warming

---

[5] This is mainly renewable energy, but the increase might affect the potential for export of renewable energy.

conditions has been observed elsewhere, such as some skiers preferring to switch activities rather than to travel further to access snow in the face of warming conditions (Landauer et al. 2009, 2012).

There is an uneasy and complex relationship between farming and tourism. While many farmers are opposed to unlimited tourism development and development that changes the traditionally rural landscape around Beitostølen, they are also dependent on the tourism sector for off-farm incomes that can sustain investments in farming activities and an acceptable material standard of living in the face of declining relative incomes from agriculture.[6] Importantly, although the tourism-led development pathways provide income diversification, they may also undermine important features of local adaptive capacity, in particular local knowledge and networks of collaboration critical to managing climatic variability as people and labor are drawn from agriculture to other tourism-related activities (Eriksen and Selboe 2012). Building cabins, hotels, and homes, tourist and ski resorts on high-quality arable land may limit long-term local agricultural production, a particular problem if global food shortages driven by economic and climatic change increase the need to rely on local food production. The pressure for constructing tourism facilities around Beitostølen has weakened farmer access to key grazing areas, and several instances of conflicts related to customary grazing rights have developed.

Lifestyle farmers try to resist the hegemony of tourism-led development and instead promote local grazing rights, small-scale farming interests, and restriction of tourism development. Several farmers voiced their dissent in public meetings, in committees, through engagement in farmer organizations and other pressure groups, and in newspapers. One farmer recounted how local resistance had reduced the scale of an ongoing residential building development scheme in an area valued for grazing in Beitostølen. Nevertheless, many lifestyle farmers lamented that the tourism industry had most clout in decision making and that there was little support for traditional agricultural practices where these practices might restrict existing or potential tourist activities or development.

The case of Øystre Slidre illustrates how negotiations and power imbalances between actors holding different values and interests may lead to a community being locked into particular development pathways. While resulting in short- and medium-term economic growth and employment, they also transform the local community and may be potentially unsustainable in terms of increasing emissions and vulnerability in a long-term perspective. What is the space for contesting current pathways and pursuing alternative pathways that would transform the local community in a different, and potentially more sustainable, direction?

---

[6] Relative to other economic activities and the dramatic growth in incomes and material standards of living that Norway has experienced over the past few decades.

Conclusions: Messy Adaptation and Unintended Transformation?

The case of Øystre Slidre illustrates that values, aspirations, and perceptions of threats and opportunities are important for how people interpret and act on climate change. Importantly, even if people do not think that climatic conditions affect them, it is the daily decisions by farmers, households, businesses, and the municipal administration that drive the direction of development toward more or less vulnerability and sustainability.

This study highlights several adaptive challenges. O'Brien (2009, 177) suggests that it is important to "identify adaptation strategies that acknowledge and address a spectrum of values." If values are not overtly acknowledged, value conflicts and impacts on what people value may be ignored, and such an inability to respond to different value priorities may represent a barrier to adaptation (Wolf et al. 2013). Findings in Øystre Slidre suggest that deliberate actions inspired by varying, but largely covert, values and aspirations may foster unintended transformations. At the level of individuals and households, there is agency, creativity, and evidence of people actively maintaining local knowledge to manage climate variability and create new income opportunities. There are also deliberate attempts by the local authorities and the tourism industry to transform Beitostølen into a thriving tourism center. However, the development pathway resulting from negotiations and power imbalances between these different interests also threatens to undermine future adaptive capacity and increase GHG emissions.

This case illustrates the adaptive challenges created by a development pathway characterized by Pelling (2011) as "trying to grow out of climate change problems," in effect increasing the problems in the process. The massive changes that Øystre Slidre has undergone over the past decades in terms of farming systems and economic development have resulted in overall unintended structural transformations, despite deliberate actions. These changes are formed by growth opportunities in tourism, national development, and policies and by the interests of powerful groups and individuals. In particular, the combination of growing dependence on tourism, a perception that climate change is insignificant in Øystre Slidre, and a fear that policies imposed by the central government will restrict tourism development may limit local responses to climate change.

The dominant development discourse may make the current pathway appear unavoidable, given the need for economic growth to maintain social services and population numbers, growing national consumption levels and demand for energy-intensive types of tourism activities, and the need for off-farm incomes to bolster farm incomes. The increase in leisure transport and consumption can also be seen as part of a national development pathway based on energy-intensive consumption, where increasing leisure consumption is required to drive continued economic growth, hence becoming part of the climate change problem. This national pathway is reinforced by local interests in expanding tourism incomes and growth, as such development is considered likely to generate the most local business and employment (Aall et al. 2011).

At the same time, however, some spaces for alternative development pathways may exist that are currently locked out by the dominant development discourse. The hegemonic discourse focuses on never-ending economic growth and consumption as an unquestioned aspiration, ignoring local aspirations related to lifestyle, cultural identity, and social networks. These more diverse aspirations could foster a wider range of options for tourism and economic development, for example, that are based on environmental stewardship and related aspects of local cultural identity, potentially generating lower rates of economic growth in the shorter term but more in line with diverse aspirations such as farming as a lifestyle and supporting long-term adaptive capacity in the face of climate change. People's actions and knowledge, even if not always deliberately aimed at climate change, also embody a potential for more sustainable transformations. There is a strong local tradition of entrepreneurship and creativity. These are also values appreciated across all three archetypes of community members. Hence the local ecological knowledge and pride in management of the cultural landscape observed in Øystre Slidre provide a potential for transformation *and* adaptation to climate change. The ability to manage climatic variability and change is central to local identity, and although it may contribute to a tendency to reject climate change as a threat, if built on, it may also contribute to more climate resilient pathways.

Aall et al. (2011) suggest that increased energy efficiency in the tourism sector and stronger regulations limiting the type and extent of tourism development, in particular number and size of cabins as well as other government planning actions, may be required to reduce rising energy use and environmental impacts in outdoor recreation. However, our study indicates that government regulations and planning may not be enough in the context of adaptation. A main obstacle to realizing the spaces for alternative pathways and adaptation potential may lie in ignoring value conflicts between different actors. A stronger overall vision for development is required that takes account of diverse aspirations and values and makes prioritization between them transparent. At present, there is a disconnect between climate responses, which are perceived as the domain and responsibility of government authorities and municipal planning, and individual farmer, tourism, and business aspirations and efforts to manage a multitude of threats. As a result, policies, and sometimes the recognition of climate change, are regarded as a potential threat to many of the strategies that actually constitute local adaptive capacity. This illustrates the dangers of depoliticizing climate change, and climate change adaptation in particular, making it a question of neutral policy making while ignoring how values and goals are contested and negotiated between different actors (Manuel-Navarrete 2010; Nightingale 2009; Shove and Walker 2007; Swyngedouw 2007, 2013). Such depoliticization often contributes to particular development goals being able to dominate (Eriksen and Marin 2014; Shove 2010); in addition, it blocks local adaptation potential and spaces for alternative, more sustainable development pathways.

Indeed, the case of Øystre Slidre illustrates how climate responses take place through a messy process of negotiation and cooperation, with tensions between different values, interests, and understandings. The direction of this process is often driven by implicit prioritization of values and unequal power relations. The result can be quite opposite to the deliberate types transformation that, according to O'Brien (2012), are required to respond to the climate change problem. In particular, the study illustrates that giving space to diverse values, interests, and aspirations in decision-making processes is critical for generating deliberate transformation. These observations support previous suggestions that adaptation needs to be an empowering process rather than a technical measure (Manuel-Navarrete 2010; Moser 2013; O'Brien et al. 2015).

This study illustrates the importance of research regarding how both adaptation and transformation are driven by values, both among different groups in a local community and among the national population that consumes goods and services, including tourism in locations like Øystre Slidre. Furthermore, the study points to a need to understand how national processes may create barriers or opportunities for alternative development pathways at the local level, and how these may in turn drive transformations at the national level. New ways of creating climate responses are needed where plurality in values, aspiration, local action, and creativity becomes integral to realizing rather than undermining the local adaptation potential and spaces for alternative development pathways. Of course, attempts at achieving any common sustainability goals are subject to the same types of politics in terms of power relations and negotiations between conflicting interests (Shove and Walker 2007). Thus moving toward more sustainable development pathways involves a process of interactions and negotiations between multiple acknowledged, rather than implicit or silenced, goals and aspirations. The case of Øystre Slidre illustrates that for local climate responses to move toward more sustainable and potentially less conflictual pathways, decision-making processes must be put in places where value differences are overtly addressed.

Acknowledgments

We would like to thank all the participants in the study for their time and willingness to share. We are also grateful to Inger Hanssen-Bauer and met.no as well as to Bioforsk Løken (Løken Research Station), and Håkon Skarstad and Tor Lunnan, in particular, for providing local meteorological data and insights regarding climatic trends. Thanks to Ian Harris at the Climatic Research Unit, UEA, Norwich, for assisting in creating Figure 7.2. We are very grateful to Josie Teurlings (Noragric, Norwegian University of Life Sciences) for helping to develop a map of Øystre Slidre (Figure 7.1). We would also like to thank Lars Otto Næss (IDS, Brighton) and two anonymous reviewers for their valuable comments on an earlier draft. The study described in this chapter forms part of the "Potentials for and Limitations to Adaptation to Climate Change in Norway" (PLAN) project, led by Karen O'Brien at the Department of Human Geography and Sociology at the University of Oslo and funded by the Research Council of Norway. The findings in this chapter remain the responsibility of the authors, however.

References

Aall, C. 2012. "The early experiences of local climate change adaptation in Norwegian compared with that of local environmental policy, Local Agenda 21 and local climate change mitigation." *Local Environment: The International Journal of Justice and Sustainability* 17 (6–7): 579–95.

Aall, C., Klepp, I. G., Engset, A. B., Skuland, E. E. and E. Støa. 2011. "Leisure and sustainable development in Norway: part of the solution and the problem." *Leisure Studies* 30 (4): 453–76.

Adger, N. W., Barnett, J., Brown, K., Marshall, N. and K. O'Brien. 2013. "Cultural dimensions of climate change impacts and adaptation." *Nature Climate Change* 3 (2): 112–17.

Adger, N. W., Barnett, J., Chapin III, F. S. and H. Ellemor. 2011. "This must be the place: Underrepresentation of identity and meaning in climate change decision-making." *Global Environmental Politics* 11 (2): 1–25.

Adger, N. W., Dessai, S., Goulden, M., Hulme, M., Lorenzoni, I., Nelson, D., Naess, L., Wolf, J. and A. Wreford. 2009. "Are there social limits to adaptation to climate change?" *Climate Change* 93 (3): 335–54.

Agrawal, A. and N. Perrin. 2009. "Climate adaptation, local institutions and rural livelihoods." In *Adapting to Climate Change: Thresholds, Values, Governance*, edited by Adger, N., Lorenzoni, I. and K. O'Brien, 350–67. Cambridge: Cambridge University Press.

Benestad, R. E. 2002. "Empirically downscaled temperature scenarios for Northern Europe based on a multi-model ensemble." *Climate Research* 21 (2): 105–25.

Bjørkhaug, H. and C. A. Richards. 2008. "Multifunctional agriculture in policy and practice? A comparative analysis of Norway and Australia." *Journal of Rural Studies* 24 (1): 98–111.

Commission of the European Union. 2009. "Adaptation to climate change: A framework for European Action. COM(2009) 147 final." http://eur-lex.europa.eu/legal-content/EN/TXT/PDF/?uri=CELEX:52009DC0147&from=EN.

Cote, M. and A. J. Nightingale. 2012. "Resilience thinking meets social theory: Situating social change in socio-ecological systems (SES) research." *Progress in Human Geography* 36 (4): 475–89.

Dannevig, H., Rauken, T. and T. Hovelsrud. 2012. "Implementing adaptation to climate change at the local level." *Local Environment* 17 (6–7): 597–611.

Dasgupta, P., Morton, J. F., Dodman, D., Karapinar, B., Meza, F., Rivera-Ferre, M. G., Toure Sarr, A. and K. E. Vincent. 2014. "Rural areas." In *Climate Change 2014: Impacts, Adaptation, and Vulnerability. Part A: Global and Sectoral Aspects. Contribution of Working Group II to the Fifth Assessment Report of the Intergovernmental Panel on Climate Change*, edited by Field, C. B., Barros, V. R., Dokken, D. J., Mach, K. J., Mastrandrea, M. D., Bilir, T. E., Chatterjee, M., et al., 613–57. Cambridge: Cambridge University Press. http://ipcc-wg2.gov/AR5/images/uploads/WGIIAR5-Chap9_FGDall.pdf.

Denton, F., Wilbanks, T., Abeysinghe, A., Burton, I., Gao, Q., Lemos, M. C., Masui, T., et al. 2014. "Climate-resilient pathways: Adaptation, mitigation and sustainable development." In *Working Group II Contribution to the Intergovernmental Panels on Climate Change Fifth Assessment Report Climate Change 2014: Impacts, Adaptation and Vulnerability*, chapter 20. http://ipcc-wg2.gov/AR5/images/uploads/WGIIAR5-Chap20_FGDall.pdf.

Eakin, H. 2006. *Weathering Risk in Rural Mexico: Climatic, Institutional, and Economic Changes.* Tucson: University of Arizona Press.

Eriksen, S. 2013. "Understanding how to respond to climate change in a context of transformational change: The contribution of sustainable adaptation." In *Changing Environment for Human Security: Transformative Approaches to Research, Policy and Action*, edited by Sygna, L., O'Brien, K. and J. Wolf, 363–74. London: Earthscan by Routledge.

Eriksen, S. and A. Marin. 2014. "Sustainable adaptation under adverse development? Lessons from Ethiopia." In *Adaptation to Climate Change: Development as Usual Is Not Enough*, edited by Inderberg, T. H., Eriksen, S., O'Brien, K. and L. Sygna, 178–99. London: Routledge.

Eriksen, S. and E. Selboe. 2012. "The social organisation of adaptation to climate variability and global change: The case of a mountain farming community in Norway." *Applied Geography* 33: 159–67.

Eriksen, S., Aldunce, P., Bahinipati, C. S., Martins, R. D'A., Molefe, J. I., Nhemachena, C., O'Brien, K., et al. 2011. "When not every response to climate change is a good one: Identifying principles for sustainable adaptation." *Climate and Development* 3 (1): 7–20.

Fresque-Baxter, J. A. and D. Armitage. 2012. "Place identity and climate change adaptation: A synthesis and framework for understanding." *WIREs Climate Change* 3 (3): 251–66.

Gjønnes, K. 1998. "Norsk landbruk igår, idag og imorgen" [Norwegian agriculture yesterday, today and tomorrow]. In *The Role of Free Market or Market Interventions in the Agricultural Policy*, edited by Romarheim, H. and A. Haglerød, 13–22. Oslo: Norwegian Agricultural Economics Research Institute.

Graham, S., Barnett, J., Fincher, R., Hurlimann, A., Mortreux, C. and E. Waters. 2013. "The social values at risk from sea-level rise." *Environmental Impact Assessment Review* 41: 45–52.

Handmer, J. and S. Dovers. 2009. "A typology of resilience: Rethinking institutions for sustainable development." In *The Earthscan Reader on Adaptation to Climate Change*, edited by Schipper, E. L. F. and I. Burton, 187–210. London: Earthscan.

Hanssen-Bauer, I. 2010. "Klima i Norge gjennom det 20. og 21. århundre. Fokus: Øystre Slidre" [The climate in Norway through the 20th and 21st centuries. Focus: Øystre Slidre]. Presentation at an open meeting in Øystre Slidre, March 18.

Hanssen-Bauer, I., Drange, H., Førland, E. J., Roald, L. A., Børsheim, K. Y., Hisdal, H., Lawrence, D., et al. 2009. *Klima i Norge 2100: Bakgrunnsmateriale til NOU Klimatilpasning.* [Norwegian climate in 2100: Background material for Official Norwegian Report on climate change adaptation]. Oslo: Norsk klimasenter.

Hanssen-Bauer, I., Førland, E. J., Haugen, J. E. and O. E. Tveito. 2003. "Temperature and precipitation scenarios for Norway: Comparison of results from dynamical and empirical downscaling." *Climate Research* 25: 15–27.

Heyd, T. and N. Brooks. 2009. "Exploring cultural dimensions of adaptation to climate change." In *Adapting to Climate Change: Thresholds, Values, Governance*, edited by Adger, W. N., Lorenzoni, I. and K. L. O'Brien, 269–82. Cambridge: Cambridge University Press.

IPCC. 2012. "Summary for policymakers." In *Managing the Risks of Extreme Events and Disasters to Advance Climate Change Adaptation*, edited by Field, C. B., Barros, V.,

Stocker, T. F., Qin, D., Dokken, D. J., Ebi, K. L., Mastrandrea, M. D., et al., 1–19. Cambridge: Cambridge University Press.
Iversen, T., Benestad, R., Haugen, J. E., Kirkevåg, A., Sorteberg, A., Debernard, J., Grønås, S., et al. 2005. "RegClim. Norges klima om 100 år. Usikkerhet og risiko." [RegClim. Norwegian climate in 100 years. Uncerntainties and risks]. Meteorologisk Institutt/Institutt for geofag/Bjerknessenteret for klimaforskning. http://regclim.met.no.
Juhola, S., Haanpää, S. and L. Peltonen. 2012. "Regional challenges of climate change adaptation in Finland: Examining the ability to adapt in the absence of national level steering." *Local Environment* 17 (6–7): 629–39.
Landauer, M., Probstl, U. and W. Haider. 2012. "Managing cross-country skiing destinations under the conditions of climate change – Scenarios for destinations in Austria and Finland." *Tourism Management* 33: 741–51.
Landauer, M., Sievanen, T. and M. Neuvonen. 2009. "Adaptation of Finnish cross-country skiers to climate change." *Fennia* 187 (2): 99–113.
Leichenko, R. M. and K. L. O'Brien. 2008. *Environmental Change and Globalization: Double Exposures*. Oxford: Oxford University Press.
Manuel-Navarrete, D. 2010. "Power, realism, and the ideal of human emancipation in a climate of change." *WIREs Climate Change* 1 (6): 781–85.
Mittenzwei, K. and M. Svennerud. 2010. "Importvern for norsk jordbruk: Status og utviklingstrekk" [Import protection in Norwegian agriculture: Status and trends]. Notat 2010-12. NILF, Norsk institutt for landbruksforskning. http://ww.nilf.no/publikasjoner/Notater/2010/N201012Hele.pdf.
Moser, S. 2013. "Individual and community empowerment for human security." In *A Changing Environment for Human Security: Transformative Approaches to Research, Policy and Action*, edited by Sygna, L., O'Brien, K. and J. Wolf, 279–93. London: Earthscan.
Næss, S. 2011. "Språkbruk, definisjoner." In [Language, definitions]. *Livskvalitet: forskning om det gode liv* [Quality of life: research on the good life], edited by Næss, S., Moum, T. and J. Eriksen, 15–51. Bergen: Fagbokforlaget.
Nersten, N. K., ed. 2001. *Norwegian Agriculture: Status and Trends 2001*. Oslo: Norwegian Agricultural Economics Research Institute.
Nightingale, A. 2009. "Warming up the climate change debate: A challenge to policy based on adaptation." *Journal of Forest and Livelihood* 8 (1): 84–90.
NILF. 1990. *Konsekvenser for jordbruksproduksjonen av økte klimagassutslipp*. [The consequences of increased GHG emissions on agricultural production]. Oslo: Norwegian Agricultural Economics Research Institute.
Norgaard, K. M. 2006. "'We don't really want to know': Environmental justice and socially organized denial of global warming in Norway." *Organization and Environment* 19 (3): 347–70.
Norgaard, K. M. 2011. *Living in Denial: Climate Change, Emotions and Everyday Life*. Cambridge, MA: MIT Press.
Norwegian Meteorological Institute. 2014. "Norge fra 1900 til i dag" [Norway from 1900 until today]. http://met.no/Klima/Klimautvikling/Klima_siste_150_ar/Hele_landet/.
O'Brien, K. 2009. "Do values subjectively define the limits to adaptation?" In *Adapting to Climate Change: Thresholds, Values, Governance*, edited by Adger, W. N., Lorenzoni, I. and K. O'Brien, 164–80. Cambridge: Cambridge University Press.
O'Brien, K. 2011. "Responding to environmental change: A new age for human geography?" *Progress in Human Geography* 35 (4): 542–49.
O'Brien, K. 2012. "Global environmental change II: From adaptation to deliberate transformation." *Progress in Human Geography* 36 (5): 667–76.
O'Brien, K. L. and R. M. Leichenko. 2000. "Double exposure: Assessing the impacts of climate change within the context of economic globalization." *Global Environmental Change* 10 (3): 221–32.

O'Brien, K. L. and J. Wolf. 2010. "A values-bases approach to vulnerability and adaptation to climate change." *WIREs Climate Change* 1 (2): 232–42.

O'Brien, K., Eriksen, S., Inderberg, T. H. and L. Sygna. 2015. "Climate change and development: Adaptation through transformation." In *Climate Change Adaptation and Development: Changing Paradigms and Practices*, edited by Inderberg, T. H., Eriksen, S. H., O'Brien, K. and L. Sygna, 273–89. London: Routledge.

Øystre Slidre kommune. 2009. "Energi- og klimaplan 2009–2013" [Energy and climate plan 2009–2013]. http://www.oystre-slidre.kommune.no/Filnedlasting.aspx?MId1=1052&FilId=1220.

Øystre Slidre kommune. 2011a. "Årsmelding" [Annual report]. Heggenes, Norway: Øystre Slidre kommune.

Øystre Slidre kommune. 2011b. "Statistikk for landbruket i Øystre Slidre" [Agricultural statistics, Øystre Slidre]. http://www.oystre-slidre.kommune.no/Filnedlasting.aspx?MId1=26&FilId=688.

Øystre Slidre kommune. 2014. "Om Øystre Slidre" [About Øystre Slidre]. https://www.oystre-slidre.kommune.no/om-oss/.

Pelling, M. 2011. *Adaptation to Climate Change: From Resilience to Transformation*. Abingdon, UK: Routledge.

Porter, J. R., Xie, L., Challinor, A. J., Cochrane, K., Howden, S. M., Iqbal, M. M., Lobell, D. B. and M. I. Travasso. 2014. "Food security and food production systems." In *Climate Change 2014: Impacts, Adaptation, and Vulnerability. Part A: Global and Sectoral Aspects. Contribution of Working Group II to the Fifth Assessment Report of the Intergovernmental Panel on Climate Change*, edited by Field, C. B., Barros, V. R., Dokken, D. J., Mach, K. J., Mastrandrea, M. D., Bilir, T. E., Chatterjee, M., et al., 485-533. Cambridge: Cambridge University Press. http://ipcc-wg2.gov/AR5/images/uploads/WGIIAR5-Chap7_FGDall.pdf.

Reid, P. and C. Vogel. 2006. "Living and responding to multiple stressors in South Africa – Glimpses from KwaZulu-Natal." *Global Environmental Change* 16 (2): 195–206.

Rognstad, O. and T. A. Steinset. 2010. "Landbruket i Norge 2009, Jordbruk – skogbruk – jakt" [Agriculture in Norway 2009, Cultivation – forestry – hunting]. Oslo: Statistics Norway.

Shove, E. 2010. "Beyond the ABC: Climate change policy and theories of social change." *Environment and Planning A* 42 (6): 1273–85.

Shove, E. and G. Walker. 2007. "Caution! Transitions ahead: Politics, practice, and sustainable transition management." *Environment and Planning A* 39 (4): 763–70.

Skarstad, H. S., Dagstad, K., Lunnan, T. and H. Sickel. 2008. "Vilkår og tiltak for å opprettalde stølsdrift i og utanfor verneområde" [Conditions and measures to sustain mountain farming in and outside protected area]. *Bioforsk Report* 3: 174.

Statistics Norway. 2013. "Strukturen i jordbruket" [Structure of agriculture], table 06459 (municipal), http://ww.ssb.no/statistikkbanken.

Statistics Norway. 2014. "Population and population changes, Q2 2014." http://www.ssb.no/en/befolkning/statistikker/folkendrkv/kvartal/2014-08-20?fane=tabell&sort=nummer&tabell=192064.

Swyngedouw, E. 2007. "Impossible 'sustainability' and the postpolitical condition." In *The Sustainable Development Paradox: Urban Political Economy in the United States and Europe*, edited by Krueger, R. and D. Gibbs, 13–40. New York: Guilford Press.

Swyngedouw, E. 2013. "The non-political politics of climate change." *Acme* 12 (1): 1–8.

Vik, J. and A. Blekesaune. 2008. "Trender i norsk landbruk 2008. Oppland" [Trends in Norwegian agriculture. Oppland]. Notat nr 2/08. Trondheim: Centre for Rural Research.

Vittersø, G. 2012. *Ren idyll?: forbrukes betydning for bygdeutvikling med utgangspunkt i lokal mat og hytteliv*. [Pure idyll?: the significance of consumption for rural development on the basis of local food and cabin life]. PhD Thesis, University of Oslo.

Wolf, J. and S. Moser. 2011. "Individual understandings, perceptions, and engagement with climate change: Insights from in-depth studies across the world." *Wiley Interdisciplinary Reviews – Climate Change* 2 (4): 547–69.

Wolf, J., Allice, I. and T. Bell. 2013. "Values, climate change, and implications for adaptation: Evidence from two communities in Labrador, Canada." *Global Environmental Change* 23 (2): 548–62.

8

Opportunistic Adaptation

New Discourses on Oil, Equity, and Environmental Security

BERIT KRISTOFFERSEN

That Arctic nation-states might utilize climate change opportunistically first occurred to me during a conference addressing adaptation measures for Svalbard in 2008. Svalbard is an Arctic Archipelago under Norwegian jurisdiction halfway between the Norwegian mainland and the North Pole. At the conference, researchers, local authorities, and businesses came together for three days to address different aspects of future scenarios and strategies to cope with Svalbard's expected 4°C to 8°C temperature increase (NorACIA 2011, 37).[1] During the final session, our findings from the conference were summarized into positive and negative consequences of climate change for Svalbard and for Norway. The summary would then be considered in the Ministry of Foreign Affairs's (MFA) forthcoming white paper "Svalbard" (Ministry of Foreign Affairs 2009). As the conference discussion progressed, the list detailing the positive consequences was getting extensive. The "threat" of climate change was being reframed into a scenario of possibilities where the changing geography of the Arctic landscape was seen as an opportunity: the rapid warming of the region might allow for new shipping routes, increased access to natural resources such as fish and oil, civilian and military preparedness measures, and tourism. These trends in future scenarios seemed more interesting to discuss in our final session than changes in permafrost, biodiversity, and negative socioeconomic impacts for the Svalbard communities. A few months after the conference the tendency toward focusing on the benefits of climate change was reflected in the MFA's white paper on Svalbard. It stated in its introduction that climate change was "creating challenges" while "simultaneously creating possibilities and expectations of increased activity in the North" (Ministry of Foreign Affairs 2009, 7–8).

This experience triggered my own research interest in whether the perceived benefits of climate change in the Arctic were in the process of being incorporated into state

---

[1] The conference was organized by NorACIA (2005–9), the Norwegian follow-up to the Arctic Council project "Arctic Climate Impact Assessment."

policy, which in and of itself could contribute to increased activities, as expressed in the Svalbard white paper. Norway is an Arctic coastal state and a major oil and gas producer, but political elites also share clear ambitions of Norway being an international leader on global climate governance. These roles and ambitions create political contradictions and tensions that are illustrated and analyzed in this chapter. After defining the concept of opportunistic adaptation and linking it to security theory debates, the chapter examines opportunism as state policy, in particular relating to the ongoing expansion of oil and gas activities into Norwegian polar areas. The chapter then looks at how this policy direction creates tensions with Norway's desire to be at the forefront of global environmental issues, especially the equity dimension of the Kyoto Protocol, where the Global North is expected to take the greatest burden when acting on climate change (United Nations 1997).

This chapter thus demonstrates how Norway is changing its focus from mitigation (reducing greenhouse gas emissions) to economic adaptation based on state-centered interests, and especially explores how this is done via an environmental-developmental discourse. By drawing on discourse cooptation (Jensen 2012a), critical geopolitics (cf. Dodds et al. 2013), and what can be termed a global environmental security approach (cf. Dalby 2008), security theory is used to expose the tensions between state-centered interests and environmental and equity concerns. The empirical evidence is drawn from speeches, white papers, and interviews with two key politicians involved in these negotiations. The interviews show how consensus is both established and challenged among Norwegian political elites.[2] They especially expose the tensions in the agendas and ideas about how Norway can balance its roles and identities as a major oil and gas producer, on one hand, and being politically progressive when it comes to climate change, on the other.

The analyses within the chapter contribute to an emerging subfield within political geography and international relations recently coined as *critical polar geopolitics*, challenging the notion of an exceptional Arctic geopolitics (Powell and Dodds 2014, 9). The chapter can also be considered to be Norway's response to Michael Bradshaw's (2013) conceptualization of the *global energy dilemma*: how to secure the supply of reliable and affordable energy while at the same time rapidly transforming to a low-carbon, efficient, and environmentally harmless energy supply. The ability to provide secure, affordable, and equitable supplies of energy that are also environmentally sustainable plays out differently, depending on the context of economic globalization, climate change, and energy security (Bradshaw 2013).

Opportunistic Adaptation

This chapter introduces the term *opportunistic adaptation*, defined as the idea that the economic benefits of climate change should be prioritized over efforts to address

---

[2] These quotes from interviews with the politicians have been approved by them before publication.

the causes. It has geographical, political, and discursive implications, and this chapter particularly focuses on two logics of opportunistic adaptation that are part of Norway's contemporary response to the global energy dilemma. The first relates to how the government and the petroleum industry reason by focusing on the benefits of climate change as an economic advantage, justifying production from the strategic perspective of the state.[3] As such, opportunistic adaptation reflects how higher temperatures are equated with the potential for more economic activities. Politically, within this logic, mitigation strategies (to reduce greenhouse gas emissions) are kept separate from adaptation strategies. Discursively and politically, this can promote a scenario where climate change in the Arctic context is an environmental problem to be managed but not resolved.

The second and more complex logic of opportunistic adaptation that the chapter elucidates is how Norwegian petroleum exports are being rescaled from a state-to-state security of supply perspective to a global security of demand scenario. This reflects a widening and deepening of the security concept, where increased production in Norway is framed as a prerequisite for economic development in the Global South. It is thus global markets and the people of the Global South that are referenced when determining "who is being secured" when new fields are opened up in the North. This security of demand approach may be, however, problematic in that it rejects the idea that Norway must reduce greenhouse gas emissions to allow for enough carbon space in the atmosphere (cf. Khor 2010) to create an emission headroom that will allow countries in the Global South to develop without causing catastrophic climate change (cf. Bradshaw 2013). One can also question whether the Norwegian approach takes into account that it is countries in the Global South that generally suffer from low levels of energy access (and/or rely on coal) but that are also most likely to suffer from the negative environmental impacts of climate change.

This discursive and political petro-climate paradox presents two material challenges for Norway. One is to uphold production levels (see Figure 8.1). To maintain the position as a major global supplier of petroleum,[4] more oil and gas needs to be found, as production peaked a decade ago and the best reserves are expected to lie in the North.

The second challenge is to reduce national greenhouse gas emissions, as Norway is currently nowhere near the policy targets established in the early 1990s, when international climate talks began. Since then, emissions have grown by more than 25 percent in national statistics, where the oil and gas sector represents the biggest

---

[3] See Kristoffersen (2014) for a related discussion, which foremost analyzes this as a negotiation *between* the industry and the state, which is characterized by *consensus* when they engage with a resource geopolitics of the Barents Sea in an Arctic context.

[4] Norway was ranked as the seventh greatest oil exporter and the fourteenth largest oil producer in the world in 2011 numbers. Norwegian production of oil reached its peak in 2000 with 181.2 million oil equivalents (o.e.); meanwhile, in 2012, production had been more than halved (89.9 million o.e.).

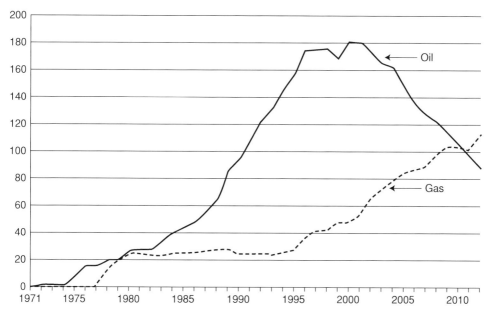

Figure 8.1. Norway's historical production of oil and gas, 1971–2012. Figure reprinted courtesy of the MIT Press from *Heating Up and Cooling Down the Petrostate: The Norwegian Experience*, by Ryggvik and Kristoffersen (2015).

increase (Statistics Norway 2012). Although there was a broad policy consensus on taking national action to stabilize emissions in the early 1990s, the objective was not addressed until the end of the decade, when it was replaced by what Hovden and Lindseth (2004) coin as the discourse of global action. Since then, Norway's main policy strategy has been to pay for emission reductions in other countries (e.g., through carbon trading or offsets), instead of reducing emissions at home. This, Hovden and Lindseth emphasize, worked in symmetry with the political wish to increase oil and gas production. The production actually more than doubled during the 1990s, and CO_2 emissions from the sector increased by almost three-quarters between 1990 and 2011 (Statistics Norway 2012).

Mitigation "versus" Adaptation

Norway is not likely to be the only fossil fuel producer that is framing the nature of global energy markets and/or the benefits of climate change in the Arctic to its own advantage. Thus we can also explore opportunistic adaptation's logics in an international context. Starting with the aforementioned intentions of changing behavior in the international climate negotiations that resulted in the Kyoto Protocol, the aim was originally to mitigate the impacts of climate change. To achieve this, nation-states would need to adjust their emissions-creating economic activities. This was the motivation behind establishing an international climate regime in the first place: as states are mutually dependent on each other, the overall objective is to reduce the total global output of greenhouse gas emissions (especially CO_2). Mitigation is thus usually negotiated as specific strategies to limit the consequences of climate change by reducing greenhouse gas emissions, whereas adaptation is a more vague concept politically and easier to fill with various meanings, depending on the context. The 2014 Intergovernmental Panel on Climate Change (IPCC) Fifth Assessment Report (AR5) defines adaptation as "the process of adjustment to actual or expected climate and its effects. In human systems, adaptation seeks to moderate harm or exploit beneficial opportunities. In natural systems, human intervention may facilitate adjustment to expected climate and its effects" (Agard et al. 2014). A UNDP report on adaptation policy discusses adaptation as a process by which strategies to "moderate, cope with and *take advantage of the consequences* of climate events are enhanced, developed and implemented" (UNDP 2005, 41, italics added).

To a great extent, then, the potential for opportunistic adaptation is closely related to the failure of nation-states to agree on how to significantly lower greenhouse gas emissions (mitigation) through binding agreements and effective policy tools. The situation opens for other political logics related to predicting global energy markets and changing geographies caused by climate change, for example, as possibilities for more intense economic activities. In this latter context, the notion of adaptation speaks to the situatedness of geophysical change and consequently to how states might incorporate territorial adaptation planning into their strategies to sustain or increase economic growth. Opportunistic adaptation thus speaks to a specific way of relating to the consequences of global warming where the gravity of change can be underplayed by the opportunities created.

Framework: Geopolitics and Environmental Security

Our quest for "security" in modern economic production is currently undermining the conditions for terrestrial habitability. This is the (environmental) security dilemma in its largest form. (Dalby 2002, 172)

This quote by Dalby is useful when setting up an analytical framework in three interrelated ways. First, it takes as the starting premise that we are now living in the era of the *Anthropocene* (Anthropo [man] + cene [new geological age]). This expression was first suggested in 2000 by Crutzen, to explain that humans are currently changing the world on a global scale and ushering in a new era in geologic time (Crutzen and Stoermer 2000; see also Dalby 2009; Hansen et al. 2008). This global human–environment interaction is an important recognition for theory development, which takes us to the second point: Dalby's environmental security dilemma takes into consideration that we are experiencing human-induced environmental change where a state's effort to secure its economic and territorial interests are undermining the conditions for terrestrial habitability. Dalby thus positions what he refers to as the environmental security dilemma as working in competition with the traditional security dilemma. States are still preoccupied with relative power, which includes gains in the energy sector, and are not responding adequately to Dalby's concerns regarding the environmental security dilemma. Despite all the rhetoric, even Norway, which proclaims a dedicated interest and prioritization of environmental concerns and awareness, is preoccupied with national (traditional) security and classical geopolitics.

The traditional security dilemma dominated analyses regarding the relations between geography and politics, where the measures taken by state A to increase its security (usually military buildup but including other forms) is perceived as a threat to state B, which in turn takes measures to increase its own security, resulting in increased perceptions of insecurity within both states. Rooted in the realist tradition of international relations, the security dilemma was mainstreamed during the Cold War, before the effects of the burning of fossil fuels on the biosphere and climate system were widely understood. Unilateral moves brought about by, for instance, interstate resource wars can set off a race among states to secure their economic, political, and territorial interests.[5] States employ knowledge, technology, and metaphors in the service of their self-interest and security through state planning toward modernization and economic development (see Scott 1998). This model for securing state interests and ultimately their power vis-à-vis each other do not take into account environmental problems like climate change. As such, the fundamental challenge of adjusting economic activities to mitigate the impacts of climate change has little chance to be adequately addressed and thus resolved within a state-centered perspective toward security.

This lack of progress on mitigation becomes particularly relevant in the context of opportunistic adaptation in the Arctic. According to Steinberg et al. (forthcoming), the focus on possibilities is now competing with the popularized discourse on increasing geopolitical tensions over resources in the Arctic. The two discourses of possibilities and resource geopolitics directly relate to state-based exercises of rescaling the Arctic region in terms of its potential for resource extraction and climate benefits. They both clearly reflect a realist understanding of security (Walt 1991) that has played a central role in the development of security conceptualization about the Arctic. This approach to polar geopolitics privileges the role of states and puts great emphasis on resources and territorial advantage, in a world frequently understood to be shaped by competition (Powell and Dodds 2014, 9). From the viewpoint of critical geopolitical scholars, the realist perspective toward security in the Arctic is suspect (cf. Dittmer et al. 2011; Gerhardt et al. 2010; Powell and Dodds 2014; Steinberg et al., forthcoming). Post–Cold War theoretical developments in security theory expose power dynamics and

---

[5] Cf. Åtland (2010) for a discussion on how this relates to Arctic geopolitics.

representations of that power. This includes the assumption that there is a separation of an "inside" of the state that equals politics and an "outside" that equals the use of force (Walker 1993), where security is defined as "the defense of a particular spatial sovereignty and the politics *within* it" (Agnew 1994, 62, italics original). This means that the generation of geopolitical knowledge is both situated and inherently value laden (Castree 2008, 425). Or in the words of Klaus Dodds et al. (2013, 6), "geopolitics in its traditional state-centric approach cannot be considered as neutral analyses of pre-given geographical facts, but rather as a deeply ideological and politicized form of analysis." The critical geopolitics approach perpetuates the process where intellectuals of statecraft construct "ideas about places, these ideas have influence and reinforce their political behaviours and policy choices, and these ideas affect how we, the people, process our own notions of places and politics" (Devetak et al. 2012, 492). The critical geopolitics perspective, then, challenges the notion of an (exceptional) Arctic geopolitics (Powell and Dodds 2014), which is the analytical starting point for the analysis of how Norwegian political elites frame geopolitics, petroleum extraction, and climate change in the Arctic.

The third important point derived from Dalby's quote relates to how we understand *environmental security*. Both Dalby and Barnett interpret the concept in terms of political interconnectedness and ecological interdependence (Barnett 2001, 2007; Dalby 2002, 2009). This is taken as a starting point for the analysis of tension that needs to be reconciled by Norwegian political elites relating to the aforementioned global energy dilemma (Bradshaw 2013) involving a widening (issues) and deepening (actors) of the security perspective. Broadening the horizon of security issues and actors involves questions about by whom, in what context, and for what political purpose security is invoked as a threat or possibility (Barnett 2001). A much-cited definition by Barnett is that environmental security is the "process of peacefully reducing human vulnerability to human-induced environmental degradation by addressing the root causes of environmental degradation and human insecurity" (129). Environmental security perspectives tend to focus on spatial unevenness of economic and political processes and its outcomes, for example, the effect Northern consumption patterns have on the Global South, rather than on being protected from dangers that originate from the aforementioned outside of the state, as reflected in a state-based security perspective (Dalby 2009; Walker 1993). A globally oriented environmental security approach focuses on the interconnectedness of human conditions, considering its varieties in different but *connected* places, rather than just taking as a starting point that *one* humanity is facing a common threat (Dalby 2008, 37, italics added). This approach, exemplified through Barnett's definition, reflects back on the discussion on mitigation where politically addressing the spatially uneven consequences of climate change is a prerequisite. Barnett and Dalby reflect specific critical approaches to human security that pay special attention to the human experience, but not necessarily for the benefit of state interests. States have also attempted to use similar human security rhetoric, not least as a form of virtuous imperialism, making decisions about who is vulnerable and how to help who is vulnerable, which corresponds their own state interests (Gjørv 2013).

Building on Dalby and Barnett's approach, and human-oriented security approaches more broadly, Schnurr and Swatuk (2012) suggest that the focus should now be on *critical environmental security* approaches that simultaneously address cause, context, and effects but also encompass environmental justice (addressing unequal power relations). This builds on radical perspectives of social injustice, underpinning the structural violence between the developed and the developing world (introduced by Galtung 1969). The added environmental justice dimension provides the justification to engage with security studies to understand the policy relevance of the empirical context: the Norwegian government strategies are legitimized through combining, coupling, or decoupling different interests and actors with regards to petroleum development and environmental and equity concerns.

Geopolitics and Opportunistic Adaptation in the High North

Climate change, greater access to natural resources and growing human activity suggest that the High North will be a region of considerable geopolitical interest. The government's overall objective is to make use of the opportunities this offers, and at the same time manage the environment and natural resources sustainably. (Ministry of Foreign Affairs 2011b)

A number of characteristics in the preceding statement are representative of how Norway presents its strategic interests in its Northern territories, and in the Arctic more broadly. Here it is noticeable how "considerable geopolitical interest" about the present and future management of the Arctic is woven into state-territorial projects (cf. Dittmer et al. 2011, 202). Climate change can be seen to provide the state with new incentives to frame the region within the spatial ontology of the state system (cf. Gerhardt et al. 2010, 999). We also clearly see how the government balances between exploiting its "natural resources" and managing "the environment." A striking feature of the preceding white paper statement is that climate change and petroleum development are discussed as two distinct factors in the same context. There is as such a decoupling, where the inconsistency between production and consumption of oil and gas, on one hand, and the reduction of greenhouse gas emissions, on the other, are overcome by treating climate change and petroleum development as different and independent variables. This resonates well with the logic adopted by the government at the time, that is, that it is not Norway's duty to resolve the paradoxical process where increased temperature in the Arctic might allow for more economic activities, such as oil and gas extraction:

introducing restrictions unilaterally in Norway or in a particular region of the world, would probably have limited, or possibly even the opposite effect; in a world with increasing demand for energy, I believe imposing such restrictions could result in energy substitution of the wrong kind, with coal and heavy oil replacing natural gas.

My point is – yes, we are facing a paradox. It is not a national or regional paradox, but a global one. We must increase our capacity to deal with challenges to economic activity. (Jonas Gahr Støre Minister of Foreign Affairs, Offshore Northern Seas conference, August 27, 2012)

It is not Norway's role to resolve this petro-climate paradox, Støre emphasizes, nor is it the role of the Arctic states. Instead, environmental change in the Arctic is to be *managed* through state capacity building. Likewise, as the earlier white paper statement also suggests, Norway contributes to an understanding of how geopolitical security challenges in the Arctic are unfolding, where Norway responds proactively. The benefits of climate change are naturalized as state interests. This is dependent on processes whereby the warming of the Arctic region is equated with the potential increase of economic activity and thereby national economic growth. Notably, Norwegian political elites do not contest the climate change science that demonstrates the linkages between climate change and human-driven (anthropogenic) factors and processes. But the question is whether, by embracing the inevitable changes, opportunistic adaptation removes attention from the relationships between the causes and effects of climate change in the Arctic.

Many policy makers and industry representatives maintain a long tradition of approaching the Arctic as a frontier for human progress through quests for commercial expansion (Strandsbjerg 2012, 1). For example, in the 1920s, Turner (2010, 297) recognized the popular framing of Alaska as a resource frontier: "her wealth of natural resources asks the nation on what new terms the new age will deal with her." Today, the imagination and articulations of the North's unknown nature, the interplay of expectations and experiences, and the writing on a supposedly empty space together create images and visions of a specific *Arcticism* (Ryall et al. 2010, x). It is the Arctic's perceived characteristic of *openness* that enables it to become a space of masculine fantasy and future adventure (Dittmer et al. 2011, 202). Referencing the instability of the nature in the Arctic through climate change, combined with this redrawing of its heroic history of national quest, Dittmer et al. (2011) employ a critical geopolitics perspective to show how realist and thus geopolitical visions are created through the notion of the Arctic as *essentially different* to other parts of the world, or what Powell (2010, in Powell and Dodds 2014, 7) analyzes as different positions toward its given scale of exceptionality.

Opportunistic adaptation accordingly speaks to these imagined geographies, where the Arctic is feminized as an open territory. This was also emphasized by politician Erling Sande (interview with the author, January 2012), who was the chair of the parliamentary committee for energy and the environment (from 2009 to 2013) and his Center Party part of the red-green coalition government (2005–13).[6] He was

---

[6] Erling Sande is now executive director for strategy and development in the renewable energy company Sogn og Fjordane Energi (SFE).

concerned with the "paradoxical process" whereby the same countries that cause global warming in turn exploit the resources that become available as a result of that same development. He emphasized that humanity has always taken advantage of the opportunities that have been available, either through natural climatic change that was the foundation for the first settlements in Norway after the last ice age or through state planning, such as the territorial expansion during colonialism, where technological progress plays a key role. So in that sense, he noted, it is not surprising that the Arctic territories were being framed in this manner, and he concluded that "we will probably be the first generation to experience an ice free Arctic Sea during summer" (interview with the author, January 2012). As such, his statements reflect on processes in the Anthropocene, where climate change is the ultimate (environmental) security dilemma (Dalby 2002). The economic legacy of states expanding their territorial and economic activities is not capable of dealing with the causes of climate change but rather intensifies them. This is clearly exemplified by what Leichenko and O'Brien (2008) call *double exposure*, where the combined drivers of climate change and globalization create positive feedbacks that facilitate increased fossil fuel extraction, increased international transport and trade, and increased net emissions of greenhouse gases, leading to further climate change.

The Global Energy Dilemma in the High North

As was reflected in the earlier statement by Minister Støre, there are continuing efforts to withstand critique that the production of oil and gas in Norway is environmentally unsustainable in the long run. The next quote is an extract from Ola Borten Moe, the Minister of Petroleum and Energy (2011–13), taken from a radio debate with an environmentalist in the context of his declaration that his first mission as a minister would be to challenge the environmental movement's depiction of reality:

I believe that we must presume that the world needs more energy, I believe that we must take into account that fossil fuels will be an important part of the energy mix globally also in the future. This means that the question is not *whether* we can carry on with oil and gas, but *how* we can do it, in the most environmentally friendly way. (NRK 2011)

Here "the world" and the "energy mix globally" have become central interventions in the initiated shift from *if* (whether to drill) to a technopolitical exercise of *how*. Drilling in the North is envisioned to contribute to emissions reductions in a global perspective (addressing mitigation) and to development in the Global South (addressing global justice). Jensen's (2012a) conceptualization of "environmental discourse cooptation" is helpful when explaining these two shifts. This empirical phenomenon is defined as how one discourse burrows into the heart of a counterdiscourse, turning its logic upside down and putting it to work to reestablish hegemony and regain political support (Jensen 2012a, 36–37). One discourse is strengthened by the addition of a new, powerful argument, whereas the other is weakened almost to the same degree. Jensen's

PhD thesis on media debates and white papers relating to oil drilling in the Barents Sea show that Norwegian political elites turned the idea of conservation on its head in the early 2000s by emphasizing the high environmental standard of Norway's drilling in the North (Jensen 2012b). Equally important in what he termed the "drilling for the environment" discourse was to set an example for Russia, where Norway has the know-how to reduce environmental risks in the Barents Sea (Jensen 2007). The discourse about the positive environmental attributes of petroleum development in the Barents Sea was thus strengthened by coopting the very foundation of the counterdiscourse, whose proponents have argued for refraining from petroleum development in the name of conservation (Jensen 2012a).

We can see cooptation taking place in Borten Moe's statement: greenhouse gas emissions from new Norwegian oil fields are more environmentally sound in a relative perspective, as they are seen as comparatively better than coming from other petroleum-producing regions of the world. This enables a hierarchy of security priorities where political tensions between state-based energy security and globally oriented environmental security concerns are managed, reworked, and resolved. Although insecurities about future energy supplies and reserves are the organizing or superior objective, climate change is structured into this reasoning in ways that are consistent with, and contribute to, upholding Norway's role as a legitimate petroleum producer and exporter of oil and gas.

This cooptation is done in part through the separation of different security logics, or discourses over *time*, with spatial implications for petroleum extraction: global warming is positioned as a subordinate threat over the next two decades due to the current and dominant global energy dependence on fossil fuels, where population growth and economic development in developing countries call for Norway to extract all available resources in a politically interconnected energy security perspective. This is the core message in *The Climate Paradox*, a book based on interviews with former Prime Minister Jens Stoltenberg (from 2005 to 2013),[7] where he elaborates on why Norway needs to develop the available resources:

> The use of fossil energy in poor countries is growing rapidly. They have a lot of coal, they are looking for oil. When the world needs more energy to fight poverty, it is not morally wrong that Norway is helping to supply that energy, that oil and gas. On the contrary, China and India are desperately dependent upon getting more oil and gas on the market so they can meet their needs. If we had a strategy to choke it [our oil and gas production], then fewer people will be raised out of poverty. That is not moral[ly right]. Secondly, it is the case that our energy is cleaner. When the world first needs to have significant amounts of oil and gas, I believe it is better that it comes from Norway than elsewhere, because we are among the countries with the lowest emissions per produced unit. (quoted in Alstadheim 2010, 71, my translation)

This energy security logic, which can be termed "drilling for global development," works in symmetry with a global environmental security approach, where political

---

[7] Stoltenberg assumed the position of secretary-general of NATO in autumn 2014.

and ecological interconnectedness is reduced to a cost-benefit analysis: emissions can and should be cut elsewhere, and Norway should sustain production levels by opening up new fields in the North to fuel economic growth in the Global South. The growing world population, often characterized as the poor and the underdeveloped, has become a rhetorical source for political intervention. This works to establish the political interconnectedness of energy security and inequality in North–South relations and brings us back to the discussion about a critical environmental security approach (addressing environmental justice), where tensions between the developed and the developing world were sought to be overcome in the Kyoto Protocol. Industrialized countries were to assume the main responsibility for climate change through cutting their own emissions before demanding the same of the developing world. However, in the two extracts from interviews with Stoltenberg and Borten Moe, the burden-sharing aspects relating to creating "carbon headroom" for the developing world are coopted by the argument of reducing the gap between the rich and the poor through energy interdependencies. At the core of this conflict is the argument of developing countries, where they claim that the industrialized countries in the Global North have a moral responsibility to reduce their emissions substantially prior to developed countries, where many of the world's poorer nations have fought hard for their right to develop (Lahn 2013).

The argument by leading figures within the Norwegian government is possible because the Kyoto mechanisms allow for different and thus *flexible* interpretations of climate policy options. In the current situation, there is no effective global cap on emissions – a mandated restraint as an upper limit on emissions – or a binding long-term international agreement "after Kyoto." It is therefore reasonable to say that these negotiations have failed, as they have not provided the tools or the means for reducing emissions or resolve the conflict of interests between developed and developing countries. The failed international efforts to act on climate change open political space for Norway to argue within a security framework to drill in its territories. The flexibility offered by the Kyoto regime enabled Prime Minister Jens Stoltenberg to argue that "we should develop more Norwegian oil fields, if we are to achieve an international climate agreement" (*Bergens Tidende* 2010). Through the flexible mechanisms of the Kyoto Protocol, Stoltenberg (who is an economist by training) figured that "it might cost 1000 NOK per ton CO_2 not to develop an oil field, to save the rainforest costs about 30 NOK per ton CO_2. So I think it would be completely meaningless to the climate if I were to say that 'we choose the absolutely most expensive initiative in the world'" (*Bergens Tidende* 2010).

Relating this back to discourse cooptation, climate measures are given meaning by turning the mitigation argument on its head, arguing that in a comparative and global perspective, opening new petroleum fields in Norway is "good for the climate" (*Bergens Tidende* 2010). A recent exposé in Norwegian media, however, based on the Oil Industry Association of Norway (OLF) annual environmental report, showed that Norwegian emissions per produced barrel are now 25 percent higher than in the Middle East and close to the average in European petroleum production.[8] The argument that Norway has "the cleanest" oil and gas production in the world has accordingly been modified to Norway being one of the greenest petroleum producers in the world.

Politician Erik Solheim (Socialist Party)[9] is particularly critical toward the rationale behind Norway's efforts to engage in what he calls "drill[ing] for climate change and the world's poorest countries" (interview with the author, April 2012). Solheim was the minister of environment and international development from 2007 to 2012 and had firsthand knowledge of processes and discussions related to the opening up of the Norwegian north for petroleum development. In the interview he emphasizes that this can be demonstrated to be "merely rhetoric" because those who promote this perspective lack an understanding of what the national response to climate change *should* be. In his view, they undermine their own efforts to drill for climate mitigation and global development when they argue that Norwegian gas can replace European coal (in power plants) while simultaneously arguing that Norwegian oil, new coal mines at Svalbard, and the two-thirds governmentally controlled oil company Statoil's tar sand projects in Canada are *also* part of global climate solutions. "You cannot say that you are concerned with climate change, [instead] you have to argue for jobs, economy and all those other considerations," says Solheim, while maintaining that the gas versus coal argument could have been "rock solid" if it had solely focused on Norwegian gas exports (Solheim, interview with the author, April 2012).

Solheim rejects key logics of opportunistic adaptation but maintains that Norwegian interests in relation to gas exports can also have a positive impact on global emission reduction and address mitigation. Another aspect of Norway's main strategy that is supported by Solheim when it comes to drilling in the North is the mapping of potential resources in the border areas toward Russia after the bilateral treaty on the maritime border came in place in 2010. The day after the treaty and offshore delimitation was officially enforced, Norway's Petroleum Directorate started mapping the seabed for potential oil and gas deposits (through seismic surveys), and the area was officially opened for oil and gas exploration in 2013. Solheim emphasized that with these tactics, Norway can avoid encountering any problems with the Russians. The agreement is considered a positive step because there is "all reason to believe that Norway will have better control and technology and systems to handle [all aspects of petroleum development], and therefore building up a parallel enterprise as a whole can lead to more responsible management of oil resources on the Russian side," according to Solheim (interview with the author, April 2012). Understood as an anticipatory logic, this suggests that if Norway is to map all its resources, the government will be in a

---

[8] OLF (2011, 30), in the newspaper Dag og Tid, January 12, 2012, based on International Association of Oil and Gas Producers (OGP) Environmental performance indicators (2011).
[9] He is now chair of the OECD Development Assistance Committee (DAC).

better position to face an uncertain future with its Eastern neighbor. This corresponds to Jensen's (2007) analysis on the role of Russia in the "drilling for the environment" discourse, where Norway should supervise Russian petroleum development in the Barents Sea based on its environmental standards and knowledge. However, and as Solheim emphasizes, there are many uncertainties and conditions that all together reflect a precautionary and restrictive approach toward contemporary political and climatic developments:

The agreement on the border demarcation with Russia was a huge breakthrough. It provides the foundation for improved relations between our two countries in the High North. However, whilst it is important for Norway to gain a better knowledge base and understanding of the situation in the border areas, we will have to move forward very gently when it comes to opening up for oil drilling or shipping. As of today, we don't have any technology to recover oil in areas with ice. And as we know, the Arctic nature is extremely vulnerable.[10]

Politician Sande, introduced earlier, held a perspective on these issues along the same lines as Solheim. Although arguing for comparative perspectives relating to greenhouse gas emissions generated on the supply side of global markets, he says that this does not necessarily imply that all of Norway's northern territories shall be opened up for commercial drilling or that all of the mapped oil and gas resources should be developed (interview with the author, January 2012). And more importantly, he says that he parts with Stoltenberg when it comes to the *relative* nature of his argument, as it is not wrong or unfounded, but we can't accept such rhetoric alone. Sande explains:

We cannot rely upon Stoltenberg's argument based on Norway's status as a clean petroleum producer as the only rhetoric that is decisive as a measure for determining which countries get to produce oil and gas. We do not allow the most "dirty" producers to produce, but the next-to-most-dirty can produce? There would be many countries with not-so-clean production that were allowed to produce, thus we would have to identify the second best, or second worst, such as Venezuela or Saudi-Arabia. This process of identifying the best and worst producers cannot function as the sole measure of how global production should continue, and more importantly, such an approach is not consistent with a global climate agreement.[11]

The interview extracts from Sande and Solheim are important when exploring the ongoing debates within governments that are not part of the official rhetoric, and how dominant approaches are – or can be – challenged. In Norway, this is reflected in the long-standing political battles between the Ministry of Climate and Environment and Ministry of Petroleum and Energy on where the geographical limits to petroleum development are being set, which have become more contested in the polar areas. However, they are also characteristic of efforts toward finding common ground between environmental and petroleum policy. As such, they expose tensions between state-centered and globally oriented approaches toward environmental security. This in turn

---

[10] This quote was added by Solheim after reading through the finished chapter draft in May 2014.
[11] This quote was added by Sande after reading through the finished chapter draft in May 2014.

raises a timely question relating to the previous theoretical discussion: shouldn't they instead be approached as conflicting interests?

Reflecting back on Leichenko and O'Brien's (2008) double exposure framework, what is missing in the opportunistic adaptation perspective is the long-term consequences of climate change, as Sande pointed out. It does not address the climate feedback loops that we are already witnessing in the Arctic, or that we are living in a time of the "Great Acceleration" in the Anthropocene (Steffen et al. 2007, in Dalby 2013, 42). Bradshaw's (2013) argument is that the global energy dilemma is context dependent in terms of how economic globalization, energy security, and climate change play out. This also brings attention to how we approach globalization. On one hand, it is (commonly) understood as multiple effects of intensifying global interconnections (Sparke 2012, 3); at the same time, globalization is an uneven economic process that includes forecasting and claims to a globalized "reality." Such claims to objectivity echo the previous discussion on claims to a geopolitical reality. Rephrased differently, globalization should not only be conceived as intensifying global interconnections; it is often simultaneously put to work through political rhetoric as an influential discourse or code word to shape policy making. This is what Sparke (2012, 5) calls (big *G*) Globalization, an instrumental term put to work in shaping and representing the growth of global interdependency. Dalby (2013) raises the problematic tendency in Anthropocene geopolitics to view globalization through Westphalian (or imperial) lenses, and there is much to be gained from an approach that does not privilege a Westphalian imagination concerning the supremacy of particular claimant or coastal states in the Arctic region (Kraska 2011, in Powell and Dodds 2014, 8). What ultimately matters in a state-centric view toward globalization is potential threats or opportunities, where environmental threats are acknowledged but often either "collapsed into problems and limits on development or ignored because predictions are impossible in the long run" (Dalby 2013, 43). The strategy then of *managing* the "global energy dilemma" through the opportunistic adaptation approach depends on how energy security can be eased into the state security discourse, with a relativistic approach toward mitigation efforts and environmental justice.

Summary and Conclusion

In the preceding analyses, I have shown that the dominant view of Norway's environmental security approach toward the expansion of petroleum activities into Norwegian Arctic territories is premised on the idea that this development does not happen at the expense of, but rather to the benefit of, insecure people of the Global South and the environment (including the biosphere). A growing world population, envisaged to be on a linear "modernization"-based (traditional) development track, is premised upon the access to fossil fuels. This modernization approach (reflecting the ways in which the Western world has developed) enforces the political agenda behind energy security and is contradictory to an environmental security perspective that connects climate change and its negative impacts to threats on human existence. A global or critical environmental security approach addresses not only global binding agreements but also national measures that reflect an awareness of how others perceive their security in light of the changing climate. As the impact from Norway in emissions reductions would be very modest on a global scale, it seems easy to argue that economic interest should be prioritized over an ambitious climate mitigation policy (Ingólfsdóttir 2014, 87). This approach, as Ingólfsdóttir points out, can weaken the agency of "small states" – such as the Nordic countries – to act as what she calls *norm entrepreneurs*[12] in international climate negotiations. To act as norm entrepreneurs requires a long-term vision and the willingness to take a higher moral ground, advocating policies that aim at supporting common interests of all states (or humanity, for that matter) rather than the self-interest or special interests of a few. To influence international discourses and thus developments, then, demands a (re)focus toward common global interests and to set an example of implementing progressive domestic policies (Ingólfsdóttir 2014). This will, for example, mean that countries like Norway listen to representatives from countries like the Philippines and enact measures that reflect an understanding about the development of climate change threats.

A main conclusion of this chapter is that there is a lot of power in defining the conditions for security. At present, the dominant security discourses provide the political foundation for a status quo approach to petroleum policies that aims at opening up the Norwegian north for extensive petroleum activities. There is thus a lot of power in what the previously quoted minister of environment and development pointed out to be "merely rhetoric": Norwegian political elites have put themselves in the position of defining the conditions of what security *is* and *whose* security is being secured, overlooking or even disregarding the perspectives of those being secured and how they themselves relate to their own insecurities. This chapter has pointed to how ethical arguments are fronted as the main arguments to drill in the High North, relating to equity and environmental concerns. By doing so the economic arguments are downplayed or hidden behind the ethics of helping the Global South and the environment. As such, an opportunistic adaptation approach can work, as it is predicated upon our own political elites deciding whose security it as risk, less so for supporting security for those who are identified as insecure, and more so for securing our own (state-centered) interests. Barnett (2001, 156) discusses this phenomenon as the double-edged sword – when climate change is made more important than other politician issues it risks state cooptation, colonization, and emptying of the environmental agenda – instead of contesting the legitimacy of prevailing (state-based interest) approaches to security. How elites respond to the accelerating changes

---

[12] Ingólfsdóttir points out that the term *norm entrepreneur* is borrowed from Christine Ingebrigtsen, who has done research on the role of the Nordic countries in world politics.

continues to matter, then, as "geopolitical framings of the future are constituted to provide the discursive context for political action" (Dalby 2013, 44).

I opened this chapter with a scene from Svalbard in 2008 and raised the question of how the politics of climate change has evolved to the point where Norwegian policy seeks to take advantage of the "opportunities" of climate change rather than focusing on mitigating the potential threats of climate change (cf. Ministry of Foreign Affairs 2009). Since that initial conference in Svalbard, I have attended a number of subsequent events where policy makers, stakeholders, and researchers have discussed the implications of climate change. It is not necessarily so that all forms of economic adaptation to a changing and new geography in the Arctic – because of climate change – are opportunistic. As time has passed and attempts to reduce overall global emissions have failed, the need to focus on adaptation has increased (Ingólfsdóttir 2014, 86). However, I have pointed to a tendency to emphasize economic adaptation in terms of taking advantage of the consequences of climate change in the High North. From a mainstream political perspective, this is not considered problematic for Norwegian policy makers, as Norway's former foreign minister Espen Barth Eide (2012–13) recently reminded me. At one conference, I had presented my thesis of opportunistic adaptation and related it to the limits of a realist perspective toward security in the Arctic, which was the main analytical focus of the conference.[13] Barth Eide acknowledged that I had raised important questions but argued that the answer to them lies at the global level. In other words, Norwegians can't take measures on their own; they have to rely on global initiatives first. The realist argument reflected in Barth Eide's statement (also emphasized earlier by for example Borten Moe) is that we have to accept the world as it is and that climate change is not a problem that "we" (the Norwegians) can fix. Barth Eide asserts that there is, or will be, a globally agreed on treaty on climate change. It is an agreement, however, that is still lacking. What environmental policies need to be enforced nationally in the petroleum sector to gain or maintain credibility as a norm entrepreneur on global environmental governance still remains unresolved, and Norway's opportunistic adaptation approach, in concert with passing the buck to the global level, moves it away from the role of norm entrepreneur. Can Norway, or other countries, wait until there is an international regime in place, and a global carbon price, before addressing climate change mitigation in domestic policies, or do they have a role to start taking globally relevant responsibility at the national level now?

Acknowledgments

I would like to thank Karen O'Brien, Kirsti Stuvøy, Gunhild Hoogensen Gjørv, Tone Huse, Hannes Gerhardt, Michael Bradshaw, Bård Lahn, Stuart Robinson, Gisle

---

[13] The conference "Security in the North: Are We Prepared?" in Bodø was organized by the Ministry of Foreign Affairs, and a summary of the conference can be downloaded at http://www.regjeringen.no (in Norwegian).

Andersen, Indra Øverland, and Fredrik Chr Brøgger for invaluable comments throughout different stages of writing this chapter. I would also like to thank all the researchers in the Norwegian Research Council's funded program "Potentials of and Limits to Climate Change Adaptation in Norway" (PLAN) for important input, from which this project got funding (under NORKLIMA).

References

Agard, J., et al. 2014. "WGII AR5 Glossary." http://ipcc-wg2.gov/AR5/images/uploads/WGIIAR5-Glossary_FGD.pdf.
Agnew, J. 1994. "The territorial trap: The geographical assumptions of international relations theory." *Review of International Political Economy* 1 (1): 53–80.
Alstadheim, K. B. 2010. *Klimaparadokset. Jens Stoltenberg om vår tids største utfordring* [The climate paradox. Jens Stoltenberg on the biggest challenge of our time]. Oslo: Aschehoug.
Åtland, K. 2010. "Security implications of climate change in the Arctic." FFI report 2010/01097. http://www.ffi.no/.
Barnett, J. 2001. *The Meaning of Environmental Security: Ecological Politics and Policy in the New Security Era*. New York: St. Martins Press.
Barnett, J. 2007. "Environmental security and peace." *Journal of Human Security* 3 (1): 4–16.
Bergens Tidende. 2010. "Nye oljefelt bra for klimaet, meiner Stoltenberg" [New oil fields are good for the climate, says Stoltenberg]. http://www.bt.no/innenriks/Nye-oljefelt-bra-for-klimaet_-meiner-Stoltenberg-1785990.html.
Bradshaw, M. J. 2013. *Global Energy Dilemmas: Energy Security, Globalization and Climate Change*. Cambridge: Polity Press.
Castree, N. 2008. "The geopolitics of nature." In *A Companion to Political Geography*, edited by Agnew, J., Mitchell, K. and G. Toal, 423–39. Malden, MA: Blackwell.
Crutzen, P. J. and E. F. Stoermer. 2000. "The Anthropocene." *Global Change Newsletter* 41.
Dag og Tid. 2012. "Myten om den reine, norske oljen" [The myth about the clean Norwegain oil]. January 13. http://dagogtid.no/nyhet.cfm?nyhetid=2192.
Dalby, S. 2002. *Environmental Security*. Minneapolis: University of Minnesota Press.
Dalby, S. 2008. "Geographies of environmental security." In *Globalization: Theory and Practices*, edited by Youngs, G. and E. Kofman, 29–39. London: Continuum International.
Dalby, S. 2009. *Security and Environmental Change*. Cambridge: Polity Press.
Dalby, S. 2013. "Realism and geopolitics." In *The Ashgate Research Companion to Critical Geopolitics*, edited by Dodds, K., Kuus, M. and J. P. Sharp, 33–47. Burlington, VT: Ashgate.
Devetak, R., Burke, A. and J. George. 2012. *An Introduction to International Relations*. New York: Cambridge University Press.
Dittmer, J., Moisio, S., Ingram, A. and K. Dodds. 2011. "Have you heard the one about the disappearing ice? Recasting Arctic geopolitics." *Political Geography* 30 (4): 202–14.
Dodds, K., Kuus, M. and J. P. Sharp. 2013. "Introduction: Geopolitics and its critics." In *The Ashgate Research Companion to Critical Geopolitics*, edited by Dodds, K., Kuus, M. and J. P. Sharp, 1–18. Burlington, VT: Ashgate.
Galtung, J. 1969. "Violence, peace, and peace research." *Journal of Peace Research* 6: 167–91.
Gerhardt, H., Steinberg, P. E., Tasch, J., Fabiano, S. J. and R. Shields. 2010. "Contested sovereignty in a changing Arctic." *Annals of the Association of American Geographers* 100 (4): 992–1002.

Gjørv, G. H. 2013. "Virtuous imperialism or a shared global objective? The relevance of human security in the global North." In *Environmental and Human Security in the Arctic*, edited by Gjørv, G. H., Bazely, D., Golovizinina, M. and A. Tanentzap, 58–80. New York: Earthscan from Routledge.

Hansen J., Sato, M., Kharecha P., Beerling, D., Berner, R., Masson-Delmotte, V., Pagani, M., Raymo, M., Royer, D. L. and J. C. Zachos. 2008. "Target atmospheric CO_2: Where should humanity aim?" *The Open Atmospheric Science Journal* 2: 217–31.

Hovden, E. and Lindseth, G. "Discourses in Norwegian climate policy: National action or thinking globally?" *Political Studies* 52 (2004): 63–81.

Ingólfsdóttir, A. H. 2014. "Environmental security and small states." In *Small States and International Security: Europe and Beyond*, edited by Archer, C., Bailes, A. J. K. and A. Wivel, 80–92. Abingdon, VT: Routledge.

Jensen, L. C. 2007. "Petroleum discourse in the European Arctic: The Norwegian case." *Polar Record* 43 (226): 247–54.

Jensen, L. C. 2012a. "Norwegian petroleum extraction in Arctic waters to save the environment: Introducing 'discourse co-optation' as a new analytical term." *Critical Discourse Studies* 9 (1): 29–38.

Jensen, L. C. 2012b. *Norway on a High in the North: A Discourse Analysis of Policy Framing*. PhD thesis, University of Tromsø.

Khor, M. 2010. "The equitable sharing of atmospheric and development space: Some critical aspects." South Centre Research Paper 33. Geneva: South Centre. http://ww.southcentre.org/.

Kristoffersen, B. 2014. "'Securing' geography: Framings, logics and strategies in the Norwegian high north." In *Polar Geopolitics? Knowledges, Resources and Legal Regimes*, edited by Powell, R. C. and K. Dodds, 131–48. Cheltenham, UK: Edward Elgar.

Lahn, B. 2013. *Klimaspillet. En fortelling fra innsiden av FNs klimatoppmøter*. Oslo: Flamme Forlag.

Leichencho, R. M. and K. L. O'Brien. 2008. *Double Exposure: Global Environmental Change in an Era of Globalization*. New York: Oxford University Press.

Ministry of Foreign Affairs. 2009. "Svalbard." White Paper 22 2008–2009. http://www.regjeringen.no/.

Ministry of Foreign Affairs. 2011a. "The High North: Vision and policy instruments." White Paper 2011–2012. http://www.regjeringen.no/.

Ministry of Foreign Affairs. 2011b. "The High North: Vision and strategies." Summary of White Paper 7 2011–2012 in English. http://www.regjeringen.no/.

NorACIA. 2011. "Climate change in the Norwegian Arctic: Consequences for life in the North." Norwegian Polar Institute Report 136. http://www.npolar.no/.

NRK. 2011. "Borten Moe mot miljøbevegelse" [Borten Moe against the environmental movement]. http://www.nrk.no/lyd/borten_moe_mot_miljobevegelsen/9B16AF088127AA04/.

OLF. 2011. "Miljørapport 2011. Fakta og utviklingstrekk" [Environmental report 2011: Facts and development patterns]. http://www.olf.no/.

Powell, R. C. and K. Dodds. 2014. "Polar geopolitics." In *Polar Geopolitics? Knowledges, Resources and Legal Regimes*, edited by Powell, R. C. and K. Dodds, 131–48. Cheltenham, UK: Edward Elgar.

Ryall, A., Schimanski, J. and H. H. Wærp. 2010. "Arctic discourses: An introduction." In *Arctic Discourses*, edited by Ryall, A., Schimanski, J. and H. H. Wærp, ix–xxii. Newcastle, UK: Cambridge Scholars.

Ryggvik, H. and B. Kristoffersen. 2015. "'Heating up and cooling down the Petrostate: The Norwegian experience." In *Ending the Fossil Fuel Era*, edited by Princen, T., Manno, J. P. and P. Martin, 249–76. Cambridge, MA: MIT Press.

Schnurr, A. S. and L. A. Swatuk. 2012. "Towards critical environmental security." In *Environmental Change, Natural Resources and Social Conflict*, edited by Schnurr, A. S. and L. A. Swatuk, 1–14. Hampshire, UK: Palgrave Macmillan.

Scott, J. C. 1998. *Seeing Like a State: How Certain Schemes to Improve the Human Condition Have Failed*. London: Yale University Press.

Sparke, M. 2012. *Introduction to Globalization: Ties and Tensions in an Unevenly Integrated World*. Oxford: Blackwell.

Statistics Norway. 2012. "Increasing climate gas emissions in 2010." http://www.ssb.no/klimagassn/.

Steinberg, P. E., Tasch, J. and H. Gerhardt. 2015. *Contesting the Arctic: Politics and Imaginaries in the Circumpolar North*. I. B. Tauris.

Strandsbjerg, J. 2012. *Territory, Globalization and International Relations: The Cartographic Reality of Space*. New York: Palgrave Macmillan.

Turner, F. J. 2010. *The Frontier in American History*. New York: Courier Dover.

UNDP. 2005. *Adaptation Policy Frameworks for Climate Change: Developing Strategies, Policies and Measures*. Edited by Lim, B. and E. Spanger-Siegfried, co-authored by Burton, I., Malone, E. and S. Huq. Cambridge: Cambridge University Press.

United Nations. 1997. "Kyoto Protocol to the United Nations Framework Convention on Climate Change." http://unfccc.int/resource/docs/convkp/kpeng.html.

Walker, R. B. J. 1993. *Inside/Outside: International Relations as Political Theory*. Cambridge: Cambridge University Press.

Walt, S. 1991. "The renaissance of security studies." *International Studies Quarterly* 35 (2): 211–39.

9

Place Attachment, Identity, and Adaptation

TARA QUINN, IRENE LORENZONI, AND W. NEIL ADGER

The adaptive challenge of climate change involves both collective and individual action. Most public discourse on climate change includes discussions on what governments should do to impose order and sustainability onto otherwise reckless and destructive behavior by individuals. We need, apparently, to be constrained by a higher order from inflicting harm on ourselves and on the global environment. Ultimately, however, governments themselves are constrained in their ability to steer action within their jurisdictions by the culture and politics of the governed. The relationship between state and citizen is an outcome of what we, as citizens, are prepared to accept and to relinquish to that higher level.

In this chapter we focus on identity as the primary means by which individuals and communities construct their decision-making frame of reference and which they seek to include in the governance of resources and lives. We show that identity and place attachment are central to well-being; they are also central to how risks, such as those posed by climate change, are framed and managed and how responses by governments and other agencies are deemed to be appropriate, legitimate, or fair. As such, identity and place attachment influence how individuals and community construct their social contract with higher levels of authority. This set of relationships between individuals, their identities, attachment to place, and the structures of governance they sit within, we assert, is central to the adaptive challenge of climate change.

We begin by reviewing the state of knowledge on identity, on place attachment, and on their role in constructing perceptions of fairness, linking to the key justice issue of climate change as an imposed harm (Adger et al. 2006). We then examine how this knowledge has been selectively applied in analyzing responses to weather-related risks and the implications for long-term planning and adaptation. Hence we make the case for the analysis of place, identity, and their manifestation in behavior, within the larger collective challenge of adapting to different risks and climates, now and in the future.

Relating Identity to Adaptation and Justice

Individuals are located within social networks, fulfilling a variety of roles depending on context. These elements of identity are not fixed: people interpret information they receive about themselves from others and consider whether this information reinforces their desired position in these networks, or whether interactions contradict their self-perceptions. Through everyday interactions, individuals form a sense of self and also a sense of difference to others (Twigger-Ross et al. 2003). It can be an uncomfortable experience when information received from external sources is inconsistent with how individuals view themselves, which in the longer term can undermine well-being. For example, individuals with a prominent environmental identity may relate strongly with the merits of a low-carbon lifestyle while continuing to fly for work-related purposes. Hence, Barr et al. (2011) show that while other aspects of individuals' lives, such as car choice and food consumption, may be shaped by an environmentalist identity, the importance of professional identity or leisure norms may prevail when making choices regarding air travel (Barr et al. 2011).

Identification with particular social groups is referred to in the psychological literature as self-categorization. Such group identification is increasingly being shown to have a bearing on risk and adaptation behavior. An individual's identification with a specific group results in ascribing the same characteristics to the group as to oneself. This widespread tendency serves to simplify the social environment, making it more predictable and manageable (see Rabinovich et al. 2011). However, identification with specific group membership can also lead to an individual's self-perception becoming depersonalized. Individuals orient their behavior toward those of the group identified with (the "in-group") and conversely become wary or nonaccepting of positions and behaviors of other groups (the "out-groups"). Such behavioral tendencies affect the perceived acceptability, legitimacy, and uptake of advice and information in the context of behavioral change. For example, a study of coffee farmers in Chiapas (Frank et al. 2011) focused on how those individuals dealt with weather-related risks to their livelihoods and farming. Frank and colleagues showed that that farmers' acquisition, internalization, and use of information about risks were strongly linked to group and social identity. Hence the farmers' personal evaluation of information sources, interpreted through the lenses of personal experience and knowledge, was a key influence on motivation to adopt new practices and deal with the risks.

Our own research with elderly people in England underscores the influence of identity on responses to climatic stresses. We studied the perceptions and identity of elderly populations at risk from heatwaves (Wolf et al. 2010). We found that some individuals in effect disassociated themselves from those perceived to be vulnerable and at risk by denying that they themselves were elderly. As such, they adopted coping strategies but were less likely to undertake proactive adaptive behaviors that would reduce their exposure to more severe heat and cold episodes. Their constructed identities revolved around being and living independently, and this manifested in their articulation of adopting "commonsense" behaviors and not wishing to relying on social support even when unable to cope alone. The research also showed how well-established networks of support may even preclude the development and adoption of long-term proactive adaptations, particularly if they refrain from sensitively challenging existing identities and associated behaviors.

These studies show the dominance, or at least central role, identity plays in affecting how individuals frame and deal with risk. Sometimes personal as well as social identities can operate antithetically or reinforce each other to decrease individual well-being. An individual's social capital (understood as social contacts and participation in networks) does not necessarily foster "positive" adaptation: if perpetuated without being questioned, it can act to hamper an individual's capacity and ability to adapt (Wolf et al. 2010).

We are not, of course, downplaying extant observations about social contacts as important determinants of resilience: Semenza et al.'s (1996) examination of heat-related mortality during the 1995 Chicago heat wave, for example, concluded that social isolation increases individual vulnerability in conjunction with other personal characteristics or environmental factors. We suggest, rather, that attention to the constructs and manifestations of identity provides a more sophisticated understanding of individual and group responses to changing environments. Benzie et al. (2012, 59) espouse this view, arguing that in the United Kingdom, vulnerability to heat "appears to have a very strong social dimension" not yet adequately reflected in policy. The Heatwave Plan designed by the Department of Health in the United Kingdom, according to Benzie et al. (2012), is based on identifying vulnerable populations through individual physiological factors, with limited recognition of wider social processes and broader factors relating to identity, place, and tenure.

By considering identity processes, we can gain insight into how adaptation options are interpreted and why decisions by policy makers are sometimes contested. Specifically, by considering identity, we may gain a better understanding of why people feel that some adaptation actions are fair as well as of how adaptation decisions may reflect on how we view ourselves or how we would like to be viewed (Holmvall and Bobocel 2008). Do such decisions align with our self-perceived identity or undermine it? And what does this mean for public acceptance of decisions in terms of fair procedures and outcomes? Fairness-based evaluations can result in strong emotional reactions as a result of what they mean for a person's identity and therefore well-being. As Clayton and Opotow (2003, 299) highlight, inclusion in the scope of justice, and the nature of the inclusion, has implications for individual identities: "the sense that a particular process or outcome is just or unjust conveys information about the identities, particularly the status, of those involved."

How individuals are included in decisions denotes implicit (or sometimes explicit) relationships of respect and power (Skitka and Crosby 2003). Conversely, being excluded can lead to feelings of marginalization and disempowerment. This can also act to delegitimize actors left outside of the decision-making process and scope of justice. An example from banking demonstrates why procedural aspects of fairness are important in determining satisfaction with decisions. Holbrook and Kulik (2001) showed that for a sample of bank loan applicants, individuals who had a long-standing relationship with the bank were more affected by the opportunity to voice their opinion and less affected by the actual outcome of the process. If people were given voice, felt they had been treated well, and that established relationships and identity roles were honored, and they were more likely to be happy with procedures even if the outcome was not particularly favorable to themselves.

Place and Climate Change

Place is described and categorized in a number of ways in the sociological, psychological, and human geography literatures; recently, attempts have sought to bring these often disparate strands together (Lewicka 2011). We review two dimensions of place that have been repeatedly discussed: place attachment and place identity.

Place attachment is often described as a positive, emotional, and cognitive attachment between individuals and community, and between individuals and place (Altman and Low 1992; Brown and Perkins 1992; Fried 2000). This bonding of people to place can lead to action to preserve emotionally significant attachments (Devine-Wright 2010). Closely linked to place attachment, and with many descriptions assigning them similar traits, is place identity.

Place identity is one element of an individual's identity that is shaped by the places that one passes through during one's life. It can motivate behavior, provide self-esteem through stability, and contribute significantly to psychological well-being (Fullilove 1996; Manzo 2003; Proshansky 1978; Proshanksy et al. 1983; Twigger-Ross and Uzzell 1996). As Manzo highlights, writing about place and identity is important because it pulls the significance of place out from internal cognitive structure and locates the relationship within "a social, historical and political milieu" (Manzo 2003, 54). Dixon and Durrheim (2000) also signal that place identity is sometimes in danger of being partitioned off as a purely cognitive process which as a result detaches it from the wider social world. They emphasize the role of dialogue in place identity formulation and discuss how places can become contested areas of collective being and belonging. By recognizing the social dimensions of place, we recognize its continuously unfolding nature and also the very political nature of decisions around places and place changes.

Increasingly, new studies are reexamining commonplace assumptions in light of the influence exerted by identity and place attachment on risk behavior, including adaptation to climate risk (Adger et al. 2013). Investigation into fisheries and agricultural livelihoods indicates that personal and occupational identities as well as a strong sense of place attachment are significant determinants of resistance to change (Daw et al. 2012). Decisions to adapt to changing risks are also driven by supporting structures, governance, household diversification, and opportunities. These new insights on the interactions between identity and underlying social structures and links to place challenge existing understandings of adaptation decisions (Daw et al. 2012).

A strong element of identity is bound up with notions of place. Hence climate risks that involve adaptations through changing places and locations have significant impacts on identity. Adger et al. (2009) highlight, for example, the negative effect of involuntary migration on place attachment. Barnett and O'Neill (2012) argue strongly that in the case of inhabitants of Pacific islands threatened by sea-level rise, involuntary migration would result in permanent and indelible erosion of identity and cultures linked to ancestral lands. Some planning for community relocation shows that collective identity and the perceptions of empowerment, autonomy, and inclusiveness are critical to how relocation is perceived. Bronen and Chapin (2013), for example, demonstrate that Alaskan settlements presently planning for relocation due to the impacts of changing climate on their village infrastructure are more empowered when issues of identity and place are incorporated into their planning processes.

The extreme events and weather-related disasters associated with climate change can forcibly alter the physical and social characteristics of a place, causing individuals in that location to reconsider their sense of place and attachment. Chamlee-Wright and Storr (2009) found that place attachment affects people's decisions to stay or relocate. Narratives of individuals who had to leave New Orleans's Ninth Ward after Hurricane Katrina revealed how their attachment to place become even stronger after displacement. For many, but not all, a return to their homes was necessary to satisfy their well-being, which was inherently connected to New Orleans. For the Ninth Ward community to rebound, it was necessary for many of the displaced individuals to be able to visualize and consider how they could be part of that regeneration by contributing actively to re-creating the sense of a place that the disaster had physically destroyed.

Climate change may manifest in significant sudden (rather than incremental) changes to particular locations, implying that large-scale adaptation will have to happen in situ or that moving to a different location will become necessary – as is now a serious consideration for low-lying small island states. This process of "loosening" ties and forming attachments to an altered or completely different place is an aspect of climate change adaptation that deserves much further attention, as the population of displaced peoples grows and the repercussions of emerging issues evolve (Agyeman et al. 2009).

Focusing on a smaller scale, for most people, their home is a secure and private domain, a safe haven with clear boundaries to the outside world. Homes are comfortable because of the familiar and the routine. Sudden, destructive events such as floods can sever these bonds and attachments, and subsequent relocation to temporary and unfamiliar cities or even temporary camps takes control away from management of living spaces. These can lead to feelings of helplessness and emotional distress. At the household level, the psychological impacts of a flood are often longer lasting than the material recovery, which in some cases may result in the development of posttraumatic stress disorder (Carroll et al. 2009; Tapsell and Tunstall 2003). Such stress has been widely reported for many years following Hurricane Katrina (Paxson et al. 2012) and has been heightened by subsequent events such as Hurricane Isaac in 2012, which made landfall on the seventh anniversary of Katrina.

The sense of disruption, distress, unease, wariness, and fear from loosening ties and reforming them when individuals get back into their homes or when communities regroup can persist in time. In other words, sense of place, through attachment and identity, has important affective implications for how an individual adapts to the impacts of climate change. This was demonstrated by recent research, based on a survey of households and individuals following major floods in 2009 in Cumbria – England and Galway – in the west of Ireland (Adger et al. 2013). We found, for example, how returning residents sometimes refurbished homes with cut-price goods and different materials in anticipation of being flooded again. The communities greeted high rainfall with a watchful eye – previously valued and secure homes now felt transient, and previously enjoyed rivers now were viewed as dangerous. Although it has been found that increased awareness of risk can facilitate adaptation (Adger et al. 2013), this study raised the question of equity and fairness in adapting to climate change, in relation to identity and place attachment. When areas are affected by a natural disaster or when climate change means that expensive adaptive measures need to be taken, those economically better off are more able to make preemptive changes, to relocate to "safer" and more secure locations (see also Box 9.1), or to return and rebuild following a sudden event. Those with greater economic or political power will be better positioned to ensure that their relationship with place and/or identity is not significantly altered.

Considering Identity and Place Attachment: Smoothing the Way for Adaptation?

Climate-induced change may be slow and incremental or may happen as sudden events or shocks that force adaptive measures to be taken immediately. Either way, this will mean change for local places. Alongside a changing climate, decision making will play a role in how places change – either through proactive management or in reactive measures following a physical disaster. Given how important place is for people, and the psychological desire for consistency, these are changes that will need to be carefully negotiated to ensure sensitive, sustainable, and fair adaptation. If adaptation

> **Box 9.1**
> **Influencing Sense of Place and Recovery: The Role of Insurance**
>
> Place attachment at the household scale may be shaped by climate change through the actions of the insurance industry (Mills 2005). The frequency and intensity of hazardous events such as flooding and hurricanes are set to increase with climate change, and one of the ways these risks are managed in everyday life is through insurance.
>
> Insurance provides security. The experience of being insured and of using insurance has been shown to have a significant impact on place attachment and identity, especially in the recovery period following a disaster. Negotiating with insurance companies for payments and new cover has often been shown to be one of the major traumas following a flood event (Carroll et al. 2009; Harries 2008; Tunstall et al. 2006). Similarly, the knowledge that coverage could be, or has been, rescinded affects how people manage their living space.
>
> Insurance companies can constrain how people replace lost items and feelings of control can be stymied by insurance company stipulations (Carroll et al. 2009). Additionally, variations in coverage can make it harder to sell houses, either if it affects the value of the property or if flood insurance is necessary for gaining a mortgage – a process that disproportionately affects poorer populations and their ability to relocate. How interactions with insurance companies are managed and whether coverage is provided shape show individuals build and experience their homes and also inform how they chose to adapt to future risks (Carroll et al. 2009).
>
> In many areas, climate-induced change will not force the choice of relocation but instead will mean slow, gradual evolution of the landscape and its communities. Yet a slow onset of change can cause a constant heightened awareness of place. Burley et al. (2007), in their analysis of a community on the Louisiana coast, found that even with coastal restoration in action, people did not feel included in the decision-making processes of their area, which added further to the fragility of their sense of place. Other areas will experience comparatively benign impacts from climate change that will influence people's sense of place to slow but immutable change. For example, in Norway, the level of snow cover is projected to decrease in low elevations, and this is affecting how places are being used; it is an especially significant change for people with more traditional values that are deeply embedded in existing surroundings and related cultural practices (O'Brien 2009).

to climate change continues to be measured in physical and financial terms, then policy makers will be overlooking the "psychological, symbolic and particularly emotional aspects of healthy human habitats" (Agyeman et al. 2009, 509).

Increased awareness of the roles that identity and place play in adaptation, we argue, is likely to be of benefit in making legitimate decisions that harness resources for community action and ensure that as far as possible decisions are fair for all parties affected. However, while place identity contributes to well-being, efforts to keep places the same may result in maladaptation where people are unwilling, or perceive

> **Box 9.2**
> **Place and Identity Promote Capacity but Constrain Radical Transformation**
>
> Many resource-intensive sectors of the economy are likely to require significant adaptation under climate change. The location of fisheries, and the suitability of whole regions for well-established agricultural and forest industries, may be in question. Hence planning for radical change or transformation of such sectors is being considered at various scales in different parts of the world (Park et al. 2012; Kates et al. 2012).
>
> Nadine Marshall et al. (2012) examined a plan for wholesale relocation of a region of peanut growing (an activity permeating and supporting an entire community) from southern Queensland (Australia) to the Northern Territories. This plan was considered through trade associations and businesses in the supply chain over a number of years until 2011, when the sector perceived its long-term future to be no longer viable due to lack of water resources. The study focused on whether place attachment and the "occupational identity" of sixty-nine surveyed farmers influenced individuals' willingness and ability to undertake incremental adaptations and their transformational capacity. The findings show a strong negative correlation between both attachment to place and attachment to job with the capacity to undertake transformational change. Inevitably individuals become attached to place and define themselves through their occupation ("I'm a peanut farmer (or logger or trawlerman), and always will be") at different points in their life course. Marshall and colleagues pinpoint how sense of place attachment and identity act as barriers to transformational change, that is, collective relocation for the benefit of the whole peanut community.

themselves as unable, to make changes. For psychological well-being, consistency in place and identity is important, but in terms of adaptation, there may be trade-offs between this desired consistency and optimal adaptation action. At the community level, strong place attachment is related to civic involvement, but at the individual level, this may prevent adaptive emigration or lifestyle changes. Such tensions are illustrated by the example of the reluctance of farmers to act collectively to, in effect, reinvent themselves in another locality (see Box 9.2). Identity and place can operate at different levels in contradictory directions regarding adaptation, and such tensions clearly need to be recognized in adaptation policy making. Both identity and place offer ways to expand and add nuance to our understanding for adaptation to climate change.

The example of the peanut farmers in Box 9.2 illustrated how a shift in place and identity undermined feelings of control and self-efficacy, which ultimately resulted in the rejection of an adaptive action. Zygmunt Bauman (2004) suggests that rather than considering identity to be a solid and unchanging structure that we should seek to keep constant, postmodernism brought with it a more fluid concept of identity, and that flexibility and openness to change are now important in identity management. Extending this perspective to adaptation, where there is no option to protect existing conditions, help will be needed for the transition to different places and roles – for example, if people are going to have to change jobs, then they need to be equipped to deal with the challenges that a such a change will bear on their identities. By being able to identify the social stressors in some forms of adaptation, individuals are able to employ mechanisms to help to deal with difficult transitions.

Conclusion

Identity and place are key elements affecting how the adaptive challenge of climate change will impact individuals' everyday lives. These are an important subset of the social dimensions to the adaptive challenge highlighted throughout this book. The scale of the challenges, and indeed the multiple levels and geographical scales at which these social processes operate, and around which values form, points to a high degree of complexity in the governance processes for adaptation.

With such multidimensional challenges in mind, this chapter emphasizes the importance of recognizing the permanence and mutability of identity and place attachment. Policies to promote pro-environmental behavior are often developed on the assumption of very limited mutability of individual identities. For instance, education and persuasion, which seek to implement correspondent and suitable behaviors, in the main work on fixed notions of what individuals perceive to be desirable or morally acceptable. We argue that such assumptions in policy are founded on a narrow view of the characteristics and influence of identity that may not be appropriate in relation to the adaptive challenge of climate change. Where people live, their recreation, and what they value are all much more dynamic than planned for in most policy interventions. A well-adapted future is likely to be diverse and heterogeneous, not something that sits well in the abounding rigid blueprints for solving the climate crisis.

References

Adger, W. N., Barnett, J., Brown, K., Marshall, N. and K. O'Brien. 2013. "Cultural dimensions of climate change impacts and adaptation." *Nature Climate Change* 3 (2): 112–17.

Adger, W. N., Dessai, S., Goulden, M., Hulme, M., Lorenzoni, I., Nelson, D., Naess, L. O., Wolf, J. and A. Wreford. 2009. "Are there social limits to adaptation to climate change?" *Climatic Change* 93 (3–4): 335–54.

Adger, W. N., Paalova, J., Huq, S. and M. J. Mace, eds. 2006. *Fairness in Adaptation to Climate Change*. Cambridge, MA: MIT Press.

Adger, W. N., Quinn, T., Lorenzoni, I., Murphy, C. and J. Sweeney. 2013. "Changing social contracts in climate-change adaptation." *Nature Climate Change* 3 (4): 330–33.

Agyeman, J., Devine-Wright, P. and J. Prange. 2009. "Close to the edge, down by the river? Joining up managed retreat and place attachment in a climate changed world." *Environment and Planning A* 41 (3): 509–13.

Altman, I. and S. M. Low, eds. 1992. *Place Attachment*. New York: Plenum.

Barnett, J. and S. J. O'Neill. 2012. "Islands, resettlement and adaptation." *Nature Climate Change* 2 (1): 8–10.

Barr, S., Shaw G. and T. Coles 2011. "Times for (un) sustainability? Challenges and opportunities for developing behaviour change policy: A case-study of consumers at home and away." *Global Environmental Change* 21 (4): 1234–44.

Bauman, Z. 2004. *Identity: Conversations with Benedetto Vecchi*. Cambridge: Polity.

Benzie, M., Harvey, A., Burningham, K., Hodgson N. and A. Siddiqi. 2012. *Vulnerability to Heatwaves and Drought: Adaptation to Climate Change*. York: Joseph Rowntree Foundation. http://www.jrf.org.uk/sites/files/jrf/climate-change-adaptation-full.pdf.

Bronen, R. and F. S. Chapin. 2013. "Adaptive governance and institutional strategies for climate-induced community relocations in Alaska." *Proceedings of the National Academy of Sciences of the United States of America* 110 (23): 9320–25.

Brown, B. B. and D. D. Perkins. 1992. "Disruptions in place attachment." In *Place Attachment*, edited by Altman, I. and S. M. Low, 279–304. New York: Plenum.

Burley, D., Jenkins, P. Laska, S. and T. Davis. 2007. "Place attachment and environmental change in coastal Louisiana." *Organization and Environment* 20 (3): 347–66.

Carroll, B., Morbey, H. Balogh, R. and G. Araoz. 2009. "Flooded homes, broken bonds, the meaning of home, psychological processes and their impact on psychological health in a disaster." *Health and Place* 15 (2): 540–47.

Chamlee-Wright, E. and V. H. Storr. 2009. "There's no place like New Orleans: Sense of place and community recovery in the Ninth Ward after Hurricane Katrina." *Journal of Urban Affairs* 31 (5): 615–34.

Clayton, S. and S. Opotow, eds. 2003. *Identity and the Natural Environment: The Psychological Significance of Nature*. Cambridge, MA: MIT Press.

Daw, T. M., Cinner, J. E., McClanahan, T. R., Brown, K., Stead, S. M., Graham, N. A. J. and J. Maina. 2012. "To fish or not to fish: Factors at multiple scales affecting artisanal fishers' readiness to exit a declining fishery." *PLoS One* 7 (2): e31460.

Devine-Wright, P. and Y. Howes. 2010. "Disruption to place attachment and the protection of restorative environments: A wind energy case study." *Journal of Environmental Psychology* 30 (3): 271–80.

Dixon, J. and K. Durrheim. 2000. "Displacing place-identity: A discursive approach to locating self and other." *British Journal of Social Psychology* 39 (1): 27–44.

Frank, E., Eakin, H. and D. Lopez-Carr. 2011. "Social identity, perception and motivation in adaptation to climate risk in the coffee sector of Chiapas, Mexico." *Global Environmental Change* 21 (1): 66–76.

Fried, M. 2000. "Continuities and discontinuities of place." *Journal of Environmental Psychology* 20 (3): 193–205.

Fullilove, M. T. 1996. "Psychiatric implications of displacement: Contributions from the psychology of place." *American Journal of Psychiatry* 153 (12): 1516–23.

Harries, T. 2008. "Feeling secure or being secure? Why it can seem better not to protect yourself against a natural hazard." *Health, Risk and Society* 10 (5): 479–90.

Holbrook, R. L. and C. T. Kulik. 2001. "Customer perceptions of justice in service transactions: The effects of strong and weak ties." *Journal of Organizational Behavior* 22 (7): 743–57.

Holmvall, C. M. and D. R. Bobocel. 2008. "What fair procedures say about me: Self-construals and reactions to procedural fairness." *Organizational Behavior and Human Decision Processes* 105 (2): 147–68.

Kates, R. W., Travis, W. R. and Wilbanks, T. J. 2012. "Transformational adaptation when incremental adaptations to climate change are insufficient." *Proceedings of the National Academy of Sciences* 109:7156–61.

Lewicka, M. 2011. "Place attachment: How far have we come in the last 40 years." *Journal of Environmental Psychology* 31 (3): 207–30.

Manzo, L. C. 2003. "Beyond house and haven: Towards a revisioning of emotional relationships with places." *Journal of Environmental Psychology* 23 (1): 47–61.

Marshall, N. A., Park, S. E., Adger, N. W., Brown, K. and S. M. Howden. 2012. "Transformational capacity and the influence of place and identity." *Environmental Research Letters* 7 (3): 034022.

Mills, E. 2005. "Insurance in a climate of change." *Science* 309 (5737): 1040–44.

O'Brien, K. 2009. "Do values subjectively define the limits to climate change adaptation?" In *Adapting to Climate Change: Thresholds, Values, Governance*, edited by Adger, W. N., Lorenzoni I. and K. O'Brien, 164–80. Cambridge: Cambridge University Press.

Park, S. E., Marshall, N. A., Jakku, E., Dowd, A. M., Howden, S. M., Mendham, E. and A. Fleming. 2012. "Informing adaptation responses to climate change through theories of transformation." *Global Environmental Change* 22 (1): 115–26.

Paxson, C., Fussell, E., Rhodes, J. and M. Waters. 2012. "Five years later: Recovery from post traumatic stress and psychological distress among low-income mothers affected by Hurricane Katrina." *Social Science and Medicine* 74 (2): 150–57.

Proshansky, H. M. 1978. "The city and self-identity." *Environment and Behaviour* 10 (2): 147–69.

Proshansky, H. M., Fabian A. K. and R. Kaminoff. 1983. "Place-identity: Physical world socialization of the self." *Journal of Environmental Psychology* 3 (1): 57–83.

Rabinovich, A., Morton T. A. and C. C. Duke. 2011. "Collective self and individual choice: The role of social comparisons in promoting public engagement with climate change." In *Engaging the Public with Climate Change, Behaviour Change and Communication*, edited by Whitmarsh, L., O'Neill S. J. and I. Lorenzoni, 66–83. London: Earthscan.

Semenza, J. C., Rubin, C. H., Falter, K. H., Selanikio, J. D., Flanders, W. D., Howe, H. L. and J. L. Wilhelm. 1996. "Heat-related deaths during the July 1995 heat wave in Chicago." *New England Journal of Medicine* 335 (2): 84–90.

Skitka, L. J. and F. Crosby. 2003. "Trends in the social psychological study of justice." *Personality and Social Psychology Review* 7 (4): 282–85.

Tapsell, S. and S. Tunstall. 2003. "An examination of the health effects of flooding in the United Kingdom." *Journal of Meteorology* 28: 341–49.

Tunstall, S. M., Tapsell, S. M., Green, C., Floyd, P. and C. George. 2006. "The health effects of flooding: Social research results from England and Wales." *Journal of Water and Health* 4: 365–80.

Twigger-Ross, C. and D. L. Uzzell. 1996. "Place and identity processes." *Journal of Environmental Psychology* 16 (3): 139–69.

Twigger-Ross, C., Bonaiuto, M. and G. Breakwell. 2003. "Identity theories and environmental psychology." In *Psychological Theories for Environmental Issues*, edited by Bonnes, M., Lee, T. and M. Bonaiuto, 203–33. Farnham, VT: Ashgate.

Wolf, J., Adger, W. N., Lorenzoni, I., Abrahamson, V. and R. Raine. 2010. "Social capital, individual responses to climate change adaptation: An empirical study of two UK cities." *Global Environmental Change* 20 (1): 44–52.

10

Values and Traditional Practices in Adaptation to Climate Change

Evidence from a Q Method Study in Two Communities in Labrador, Canada

JOHANNA WOLF, ILANA ALLICE, AND TREVOR BELL

Communities in northern Canada are experiencing the effects of a changing climate. The impacts in the north are materializing faster due to recent accelerated arctic warming and its effects, as well as increasing extreme weather events (Barber et al. 2008; Graversen et al. 2008; Kaufman et al. 2009; Min et al. 2008). These changes also have more direct consequences for communities in the Canadian north than for their southern counterparts. Vulnerability studies have characterized the origins of risk in many communities across the Canadian arctic, and implications for adaptation policy and practice have been outlined (Ford et al., 2010). The studies suggest that communities are particularly vulnerable to the effects of climate change due to underlying socioeconomic stresses, pervasive inequalities, and the magnitude of climatic change locally (see, e.g., Ford et al. 2006; Laidler et al. 2009). Recently, however, an emerging literature points to the importance of intangible and subjective impacts of the changing climate, for example, on places (Adger et al. 2011), values (O'Brien and Wolf 2010), beliefs and perceptions of risk (Kuruppu 2009; Mortreux and Barnett 2009), culture (Adger et al. 2012), and the affective dimensions of climate change impacts in the north (Cunsolo Willox et al. 2013).

Research that examines climate change in Labrador is limited but growing. Furgal et al. (2002) document Inuit knowledge of climate change from residents of Nain, Labrador, and explore the health impacts of these changes. A conference held in North West River, Labrador, in 2008 highlighted public interest in and need for further research and community engagement on issues related to environmental and climate change in Labrador (Bell et al. 2008). Demarée and Ogilvie (2008) have examined historical weather observations by the Labrador Moravian missionaries (*Unitas Fratrum*, also called Moravian Brethren), who documented thermometric (and some barometric) readings from 1771 onward on the Labrador coast. Some evidence from the region suggests that increasing variability in precipitation may result in health impacts through waterborne gastrointestinal illness (Harper et al.

2011). Recent research suggests there are mental health impacts of extreme weather and climatic variability in Rigolet (Cunsolo Willox et al. 2013).

This chapter draws on results from 42 Q sorts (see Methods section for an introduction to Q methodology) conducted in two Labrador coastal communities to examine how participants' values relate to their experience of climate change and adaptation to it. In particular, we discuss four distinct discourses on the perception of and adjustments to climate variability and change and their associated values that emerge from the analysis of the Q sorts. The study was conducted following an unusually mild winter in 2009–10 during which freeze up of sea ice and freshwater ponds in Labrador occurred later than expected, breakup was weeks earlier than expected, and temperatures during the winter were milder than normal. Less snowfall and unseasonal rain together with milder temperatures contributed to poor sea and freshwater ice and poor winter traveling conditions on the land. The chapter argues that diverse views exist both within and among communities in close geographical proximity that also partially share history and culture and that these views are underpinned by diverse and at times competing values that support distinct goals for adaptation.

The chapter first reviews literature on values and how they have been considered in the context of climate change. Next, we outline the setting of this research in Labrador in northeastern Canada and explain the use of Q methodology. We then outline the results in the form of four discourses that emerge from the analysis of the Q sorts. The discussion examines the values that underpin the four discourses and considers the findings in the context of adaptation literature.

What Are Values?

Many disciplines, including sociology, anthropology, and psychology, discuss values, and a diverse array of conceptions and definitions exists. Values have been understood as lasting beliefs that support specific actions because these actions, or their outcomes, are perceived as preferable to potential alternatives (Rokeach 1973). In this way, values serve as standards that guide choices, judgments, and arguments, and they underlie attitudes and ways of evaluating situations, choices, and information (Rokeach 1979). Five features common to all definitions of values have been identified: values are (1) concepts or beliefs about (2) desirable actions or outcomes that (3) transcend specific situations, (4) guide selection of behavior, and (5) are ordered according to their relative importance (Schwartz and Bilsky 1987).

In the context of climate change, values have recently been examined as elements that shape adaptation outcomes. Cultural values in Kiribati (tropical Pacific Ocean) have been found to shape perception of risk and adaptation in the water sector (Kuruppu 2009). Also in the context of adaptation, O'Brien (2009) argues that limits to adaptation are subjectively defined through values that underpin adaptation decisions and actions. In northern Canada, research that characterizes vulnerability has pointed to cultural values as underpinning historical adaptability of the Inuit (Ford et al. 2006; Laidler et al. 2009; Wenzel 2009) and implicitly refers to cultural values as they are embedded in the concept of *Inuit Qaujimajatuqangit* (IQ), Inuit traditional knowledge (Gearheard et al. 2010; Laidler 2006). Wolf et al. (2013) argue that values shape how the effects of the changing climate are felt and further underpin which adaptations are perceived as useful and effective.

Values can be organized in systems, defined as "an organized set of preferential standards that are used in making selections of objects and actions, resolving conflicts, invoking social sanctions, and coping with needs or claims for social and psychological defenses of choices made or proposed" (Williams 1979, 20). One such system has been identified by Schwartz (1994), who elaborates ten basic universal values – security, tradition, conformity, power, achievement, hedonism, stimulation, self-direction, universalism, and benevolence – and the characteristic motivations that organize them into value systems. Two orthogonal dimensions represent these motivations; self-enhancement *versus* self-transcendence and conservation *versus* openness to change (Schwartz 2006). A similar conceptualization of values emerges from self-determination theory, which distinguishes between intrinsic values (such as personal growth, social connection, societal contribution) and extrinsic values (such as success, popularity, beauty) (Deci and Ryan 2002). Value conflicts emerge from the diverging dimensions between intrinsic values (or self-transcendence and associated values) and extrinsic values (self-enhancement and associated values). Cultural values developed by Schwartz (1999) include conservatism, intellectual autonomy, affective autonomy, hierarchy, egalitarianism, mastery, and harmony. Another value system is represented in the results of the World Values Survey (WVS), which asserts that there are two major dimensions of values: traditional versus secular-rational values and survival versus self-expression values (Inglehart and Welzel 2005). Analysis of WVS data shows that socioeconomic development tends to be associated with a shift from survival values to self-expression values. Schwartz's values were used in this study for three reasons: they have been developed to represent universal dimensions in values, they include specific values that go beyond the two dimensions of the WVS, and they include tradition as a specific value considered important in an aboriginal setting. The individual and cultural values elaborated by Schwartz are shown in Tables 10.1 and 10.2.

Setting

Labrador in northeastern Canada is a vast subarctic and very sparsely populated territory that represents the mainland portion of the province of Newfoundland and Labrador. Two regional centers, Labrador City and Goose Bay, are service hubs for small communities, many of which are situated on Labrador's coastline. This study was conducted in 2010–11 in two coastal communities in Labrador, Rigolet, and

Table 10.1. *Schwartz's cultural values*

Defining the relationship between the individual and the group:
Conservatism: Maintenance of the status quo, propriety, and restraint of actions or inclinations that might disrupt the solidarity group or the traditional order (social order, respect for tradition, wisdom).
Intellectual autonomy: individuals independently pursuing their own ideas and intellectual directions (curiosity, broadmindedness, creativity).
Affective autonomy: individuals independently pursuing affectively positive experience (pleasure, exciting life, varied life).

Guaranteeing responsible behavior that will preserve the social fabric:
Hierarchy: Legitimacy of an unequal distribution of power, roles and resources (social power, authority, humility, wealth).
Egalitarianism: Transcendence of selfish interests in favor of voluntary commitment to promoting the welfare of others (equality, social justice, freedom, responsibility, honesty).

Relationship of humankind with the natural world and the social world:
Mastery: Getting ahead through active self-assertion (ambition, success, daring, competence).
Harmony: fitting harmoniously into the environment (unity with nature, protecting the environment, world of beauty).

Source: Adapted from Schwartz (1999).

St. Lewis. The communities were chosen for their setting in remote Labrador, where very few studies have examined climate change adaptation. Rigolet is a town of 200 in the Inuit Settlement Region of Nunatsiavut that provides Inuit regional and local self-government over many aspects of natural resources, education, and health and

Table 10.2. *Schwartz's individual values*

Self-Direction: Independent thought and action; choosing, creating, exploring.
Stimulation: Excitement, novelty, and challenge in life.
Hedonism: Pleasure and sensuous gratification for oneself.
Achievement: Personal success through demonstrating competence according to social standards.
Power: Social status and prestige, control or dominance over people and resources.
Security: Safety, harmony, and stability of society, of relationships, and of self.
Conformity: Restraint of actions, inclinations, and impulses likely to upset or harm others and violate social expectations or norms.
Tradition: Respect, commitment, and acceptance of the customs and ideas that traditional culture or religion provide the self.
Benevolence: Preserving and enhancing the welfare of those with whom one is in frequent personal contact (the 'in-group').
Universalism: Understanding, appreciation, tolerance, and protection for the welfare of all people and for nature.

Source: Adapted from Schwartz (2006).

social development. St. Lewis is a town of about 200 primarily Inuit-Métis on the Labrador south coast outside of the Nunatsiavut land claim area and is therefore under provincial jurisdiction of the Government of Newfoundland and Labrador. The communities share a history of a declining commercial fishery, socioeconomic stress, and decreasing population from a lack of employment opportunities during the past several decades. Their recent histories, however, are more distinct. The Nunatsiavut land claim agreement arguably helped shape Inuit identity in the communities of northern Labrador, including Rigolet, because it formalized the status of many descendants of Inuit-European settler unions in the land claim area as Inuit. While Inuit ancestry is common in St. Lewis, it is not part of the land claim agreement, and it has been argued that identity on the south coast of Labrador was and is still affected by this exclusion (Plaice 2008). Many Labradorians strongly distinguish their identity from that of residents of Newfoundland, the island portion of the province, and this can be seen also in their suspicious attitudes toward the provincial government situated in the capital city of St. John's.

Land-based activities, including hunting, fishing, and ice fishing, have historically been and remain important aspects of life in both Rigolet and St. Lewis. These activities continue to support not only livelihoods but also identity, well-being, and health. Many changes to daily life common in northern aboriginal communities are putting pressure on these activities. The introduction of the cash economy has altered life in the north (Chabot 2003; Wenzel 1991). Many families interviewed for the present study found it difficult to find the time needed "to go off on the land," a phrase used to describe traditional activities such as hunting, fishing, trapping, and berry picking outside of the settlement. As a result, many households report they can at best spend weekends on the land. Since the introduction of compulsory schooling in Newfoundland and Labrador in 1942 (Phillips and Norris 2001), the children's school schedule has to be considered when planning trips. To some, the cost of a snowmobile, or indeed fuel for it, is prohibitive. St. Lewis was connected by road to neighboring communities in 2001 and via the new extension of the Trans-Labrador Highway to Happy Valley-Goose Bay in 2010. Rigolet has no road access. It is accessible in summer by ferry and air from central Labrador and in winter by air or snowmobile.

Food sharing remains an important part of life in both communities, as it is in other arctic communities (Berkes and Jolly 2001; Collings et al. 1998). In Rigolet, food sharing is practiced both within and outside of familial ties. In St. Lewis, food-sharing practices are somewhat more limited to sharing within familial relations. Another dimension of sharing in the communities relates to firewood. In Rigolet, firewood is shared between households, and especially with elders in the community. In St. Lewis, this practice is more limited due to the community not being within an aboriginal land claim area, and provincial forestry regulations are perceived as prohibiting wood sharing.

Method

This study used Q methodology, a systematic analysis of subjective views on an issue first developed by Stephenson (1935). The method can be used with qualitative statements or imagery and is gaining increasing recognition as an ideal tool for investigating an issue through discourses (Barry and Proops 2000), understood here as shared meanings that form distinct story lines (Dryzek 2005). The discourses that Q reveals embody personal values, beliefs, and attitudes toward the issue in question (Darier and Schüle 1999). Q methodology has been used in examining discourses of ecological citizenship in western Canada (Wolf et al. 2009), salience and personal efficacy in relation to climate change in the United Kingdom (Lorenzoni et al. 2007), and the role of deliberation in building adaptive capacity (Hobson and Niemeyer 2011).

The empirical component of the study started with qualitative interviews with fifty-three participants, twenty-nine from Rigolet and twenty-four from St. Lewis, conducted in summer 2010. Participants were recruited through a local community research assistant in each community working with the research team.[1] The interviews explored what participants value in their way of life, their perception of change in the environment and their community, their responses to these changes, and their perceptions of the unusually mild winter of 2009–10. The sample of participants aimed to represent the community at large and included current and former natural resource officers, fishermen, heavy machine operators, teachers, municipal employees and decision makers, translators, government employees and decision makers, youth coordinators, health care workers, and social care coordinators. The sample was as close to gender balanced as possible (twenty-seven male, twenty-six female), and participants ranged in age from sixteen to eighty in St. Lewis and nineteen to sixty-three in Rigolet. Our sample included youths and those who do not spend a significant amount of time on the land so as to broaden the sample beyond the type of participants that vulnerability studies in the arctic have considered. Figure 10.1 illustrates the process of data collection and analysis.

A large sample of Q statements was developed from the qualitative analysis of the interviews. These statements were supplemented with some drawn from other sources, including local media and participants' informal conversations with the researchers, to create an inclusive overall statement set that represents the broadest possible spectrum of views held on the issue (called "concourse" in Q; see Stephenson 1978; Brown 1999). This large set was reduced to thirty-two statements through an iterative process of grouping statements into emergent categories to eliminate overlaps. The initial set of thirty-two statements was piloted to determine the balance and inclusivity of the statement set, and feedback from pilot participants was used to adjust the set. The

---

[1] All participants opted to conduct the interview in English.

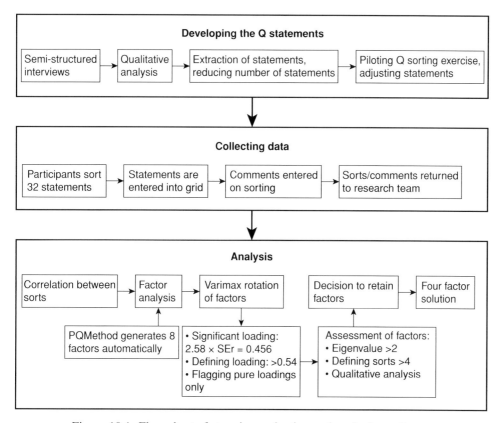

Figure 10.1. Flow chart of steps in conducting and analyzing a Q sort.

research team then matched each of the statements with one of Schwartz's individual or cultural values. The use of Q methodology with Schwartz's values represented by statements is a novel approach.

Participants who indicated interest in continued participation after the initial interview completed the Q sort with the community research assistant and the researcher (in February 2011). Forty-two participants completed the Q sort with an identical set of thirty-two statements, twenty in St. Lewis and twenty-two in Rigolet. To complete a Q sort, participants were asked to sort the thirty-two statements printed on individual cards into a fixed pattern on a scale from −3 (I strongly disagree) to +3 (I strongly agree). Figure 10.2 shows the sorting grid.

This sorting process demands that participants prioritize statements over each other to fit them into the given sorting grid in which only two statements each are allowed under "I agree strongly" and "I disagree strongly." While Q sorting is typically completed by participants alone, the setting of this research required that assistance be available to participants in case they had questions, needed help reading the statements, or wanted help filling out the sorting grid. Some participants asked for clarification on how to sort the statements into the grid, and a few struggled initially

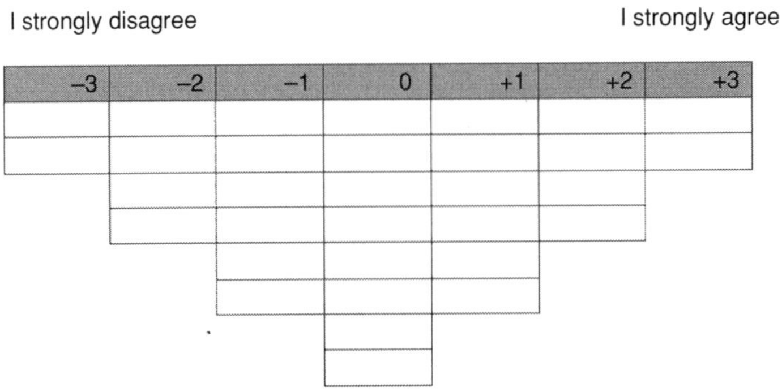

Figure 10.2. Q sorting grid.

with prioritizing the statements so that only the required number was placed under each category of agreement/disagreement.

As is typical in Q, the completed sorts were analyzed with correlation and principal component factor analysis (PCA) and were followed by a varimax rotation (Brown 1980), all completed using a Q method specific software program called PQ Method. The correlation and PCA examine relationships between the rankings of statements among sorts to characterize the main dimensions within the data. The factors that emerge are calculated based on the relationship between the sorts, based on the ranking of the statements in the sorting grid. Each calculated factor represents a typified response to the issue in question and is shaped like a Q sort, with statements ranked as described for the sorts above. After varimax rotation, the extent to which each participant's sort corresponds with the computed factors is calculated; this is called loading. To draw out distinct factors from the eight generated automatically by PQ Method, five procedures and criteria were used: only "pure" sorts (those which loaded significantly only on one factor) were used to define any factors (Addams and Proops 2000); the eigenvalue of the factors had to be greater than 1 to be selected (Brown 1980); the number of defining sorts per factor after rotation had to be equal to or greater than 2 (Sexton et al. 1998); and only those sorts whose squared loading was greater than half of the variance were used to define factors (Fürntratt 1969). Most importantly, the factors were evaluated for their qualitative content (Schlinger 1969). These measures resulted in four factors.

The analysis of the Q sorts conducted in both communities allows insight into how issues and perspectives are prioritized. The factors that emerge from factor analysis of the statements are typified expressions of views with specific priorities based on the statistical relationships between statements and sorts. Each factor therefore represents a story line, here called a discourse, that emerges from the statements that are ranked at +3 and −3 and +2 and −2, that is, those statements that are prioritized. The discourses that emerge here are correlated, as discussed later, but are statistically separate as each contributes significantly to the explained variance, and they are qualitatively distinct.

Interested participants were invited to a focus group held in each community (Rigolet in March 2011 and St. Lewis in April 2011) to present the key results of the interviews and the Q sort and seek their input and feedback to validate the analysis and, in particular, the four discourses that emerged from the Q analysis.

Results: Four Discourses on the Changing Climate in Labrador

The analysis of the Q sorts revealed four distinct discourses, and the focus groups confirmed that participants felt these represented their collective views. Together, the factors explain 59 percent of the variance in the data, factors 1–4 each accounting for 17 percent, 15 percent, 12 percent, and 15 percent of the variance, respectively. This suggests that no one factor is dominant, as each contributes similarly to the overall explained variance. The analysis of who defined and loaded significantly on the four discourses highlights that values differ both between and within the two communities. In this section, we first discuss which statements in each of the four discourses were ranked at the extremes of the grid, +3, +2 and −2, −3 (Table 10.3). We discuss those statements that set the factors apart from each other, in Q called "distinguishing statements." We then turn to those statements on which the discourses shared views and that do not distinguish between the discourses, in Q called "consensus statements."

Exclusively eight sorts completed in Rigolet define the first discourse, and a further four sorts from Rigolet loaded significantly on it. Notably, no sort from St. Lewis loaded significantly on this first discourse. The view expressed by this discourse emphasizes that the changes in weather are changing culture and traditional lifestyle (statement (S) 27, ranked +3). The discourse represents sentiments of missing the old Labrador winters with lots of ice, snow, and cold (S23, +3). The freedom to get out on the land is seen as the most important part of living in Labrador (S30, +2). Riding one's snowmobile is seen as so enjoyable that it would not be given up (S26, +2). The ranking of S21 (+2) supports this sentiment and suggests that being able to spend time on the land in the winter contributes to feeling happier and healthier. The recent mild winters were therefore not enjoyable and there is concern they may recur (S2, −3). Although this concern also emerges in the other discourses, this discourse explicitly links the effects of the mild winter to a lack of enjoyment and to negative effects on the way of life and culture. In the context of the changing climate, these items together suggest that if conditions were to prevent access to the land in winter, as they did in 2009–10, sadness, a sense of loss, and the loss of freedom and enjoyment would result. We call this discourse "losing winter, losing culture."

In contrast to discourse 1, which emerged exclusively from Rigolet, exclusively five sorts completed in St. Lewis define the second discourse, and a further seven sorts from St. Lewis loaded significantly on it. Again, notably, no one from Rigolet

Table 10.3. *List of Q statements and associated Schwartz values in increasing order of disagreement between discourses*

Statement	Discourse 1	2	3	4
9. Changes in weather and land will mean that I can grow more local foods. [Intellectual autonomy]	0	0	−1	0
11. Traveling by snowmobile out on the ice is as safe nowadays as it used to be. [Security]	−2	−3	−2	−3
16. Our community should have more control over what happens on our land. [Power]	1	2	2	1
12. When we're bored and stuck at home because of the weather, it would help to have more community activities to take part in. [Benevolence]	1	0	0	0
8. You can't fight the weather so we just have to get used to changing our daily activities. [Harmony]	0	1	0	1
31. I'll just sell my snowmobile and get a four-wheeler if these mild winters keep happening. [Self-direction]	−1	−2	−1	−1
32. The mild winters have made it easier to get out on the land. [Self-direction]	−1	−3	−3	−2
25. I feel a sense of freedom in the community because it is safe and connected. [Security]	1	2	0	1
10. Having a road opens the community to new problems from outside. [Conservatism]	0	1	0	0
22. Changes in weather might bring new animals that we can hunt and fish. [Intellectual autonomy]	−1	−1	0	0
7. The changing weather will make the ferry season longer so I can enjoy getting out of town more often. [Stimulation]	0	0	1	−1
5. You don't come across country foods as much anymore, but that's okay with me. [Tradition]	−2	−1	−3	−3
20. If everyone keeps their by-catch when salmon fishing, I might as well do it too. [Conformity]	−1	1	0	−1
24. I am going to move away if these mild winters keep happening. [Intellectual autonomy]	−2	−2	−1	−2
2. The last couple of winters likely won't happen again so we don't need to worry about having milder winters in the future. [Conservatism]	−3	−2	−1	0
17. We need better communication technology and rescue response, especially for youth, to deal with the dangerous conditions on the land and ice. [Benevolence]	1	2	1	0
13. Road access to our community is an important alternative in order to get around. [Intellectual autonomy]	0	2	1	2
26. I really enjoy riding my snowmobile and would never want to give it up. [Hedonism]	2	0	0	0
1. Our government's hunting and fishing regulations help our community to feed itself. [Universalism]	0	−1	−1	−2
28. I've really enjoyed the milder winters we've recently had because it meant a lot less shoveling. [Hedonism]	−3	0	−2	−1
21. I feel healthier and happier when I am able to get out on the land or get to the cabin during the winter. [Affective autonomy]	2	−1	2	2

Table 10.3 (continued)

Statement	Discourse 1	2	3	4
18. Our government should make sure that our seniors and other people who cannot hunt or collect firewood get enough food and wood for the year. [Egalitarianism]	0	1	3	−1
29. I identify first as Labradorian. [Hierarchy]	1	1	0	3
23. I miss the old Labrador winters we used to have with lots of snow and ice and cold. [Harmony]	3	−1	1	1
30. The freedom to get out on the land is the most important part of living in Labrador. [Self-direction]	2	0	1	3
19. I'm not concerned about the changes in weather because people from around here have always dealt with changes. [Mastery]	−2	0	−2	2
6. Sharing country foods with other people is an important part of my life. [Benevolence]	2	−1	3	2
27. The changes in weather are changing our culture and our traditional lifestyle. [Tradition]	3	1	1	−1
3. My role in the community will change if I can no longer access the land and its resources. [Achievement]	−1	−2	2	1
14. It's the government's responsibility to create new jobs here in order for our community to survive. [Egalitarianism]	0	3	2	0
15. I can't use the information and knowledge I was taught by my parents and grandparents because the weather is so different now. [Tradition]	1	0	−2	−2
4. The biggest problem facing the community is that too many people are leaving and there aren't enough young people coming back. [Universalism]	−1	3	−1	1

loaded significantly on the second discourse. The views represented by this discourse suggest that the biggest problem facing the community is that too many people are leaving and not enough young people are coming back (S4, +3). It is seen as government's responsibility to create jobs in the community to help the community survive (S14, +3). Participants who loaded significantly on this discourse feel that their role in the community will not be affected if they can no longer access the land and its resources (S3, −2). This discourse emphasizes feelings of safety and connectedness in the community (S25, +2); some participants demonstrated in interviews and focus groups that they also draw a sense of freedom from these feelings. The views represented by this discourse, therefore, do not identify changes in weather or climate as the most important issue in the community. Rather, they place significant emphasis on demographic change and outmigration from the community and government responsibility to help address this problem. We title this discourse "securing community viability."

Five sorts completed in Rigolet define the third discourse, and four sorts from St. Lewis loaded significantly on it. The discourse emphasizes that government has a responsibility to ensure that seniors and others who cannot hunt or collect firewood have enough food and wood for the year (S18, +3). Similar to discourse 2, it is seen as governments' responsibility to create new jobs in the community to help it survive. Yet, contrary to discourse 2, there is concern that people's role in the community could change if they can no longer access the land and its resources (S3, +2). From interview data and the focus groups, this is primarily relevant in connection to food and wood sharing. This discourse emphasizes the role country foods play in the local way of life (S6, +3) and perceives effects of changing weather conditions on traditional food practices (S5, −3). Together these rankings suggest that individuals' social roles in the community are perceived as at risk if access to the land is constrained and that it is then government's responsibility to provide for those who are most likely to be affected. This discourse is called "securing traditional foods."

Three sorts completed in Rigolet and two in St. Lewis define the fourth discourse. Another two sorts from Rigolet and six from St. Lewis loaded significantly on it. This discourse emphasizes a Labradorian identity over any other (S29, +3) and feels that the changes in weather are no cause for concern because the people of Labrador have always dealt with changes (S19, +2). Similarly to discourse 1, the freedom to get out on the land is seen as the most important part of living in Labrador (S30, +3). This discourse emphasizes the connection to the land and the critical role the land plays in well-being (S21, +2). Discourse 4 therefore expresses a strong belief in people's ability to deal with changes and applies this to potential changes in weather. Considering these rankings together, adaptability, identity, and freedom to access the land feature prominently in this discourse. We title this discourse "Labradorians adapt naturally." Table 10.4 summarizes the results of the Q sort.

The four discourses have certain viewpoints in common. Statements that do not statistically distinguish between any of the discourses, in Q called "consensus statements," include 8, 9, 11, 12, 16, and 31, and they explain the relationship between discourses, which in the case of this study was considerable. All discourses feel that the communities should have more control over what happens on their land (S16) and find that traveling on the ice is not as safe now as it used to be (S11). The statement embodying an acceptance toward changes in weather (S8) that emerged from interviews and the statement pointing to the possibility of growing more food (S9) are ranked neutrally by all discourses. Having more community activities in which to take part to prevent feeling isolated in mild winters (S12) is viewed neutrally by all discourses, while the possibility of selling the snowmobile and buying a four-wheeler (S31) is met with general disagreement. Furthermore, statement 24, "I am going to move away if these mild winters keep happening," while not a statistically significant consensus statement, was ranked with disagreement by all four discourses, suggesting participants feel significant attachment to the land surrounding the communities,

Table 10.4. *Four discourses based on pooled data from Rigolet and St. Lewis*

Discourse	−3 Statements (Strongly disagree)	−2 Statements	+2 Statements	+3 Statements (Strongly agree)
1 – Losing winter, losing culture Rigolet: 15 significant loadings, 8 defining sorts St. Lewis: 0 significant loadings, 0 defining sorts Var=17%	2. The last couple of winters likely won't happen again so we don't need to worry about having milder winters in the future. [Conservatism]	11. Traveling by snowmobile out on the ice is as safe now as it used to be. [Security] 5. You don't come across country foods as much anymore but that's ok with me. [Tradition]	6. Sharing country foods with other people is an important part of my life. [Benevolence] 21. I feel healthier and happier when I am able to get out on the land or get to the cabin during the winter. [Affective autonomy]	27. **The changes in weather are changing our culture and our traditional lifestyle.** [Tradition]
	28. I have really enjoyed the milder winters we've recently had because it meant a lot less shoveling. [Hedonism]	19. I'm not concerned about the changes in weather because people here have always dealt with changes. [Mastery] 24. I am going to move away if these mild winters keep happening. [Intellectual autonomy]	26. **I really enjoy riding my snowmobile and would never want to give it up.** [Hedonism] 30. **The freedom to get out on the land is the most important part about living in Labrador.** [Self-direction]	23. I miss our old Labrador winters we used to have with lots of ice and snow and cold. [Harmony]
2 – Securing community viability St. Lewis: 12 significant loadings, 5 defining sorts Rigolet: 0 significant loadings, 0 defining sorts Var=15%	11. Traveling by snowmobile out on the ice is as safe now as it used to be. [Security]	2. The last couple of winters likely won't happen again so we don't need to worry about having milder winters in the future. [Conservatism] 3. **My role in the community will change if I can no longer access the land and its resources.** [Achievement]	13. Road access to our community is an important alternative in order to get around. [Intellectual autonomy] 16. Our community should have more control over what happens on our land. [Power]	4. **The biggest problem facing the community is that too many people are leaving and there aren't enough young people coming back.** [Universalism]
	32. The mild winters have made it easier to get out on the land. [Self-direction]	24. I am going to move away if these mild winters keep happening. [Intellectual autonomy] 31. I'll just sell my snowmobile and get a four-wheeler if these mild winters keep happening. [Self-direction]	17. We need better communication technology and rescue response, especially for youth, to deal with the dangerous conditions on the land and ice. [Benevolence] 25. I feel a sense of freedom in the community because it is safe and connected. [Security]	14. **It's the government's responsibility to create new jobs here in order for our community to survive.** [Egalitarianism]

(*continued*)

Table 10.4 (continued)

Discourse	−3 Statements (Strongly disagree)	−2 Statements	+2 Statements	+3 Statements (Strongly agree)
3 – Securing traditional foods St. Lewis: 4 significant loadings, 0 defining sorts Rigolet: 5 significant loadings, 5 defining sorts Var=12%	5. You don't come across country foods as much anymore but that's ok with me. [Tradition] 32. The mild winters have made it easier to get out on the land. [Self-direction]	11. Traveling by snowmobile out on the ice is as safe now as it used to be. [Security] 15. I can't use the information and knowledge I was taught by my parents and grandparents because the weather is so different now. [Tradition] 19. I'm not concerned about the changes in weather because people here have always dealt with changes. [Mastery] 28. I have really enjoyed the milder winters we've recently had because it meant a lot less shoveling. [Hedonism]	**3. My role in the community will change if I can no longer access the land and its resources. [Achievement]** **14. It's the government's responsibility to create new jobs here in order for our community to survive. [Egalitarianism]** 16. Our community should have more control over what happens on our land. [Power] 21. I feel healthier and happier when I am able to get out on the land or get to the cabin during the winter. [Affective autonomy]	6. Sharing country foods with other people is an important part of my life. [Benevolence] **18. Our government should make sure that our seniors and other people who cannot hunt or collect firewood get enough food and wood for the year. [Egalitarianism]**
4 – Labradorians adapt naturally St. Lewis: 8 significant loadings, 2 defining sorts Rigolet: 5 significant loadings, 3 defining sorts Var=15%	5. You don't come across country foods as much anymore but that's ok with me. [Tradition] 11. Traveling by snowmobile out on the ice is as safe now as it used to be. [Security]	1. Our government's hunting and fishing regulations help our community to feed itself. [Universalism] 15. I can't use the information and knowledge I was taught by my parents and grandparents because the weather is so different now. [Tradition] 24. I am going to move away if these mild winters keep happening. [Intellectual autonomy] 32. The mild winters have made it easier to get out on the land. [Self-direction]	6. Sharing country foods with other people is an important part of my life. [Benevolence] 13. Road access to our community is an important alternative in order to get around. [Intellectual autonomy] **19. I'm not concerned about the changes in weather because people from around here have always dealt with changes. [Mastery]** 21. I feel healthier and happier when I am able to get out on the land or get to the cabin during the winter. [Affective autonomy]	**29. I identify first as Labradorian. [Hierarchy]** **30. The freedom to get out on the land is the most important part of living in Labrador. [Self-direction]**

Note: Total variance explained = 59% (significant sorts counted include defining sorts). [Conservatism] = Schwartz's cultural values – [Self-direction] = Schwartz's individual values. **Boldface** indicates statements that distinguish between factors.

and this likely ties people to the places even during difficult times such as mild winters.

While discourses 1 and 2 clearly set apart Rigolet from St. Lewis, participants from both communities defined and loaded significantly on discourses 3 and 4, suggesting there is variation of views not only between but also within the communities. The approach taken here reveals that there are substantively distinct views within the two communities on issues that are priorities to people. One example of this is the ranking of statement 13 in discourses 1 and 4. While the dominant view emerging from Rigolet on the issue of a potential road connection to that community is that of discourse 1 (ranked at 0), discourse 4 suggests the road would be an important transportation link to the regional hub of Happy Valley–Goose Bay. Because five participants from Rigolet loaded significantly on discourse 4, and three from the community defined the discourse, these findings suggest that some participants in Rigolet view a potential road to their community favorably, while those participants whose sorts shaped discourse 1 do not attribute it much importance or view it with ambivalence. This result is supported by findings from the interviews that suggest that some participants in Rigolet indeed are in favor of a road to the community, while others view it with caution (Wolf et al. 2013). The results from the Q sort, however, point out that those in Rigolet who view a road favorably identify as Labradorians who have always dealt with changes and therefore see themselves as resilient in the face of changing weather and climate conditions. They may also view Labrador more as a whole, rather than focusing on separate Nunatsiavut communities. Those who view the road neutrally in discourse 1 emphasize the sense of loss to lifestyle and culture experienced as a result of the changing conditions rather than focusing on response strategies such as a road connection.

Discussion: Schwartz's Values in the Four Discourses on Climate Change and Adaptation

Evidence from both the Q sort and the interviews points to different views both between and within the two communities on what the changing climatic conditions mean and how best to adapt to them. This section discusses the findings displayed in Table 10.4 and examines the values represented in the statements ranked at -3, -2, $+2$, and $+3$.

Losing Winter, Losing Culture

Discourse 1 clearly points to a sense of loss of culture and traditional way of life in the face of potential future climate change. This discourse did not rank highly any of the statements that suggest adaptive responses. The ranking of "The freedom to get out on the land is the most important part of living in Labrador" (S30, +2), however, suggests that the sense of freedom derived from having access to the land could be *a goal* for adaptation. This is indeed supported by interview results. Land-based activities are being protected with reactive adaptive strategies by shifting the timing of practices, shifting species hunted, and shifting location of the hunt, all to ensure traditional practices continue whenever possible. In the process of prioritizing statements, however, such specific adaptations were seen as less important by the sorts that shaped discourse 1 than statements that highlight the sense of loss of culture, traditional way of life, and the old Labrador winters. This discourse supports findings from Rigolet that suggest there are mental health implications of changing climatic conditions in the community that relate to a sense of loss (Cunsolo Willox et al. 2013).

The individual values prioritized by discourse 1 that distinguish it from the other discourses span across three value dimensions: one distinguishing statement represents conservation (tradition, S27, "The changes in weather are changing our culture and our traditional lifestyle"), one self-enhancement (hedonism, S26), and one openness to change (self-direction, S30, "The freedom to get out on the land is the most important part about living in Labrador"). In terms of tradition, this discourse emphasizes the effects of changes in weather on culture and traditional way of life (S27). The enjoyment of riding a snowmobile (S26) and the freedom to access the land (S30) together suggest that these are integral components of what participants feel is a valued way of life. The coexistence of both conservation and openness-to-change values between statements 27 and 30 may stem from the connection between the changes in weather and climate that alter the ability to access the land and the perceived consequences of this on the traditional aspects of the way of life. This finding is supported by interview results that suggest many participants in Rigolet mourn the loss of access to the land during unusually mild winters and the effects this has on the local lifestyle.

Harmony and affective autonomy emerge as priority cultural values in discourse 1. Statements 21 and 23 emphasize that the connection with the land contributes to health and well-being and identify a sense of longing for the old Labrador winters that reliably enabled access to the land. Based on the distinguishing statements, the cultural value of harmony sets this discourse apart from the others.

Securing Community Viability

Discourse 2 emphasizes the need for communication technology and rescue response to deal with dangerous conditions on the land and sea ice (S17) as ways to adapt. This discourse disagrees with simply selling the snowmobile and buying a four-wheeler in its place as a meaningful adaptation strategy. Together these items point to specific preferences for adaptation using certain technologies but not others. Implied here is that the meaning of the activity would change if the means of the activity changed. This suggests there are potentially important cultural implications of using technologies to adapt to the changing conditions.

The distinguishing statements for discourse 2 represent both self-enhancement (achievement, ranked negatively) and self-transcendence (universalism), and it is the only discourse that emphasizes universalism values. The coexistence of both self-enhancement and self-transcendence values within this discourse may be explained by examining the ranking of statement 3 at −2. The ranking suggests that participants loading on this discourse do not feel their roles in the community will be affected by changes in access to the land. According to Schwartz's definition of achievement (see Table 10.1), this implies that personal success demonstrated through competence is not (solely) tied to the land. According to the distinguishing statements for this discourse, this emphasis on universalism sets it apart (S4, +3).[2] The findings from Labrador indicate that values such as universalism that produce feelings of broader social commitments to the community may lead to prioritizing nonclimatic social issues, in our case outmigration by youth and young families, over the effects of climate change.

Discourse 2 prioritizes the cultural values of egalitarianism and intellectual autonomy. These emerge from the statements 14 ("It's the government's responsibility to create new jobs here in order for our community to survive") and 13 ("Road access to our community is an important alternative in order to get around"). Statement 14 is also a distinguishing statement of this discourse, suggesting the egalitarian view it represents is a distinct feature of this discourse compared to the others.

Securing Traditional Foods

Discourse 3 centers on various aspects of country foods. The views represented in this discourse emphasize the importance of sharing country foods and see any difficulties in accessing these foods as problematic. Perceptions of individuals' social roles in the community also reflect a strong connection to country foods. The discourse points to the role of government in securing the well-being of those who are most vulnerable to the effects of the changing conditions. It also points to using traditional knowledge in the face of changing weather as a way to adapt.

The distinguishing statement that sets apart discourse 3 represents self-enhancement values (achievement), in addition to one representing egalitarian cultural values. Contrary to discourse 2, this discourse conveys that changes in access to the land will affect the social roles of individuals in the community. The interviews support this finding as they suggest some participants feel it is very important to provide country foods and wood to other members in the community, which of course hinges on the

---

[2] Such social commitments feature prominently also in findings from Kiribati, where financial contributions to the church reduce available funds for adaptation (Kuruppu 2009). These contributions represent a commitment to collective activities that is supported by I-Kiribati cultural values (Kuruppu 2009).

ability to access the land. Changes in weather are seen to have a direct effect on the social roles of people in the community.

Discourse 3 includes the cultural values of egalitarianism and affective autonomy. Two distinguishing statements (S14 and S18, ranked at +2 and +3, respectively) both represent egalitarianism as a value that supports the survival of the community in the face of a changing demographic and outmigration. The discourse also places emphasis on protecting those most at risk from the effects of changing conditions, again representing egalitarian values. In discourse 2 from St. Lewis, however, the egalitarian values translate into concern for those most at risk, while the responsibility for protecting such groups is seen to reside with government.

Labradorians Adapt Naturally

Discourse 4 emphasizes the freedom of accessing the land and the sense of health and happiness derived from it, similar to discourse 1. Unlike discourse 1, however, it is not concerned about the changes in weather because it feels that people in Labrador have always dealt with changes. It also does not emphasize a sense of loss, as in discourse 1. This finding points to a belief in an inherent adaptability that can be relied on in the face of change rather than a need for proactive adaptation.

The distinguishing statements for discourse 4 reflect the individual value of openness to change (self-direction, S30, "The freedom to get out on the land is the most important part of living in Labrador"), viewed as the freedom to get out on the land, and this sentiment is tied to a strong sense of Labradorian identity. Together with two cultural values from distinguishing statements 19 and 29 (hierarchy and mastery, discussed later), this suggests that the focus on a strong sense of independence emerging from ability to travel on the land is what separates this discourse's values from the others. While discourse 1 shares the priority placed on self-direction, it does so in the context of tradition and hedonism, values that are not prioritized in discourse 4.

Discourse 4 includes four cultural values: mastery, hierarchy, affective autonomy, and intellectual autonomy. Based on the distinguishing statements, mastery and hierarchy are prioritized. Mastery is represented by the sentiment of resilience in the face of changes (S19) and leads this discourse to be unconcerned about changes in weather. This is the opposite view of discourses 1 and 3, both of which disagreed with this statement (−2). Discourse 4 clearly prioritizes a Labradorian identity over any others (S29), and the sense of hierarchy embedded in this notion of identity is unique to this discourse. In St. Lewis, eight participants ranked statement 29 at +3, whereas five in Rigolet ranked it at +3. The interviews suggest that most participants in Rigolet identify most as Inuit and second as Labradorians, whereas most in St. Lewis identify as Inuit-Métis and second as Labradorians. A third group in both communities identifies most as Labradorian and second as either Inuit or Inuit-Métis. Discourse 4 appears to reflect this perception of identity. Some participants who loaded significantly on this discourse associated being Labradorian with a sense of pride of having been born in Labrador or refer to Labrador as "in my blood" (Rigolet participant 29). Discourse 4 resonates with these findings because of its emphasis on inherent adaptability of Labradorians.

The discourses emerging from this analysis and the values associated with them underscore that the effects of the changing weather on local culture and way of life are not only seen as important but are meaningful in diverse ways to people in the remote northern communities studied. Rather than characterized as vulnerable communities (cf. Ford et al. 2006, 2010), the values-based approach taken here shows that a rather more differentiated set of views exists both within and between the communities on changes in weather and climate and any potential responses to them.

Conclusion

This chapter has examined how participants' values relate to the experience of climate change and adaptation to it. In particular, we have discussed four distinct discourses on the perception of and adjustments to climate variability and change and their associated values that emerge from the analysis of the Q sorts. The four discourses, "losing winter, losing culture," "securing community viability," "securing traditional foods," and "Labradorians adapt naturally," demonstrate the distinct views on what the changing conditions mean and how best to respond to them. The differences encompass what is understood as the main effect or outcome of changing conditions, how this is seen to affect the communities, who or what aspect is most at risk, and whether and how adaptive adjustments might take place. These findings support those by Roman et al. (2010) that point out that different actors have diverse goals for adaptation.

The values associated with the four discourses point to different ways to adapt. The values of the discourse "losing winter, losing culture" (tradition, harmony, self-direction, and hedonism) point to adaptation only indirectly, and only as a means to preserve traditional activities. The strong sense of loss, however, suggests there may be limits perceived to the extent to which adaptation can preserve these activities (cf. Adger et al. 2009). The discourse "securing community viability" with values of universalism, egalitarianism, and security sees adaptive responses to a socioeconomic threat to the community. In this nonclimatic context, it points to the responsibility of government to maintain community viability and prevent outmigration. The discourse identifies technological responses to improve winter travel safety to respond to changing weather conditions. Because the main threat perceived to the community is outmigration, however, adaptation to climate change is seen primarily in the context of preserving the community in the face of other pressures and stresses (Leichenko and O'Brien 2008). The values of the discourse "securing traditional foods," egalitarianism and achievement, conversely, suggest adaptation should be targeted toward those most at risk in the community and perceive a lack of access to the land as impeding an important social obligation to the community – to provide country food. Adaptation here is based on perception of differentiated risk within the community and is closest to what has been conceptualized as adaptation through vulnerability reduction (Adger et al. 2005). The discourse "Labradorians adapt naturally," underpinned by self-direction, hierarchy, and mastery, perceives adaptation as a process that has and will always take place in Labrador and exhibits strong beliefs that Labradorians and their way of life are inherently adaptable. This view implies that adaptation arises out of local circumstances and the perceived effects of changing conditions and therefore likely yields reactive adaptive measures.

The need for and type of adaptation that is perceived to be effective and useful clearly depends on how the issue of climate change is perceived, and further hinges on the values that underpin these perceptions. The different emphases placed on certain aspects of changing weather conditions, their impacts, and any responses to them, such as a sense of loss compared to Labradorian adaptability, highlight how distinct the views both within and between otherwise similar communities can be. There are at least three implications of this for adaptation planning. First, the very beginning of adaptation planning should best arise from a concrete local need identified by the community. This need should best span across at least some values in the community so as to avoid value-based conflict later. Second, in some cases, adaptation planning may not be perceived as necessary by a community, perhaps even despite scientific assessments to the contrary. Such views may arise out of what is represented in discourse 4, a strong belief in an inherent adaptability. Adaptation planning efforts, especially those largely directed from outside the community, ought to emphasize respect for local perceptions rather than attempt to challenge them. Third, adaptation planning that considers values needs to find ways to reconcile competing goals and values and build common ground to move forward. Given the diverse actors in adaptation, further research is needed on how mediation and conflict resolution may best be used in adaptation planning.

References

Addams, H. and J. Proops, eds. 2000. *Social Discourse and Environmental Policy*. Cheltenham, UK: Edward Elgar.

Adger, W. N., Arnell, N. and E. L. Tompkins. 2005. "Successful adaptation to climate change across scales." *Global Environmental Change* 15 (2): 77–86.

Adger, W. N., Barnett, J., Brown, K., Marshall, N. and K. O'Brien. 2012. "Cultural dimensions of climate change impacts and adaptation." *Nature Climate Change* 3 (2): 112–17.

Adger, W. N., Barnett, J., Chapin, F. S. and H. Ellemor. 2011. "This must be the place: Underrepresentation of identity and meaning in climate change decision-making." *Global Environmental Politics* 11 (2): 1–25.

Adger, W., Dessai, S., Goulden, M., Hulme, M., Lorenzoni, I., Nelson, D., Naess, L., Wolf, J. and A. Wreford. 2009. "Are there social limits to adaptation to climate change?" *Climatic Change* 93 (3–4): 335–54.

Barber, D. G., Lukovich, J. V., Keogak, J., Baryluk, S., Fortier, L. and G. H. R. Henry. 2008. "The changing climate of the arctic." *Arctic* 61 (1): 7–26.

Barry, J. and J. Proops. 2000. *Citizenship, Sustainability and Environmental Research – Q Methodology and Local Exchange and Trading Systems.* Cheltenham, UK: Edward Elgar.

Bell, T., Jacobs, J. T., Munier, A., Leblanc, P. and A. Trant. 2008. "Climate change and renewable resources in Labrador: Looking toward 2050." In *Proceedings and Report of a Conference Held in North West River, 11–13 March 2008*, 95 pp. North West River, Labrador, Canada: Labrador Highlands Research Group, Memorial University of Newfoundland.

Berkes, F. and D. Jolly. 2001. "Adapting to climate change: Social-ecological resilience in a Canadian western arctic community." *Conservation Ecology* 5 (2): 18–39.

Brown, S. R. 1980. *Political Subjectivity: Applications of Q Methodology in Political Science.* New Haven, CT: Yale University Press.

Brown, S. 1999. "Subjective behaviour analysis." Paper presented at "The Objective Analysis of Subjective Behavior: William Stephenson's Q Methodology," panel presented at the 25th Anniversary Annual Convention of the Association for Behavior Analysis, May 27, Chicago.

Chabot, M. 2003. "Economic changes, household strategies, and social relations in contemporary Nunavik Inuit." *Polar Record* 39 (1): 19–34.

Collings, P., Wenzel, G. and R. Condon. 1998. "Modern food sharing networks and community integration in the central Canadian Arctic." *Arctic* 51 (4): 301–14.

Cunsolo Willox, A., Harper, S. L., Edge, V. L., Landman, K., Houle, K. and J. D. Ford. 2013. "'The land enriches the soul': On climatic and environmental change, affect, and emotional health and well-being in Rigolet, Nunatsiavut, Canada." *Emotion, Space and Society* 6: 14–24.

Darier, E. and R. Schüle. 1999. "Think globally, act locally? Climate change and public participation in Manchester and Frankfurt." *Local Environment* 4 (3): 317–29.

Deci, E. L. and R. M. Ryan, eds. 2002. *Handbook of Self-Determination Research.* Rochester, NY: University of Rochester Press.

Demarée, G. and A. Ogilvie. 2008. "The Moravian missionaries at the Labrador coast and their centuries-long contribution to instrumental meteorological observations." *Climatic Change* 91 (3): 423–50.

Dryzek, J. S. 2005. *The Politics of the Earth – Environmental Discourses.* Oxford: Oxford University Press.

Ford, J. D., Pearce, T., Duerden, F., Furgal, C. and B. Smit. 2010. "Climate change policy responses for Canada's Inuit population: The importance of and opportunities for adaptation." *Global Environmental Change* 20 (1): 177–91.

Ford, J. D., Smit, B. and J. Wandel. 2006. "Vulnerability to climate change in the Arctic: A case study from Arctic Bay, Canada." *Global Environmental Change* 16 (2): 145–60.

Furgal, C., Martin, D. and P. Gosselin. 2002. "Climate change and health in Nunavik and Labrador: Lessons from Inuit knowledge." In *The Earth Is Faster Now: Indigenous Observations of Arctic Environmental Change*, edited by Krupnik, I. and D. Jolly, 266–300. Washington, DC: Arctic Research Consortium of the United States, Arctic Studies Centre, Smithsonian Institution.

Fürntratt, E. 1969. "Zur Bestimmung der Anzahl interpretierbarer gemeinsamer Faktoren in Faktorenanalysen psychologischer Daten" [Determining the number of interpretable common factors in factor analyses of psychological data]. *Diagnostica* 15: 62–75.

Gearheard, S., Pocernich, M., Stewart, R., Sanguya, J. and H. Huntington. 2010. "Linking Inuit knowledge and meteorological station observations to understand changing wind patterns at Clyde River, Nunavut." *Climatic Change* 100 (2): 267–94.

Graversen, R. G., Mauritsen, T., Tjernstrom, M., Kallen, E. and G. Svensson. 2008. "Vertical structure of recent Arctic warming." *Nature* 451 (7174): 53–56.

Harper, S. L., Edge, V. L., Schuster-Wallace, C. J., Berke, O. and S. A. McEwen. 2011. "Weather, water quality and infectious gastrointestinal illness in two Inuit communities in Nunatsiavut, Canada: Potential implications for climate change." *EcoHealth* 8 (1): 93–108.

Hobson, K. and S. Niemeyer. 2011. "Public responses to climate change: The role of deliberation in building capacity for adaptive action." *Global Environmental Change* 21 (3): 957–71.

Inglehart, R. and C. Welzel. 2005. *Modernization, Cultural Change, and Democracy: The Human Development Sequence.* Cambridge: Cambridge University Press.

Kaufman, D. S., Schneider, D. P., McKay, N. P., Ammann, C. M., Bradley, R. S., Briffa, K. R., Miller, G. H., Otto-Bliesner, B. L., Overpeck, J. T. and B. M. Vinther. 2009. "Recent warming reverses long-term arctic cooling." *Science* 325 (5945): 1236–39.

Kuruppu, N. 2009. "Adapting water resources to climate change in Kiribati: The importance of cultural values and meanings." *Environmental Science and Policy* 12 (7): 799–809.

Laidler, G. 2006. "Inuit and scientific perspectives on the relationship between sea ice and climate change: The ideal complement?" *Climatic Change* 78 (2): 407–44.

Laidler, G., Ford, J., Gough, W., Ikummaq, T., Gagnon, A., Kowal, S., Qrunnut, K. and C. Irngaut. 2009. "Travelling and hunting in a changing Arctic: Assessing Inuit vulnerability to sea ice change in Igloolik, Nunavut." *Climatic Change* 94 (3): 363–97.

Leichenko, R. and K. O'Brien. 2008. *Environmental Change and Globalization – Double Exposures.* Oxford: Oxford University Press.

Lorenzoni, I., Nicholson-Cole, S. and L. Whitmarsh. 2007. "Barriers perceived to engaging with climate change among the UK public and their policy implications." *Global Environmental Change* 17 (3): 445–59.

Min, S. K., Zhang, X. B. and F. Zwiers. 2008. "Human-induced Arctic moistening." *Science* 320 (5875): 518–20.

Mortreux, C. and J. Barnett. 2009. "Climate change, migration and adaptation in Funafuti, Tuvalu." *Global Environmental Change* 19 (1): 105–12.

O'Brien, K. L. 2009. "Do values subjectively define the limits to climate change adaptation?" In *Adapting to Climate Change: Thresholds, Values, Governance*, edited by Adger, W. N., Lorenzoni, I. and K. O'Brien, 164–80. Cambridge: Cambridge University Press.

O'Brien, K. L. and J. Wolf. 2010. "A values-based approach to vulnerability and adaptation to climate change." *Wiley Interdisciplinary Reviews: Climate Change* 1 (2): 232–42.

Phillips, L. M. and S. P. Norris. 2001. "Literacy policy and the value of literacy for individuals." In *Citizenship in Transformation in Canada*, edited by Y. V. Hebert, 209–27. Toronto: University of Toronto Press.

Plaice, E. 2008. "The lie of the land – Identity politics and the Canadian land claims process in Labrador." In *The Rights and Wrongs of Land Restitution: "Restoring What Was Ours,"* edited by Fay, D. and D. James, 67–84. New York: Routledge-Cavendish.

Rokeach, M. 1973. *The Nature of Human Values.* New York: Free Press.

Rokeach, M., ed. 1979. *Understanding Human Values: Individual and Societal.* New York: Free Press.

Roman, C. E., Lynch, A. H. and D. Dominey-Howes. 2010. "What is the goal? Framing the climate change adaptation question through a problem-oriented approach." *Weather, Climate, and Society* 3 (1): 16–30.

Schlinger, M. J. 1969. "Cues on Q-technique." *Journal of Advertising Research* 9 (3): 53–60.

Schwartz, S. H. 1994. "Are there universal aspects in the structure and contents of human values?" *Journal of Social Issues* 50 (4): 19–45.

Schwartz, S. H. 1999. "A theory of cultural values and some implications for work." *Applied Psychology – An International Review* 48 (1): 23–47.

Schwartz, S. H. 2006. "Les valeurs de base de la personne: théorie, mesures et applications" [Basic human values: Theory, measurement, and applications]. *Revue française de sociologie* 47 (4): 929–68.

Schwartz, S. H. and W. Bilsky. 1987. "Toward a psychological structure of human values." *Journal of Personality and Social Psychology* 53 (3): 550–62.

Sexton, D., Snyder, P., Wadsworth, D., Jardine, A. and J. Ernest. 1998. "Applying Q methodology to investigations of subjective judgments of early intervention effectiveness." *Topics in Early Childhood Special Education* 18 (2): 95–110.

Stephenson, W. 1935. "Technique of factor analysis." *Nature* 136 (3434): 297.

Stephenson, W. 1978. "Concourse theory of communication." *Communication* 3: 21–40.

Wenzel, G. 1991. *Animal Rights, Human Rights*. Toronto: University of Toronto.

Wenzel, G. 2009. "Canadian Inuit subsistence and ecological instability – If climate changes, must the Inuit?" *Polar Research* 28 (1): 89–99.

Williams, R. M., Jr. 1979. "Change and stability in values and value systems: A sociological perspective." In *Understanding Human Values: Individual and Societal*, edited by M. Rokeach, 15–46. New York: Free Press.

Wolf, J., Allice, I. and T. Bell. 2013. "Values, climate change, and implications for adaptation: Evidence from two communities in Labrador, Canada." *Global Environmental Change* 23 (2): 548–62.

Wolf, J., Brown, K. and D. Conway. 2009. "Ecological citizenship and climate change: Perceptions and practice." *Environmental Politics* 18 (4): 503–21.

11

Exploring Vulnerability and Adaptation Narratives among Fishers, Farmers, and Municipal Planners in Northern Norway

GRETE K. HOVELSRUD, JENNIFER J. WEST, AND HALVOR DANNEVIG

The apparent disconnect that exists between the abundance of scientific knowledge about climate change and a lack of concerted political commitment and practical effort to deal with the challenges is receiving increasing attention among scholars (Hulme 2008, 2009; Jasanoff 2010; Norgaard 2011; O'Brien and Hochachka 2010; O'Riordan and Jordan 1999; Szersznynski and Urry 2010; West and Hovelsrud 2010). It is clear that the evidence and information about anthropogenic climate change do not automatically translate into adaptive behavior, despite the substantial long-term threats that climate change poses to society. The role of narratives in reflecting or shaping adaptation and adaptive capacity to climate change within particular sectors and communities has so far received little attention in the literature, although the importance of considering worldviews and values as integral aspects of adaptation is increasingly recognized (Jasanoff 2010; O'Brien and Hochachka 2010; O'Brien and Wolf 2010). While community-based adaptation research has tended to view communities as homogeneous entities (Sabates-Wheeler et al. 2008), it is increasingly understood that communities are constituted by individuals with different knowledge, beliefs, values, social roles, goals, life opportunities, and jobs, which shape how they interpret, perceive, and act upon knowledge about climate change as well as the potential risks and opportunities associated with it (e.g., Crate 2008; Heyd 2008).

This chapter contributes to this discussion by (1) examining the different perceptions of climate vulnerability among three occupational groups (fishers, farmers, municipal officials) in three case municipalities and by (2) discussing the implications of such vulnerability narratives for future climate change adaptation. To shed light on why individuals within the same communities express different vulnerability and adaptation narratives, we draw on insights from cultural theory, which provide a conceptual bridge for understanding the divide between scientific findings and local understanding and perceptions of the same phenomenon. We conclude with some reflections on how such focus may contribute to bridging the gap between scientific knowledge on climate change, which calls for action, and the lack of urgency expressed through narratives at the local level.

Climate variability and change are projected to affect communities in northern Norway via changes in prevailing weather (i.e., frequency, duration and timing of extreme events), sea-level rise, and long-term changes in mean ocean and air temperatures and precipitation. The temperature in northern Norway is expected to increase by up to 4°C by 2100, while ocean temperatures in the Barents Sea region are expected to rise by 1°–2°C over the coming century (Førland et al. 2009; Hanssen-Bauer et al. 2009). Warming ocean temperatures are in turn expected to lead to changes in the magnitude, composition, and spatial and temporal distributions of important commercial fish stocks, such as cod (*Gadhus morhea*) and herring (*Clupea harengus*) (Drinkwater 2005; Sundby and Nakken 2008). Climate change will affect physical infrastructure and the timing, profitability, and viability of various primary production and harvesting activities locally, including in agriculture, where an increase in the growing season is expected (Hanssen-Bauer et al. 2009; Kvalvik et al. 2011).

Although climate change is projected to substantially influence fisheries, agriculture, reindeer husbandry, and municipal planning activities in northern Norway (Hovelsrud and Smit 2010; Øseth 2010), research on coastal communities in the region indicates that local perceptions of vulnerability to climate change differ from those of the researchers and that communities may not perceive climate change to be an immediate concern when compared to more locally pressing issues such as out-migration, jobs, and the social and economic viability of municipalities (Hovelsrud et al. 2010; West and Hovelsrud 2010).

Cultural theory of risks (e.g., Douglas 1982; Thompson et al. 1990) provides one useful alternative to both traditional agent-centered rationalism and structuralist-institutionalist theories for explaining human behavior, perceptions, and preferences. Whereas both these strands of social theory have failed to explain why people and institutions do not react to the threat of climate change (Hulme 2008, 2009; Jasanoff 2010; O'Brien and Hochachka 2010; O'Riordan and Jordan 1999), "cultural theory sheds light on why we as individuals find it hard to respond to 'mega' risks like Climate Change" (O'Riordan and Jordan 1999, 88). Cultural theory categorizes human interaction loosely as four *ways of life*: fatalist, hierarchist, individualist, and egalitarian. These are determined by the degree to which social regulation and social contact, or group membership, influence the individual. According to O'Riordan and Jordan (1999, 86), "people's policy choices are supportive of and rationalized on the basis of these different 'ways' or value orientations." Fatalists are described by a high degree of social regulations, but with weak or no group membership, whereas egalitarians have strong group membership and weak social regulations. Individualists are described by strong group membership but weak social regulation. Hierarchists are described by strong social regulation and a high degree of group membership

(Thompson 1990). O'Riordan and Jordan (1999) and Kahan et al. (2012) find that different ways of life are associated with different perceptions of climate change risk. *Hierarchists* tend to trust scientific authority, such as climate scientists, and will accept state intervention as long as they are appropriately legitimized. *Individualists* tend to be concerned about problems that impinge on their personal freedom, as could be expected to be the case with climate change policy, in particular with respect to mitigation measures (Kahan et al. 2012). In reality, no individuals adhere to only one way of life but have preferences and values that might fit within several ways of life. The way of life concept is closely linked to the notion of *worldviews,* defined by O'Brien and Wolf (2010, 234) as "the basic assumptions and beliefs that influence much of an individual or group's perceptions of the world, their behavior and their decision-making criteria." These approaches to understanding the context of climate change adaptation serve as the inspiration for analyzing our extensive empirical material. In these pages, we present the research context, empirical material, and methods of our study before moving on to our discussion and findings.

Research Context and Methods

Our analysis draws on research carried out with farmers, fishers, and municipal officials in Nordland and Finnmark Counties in northern Norway (see Figure 11.1)

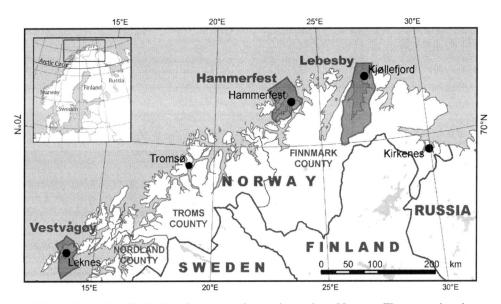

Figure 11.1. Map displaying the case study area in northern Norway. Figure reprinted with kind permission of Springer Science+Business Media from Hovelsrud, G. K., Dannevig, H., West, J. and H. Amundsen, 2010, "Adaptation in fisheries and municipalities: Three communities in northern Norway," pp. 23–62 in *Community Adaptation and Vulnerability in Arctic Regions*, edited by Hovelsrud, G. K. and B. Smit.

in the period 2007–10. This region, while boasting abundant natural resources and related activities, including fisheries, oil and gas production, and land-based extractive industries, is peripheral in Norway, a country which in a global context may also be considered peripheral (Watten 2004). The cases are all coastal municipalities: Vestvågøy in the Lofoten region in Nordland County, Hammerfest in western Finnmark, and Lebesby in eastern Finnmark. Residents across the communities face similar challenges when dealing with natural variability in weather and natural resources.

Fisheries (both fishing and fish processing activities), municipal planning, and agriculture are three important employers in the three municipalities. We have chosen to study these occupational groups as they represent sectors that are both vital to their local communities and expected to be highly exposed to climate change (Øseth 2010). Fishing and farming activities also constitute an important historical basis for livelihoods, culture, and identity across the case sites. This is clearly reflected in the vulnerability and adaptation narratives expressed within these occupational groups, which constitute an important focus of this chapter. In fact, our data analysis across the municipalities reveals important commonalities (and differences) in the vulnerability and adaptation narratives of these three occupational groups. By teasing out and describing these similarities and differences, the chapter seeks to demonstrate that there are different voices, values, and interests within a municipality that have bearings for climate adaptation even in cases where individuals share a common cultural repertoire. At the same time, there may be similarities in the dominant values, interests, and priorities espoused by particular occupational groups that cut across municipalities and create both opportunities and challenges for implementing adaptive actions beyond the communities themselves. The three sectors are briefly introduced in the following.

Fisheries

Employment in Norwegian fisheries has seen a steady decline over the past decades (West and Hovelsrud 2008), but together with land-based processing activities, fisheries remain important for the local economies and employment, for cultural identity, and for connection to place in the case communities (Amundsen 2013). Fisheries in Norway are highly regulated and managed and are differentiated by vessel length, gear type, species quota, and fishing ground. Recent and ongoing structural reforms have contributed to a reduction in the number of fishers and fish vessels, particularly in the coastal fleet. In northern Norway, the winter fishery for spawning cod is particularly important, and it is one of the world's largest fisheries in terms of landed quantum (Berge 1996). A major part of this fishery has traditionally taken place around the Lofoten islands. In recent years, warmer ocean temperatures have caused this fishery to shift northward and further out to sea (Ministry of Environment 2010).

Agriculture

Agriculture in northern Norway is an undertaking characterized by major seasonal and interseasonal climatic variations. The most important agricultural activities are dairy and sheep farming, which both require large areas for pastures and grass fields for feed production. The market for agricultural products in Norway is fully regulated, with fixed prices for products such as milk and meat, and is protected from competition from imported agricultural products through a high customs tariff. Farmers in northern Norway also depend on a variety of government subsidies. Traditionally, farming complemented fisheries to provide a varied livelihood for the rural population in northern Norway (*fiskarbonden*) (Brox 1966). Farming remains important in several communities, even though it occupies a minor share of the working population across the sites. A prolonged growing season is observed in the case study areas (Kvalvik et al. 2011).

Municipal Planning

Municipalities (430 in total at the time of writing) are the local level of government in Norway and serve as a link to national management bodies and governance institutions and are the bounded local community in terms of political administration, physical location, and region of data collection. The municipal authorities provide a number of important services for their inhabitants. These include elementary schools, local road maintenance, waste collection and management, emergency response, provision of social services, fire protection, water supply, and local and spatial planning. The services are highly regulated by national laws and regulations, and the majority of the municipal budget is spent on fulfilling mandatory obligations. Norwegian municipalities have the main responsibility for local land-use planning and for preventing damage from extreme weather events such as floods and avalanches. Other societal responsibilities are shared with other government institutions. Municipalities also have an important role to play in coordinating responses to future extreme events and damage through "rebuilding" after an event (for an overview, see Næss et al. 2005). The tools available to municipalities for managing climatic risks include the prohibition of new buildings in areas prone to flooding or slides, and the construction of landslide and avalanche protection for existing infrastructure.

Methods and Data

Our empirical data result from a number of interlinked projects[1] on adaptation to multiple stressors. Our research approach is driven by locally relevant questions

---

[1] CAVIAR (Community Adaptation and Vulnerability in the Arctic Regions), an IPY 2007–8 consortium, RCN (2006–10); NORADAPT (Community Adaptation and Vulnerability in Norway) RCN – NORKLIMA (2007–11);

Table 11.1. *Summary of research methods*

	Vestvågøy	Lebesby	Hammerfest
Semistructured interviews	21	25	20
Group interviews with (a) coastal fishers, (b) municipal officers, and (c) farmers	1 a), 1 b), 1 c)	1 a), 1 b), 1 c)	1 a), 1 b)
Feedback sessions in townhall meetings	1	2	1
Other	Attended and observed community events and meetings, collected gray literature and documents	Key-informant discussions, participant observation on board fishing vessels and at community events, collection of gray literature and documents	Attended and observed community events and meetings, collected gray literature and documents

framed under cooperation between local partners and researchers. Understanding the range of adaptation priorities and needs within municipalities requires consultation with local actors and inclusion of their knowledge, and our partners in the three municipalities have been instrumental in shaping the focus of the research and the research design, including the development of downscaled climate scenarios for the case study sites and the identification of the most salient aspects of climate and weather changes and societal challenges.

Our research began with an initial visit to the municipalities where we presented the project and invited local partners to participate. The development of the research was an iterative process in which preliminary results were presented to the municipalities for feedback and adjustments. Determining the research focus together with local stakeholders means that the particular research focus varies between the cases (Hovelsrud et al. 2010). The primary research methods consisted of semistructured and open-ended interviews, following an interview guide, and key-informant and group discussions carried out in the three municipalities in the period 2007–10 (see Table 11.1). Reviews of published scientific literature, gray literature, statistics, local media, and other relevant information supplemented with community meetings,

DAMOCLES (Developing Arctic Modelling and Observing Capabilities for Long-term Environmental Studies) EU Sixth Framework Programme (2007–11); PLAN – Klimatilpasning i Norge, RNC (2007–11).

presentations, and discussions in the three municipalities over the course of the research period.

In addition to primary data from interviews and group discussions, our analysis in this chapter draws on and further develops published and unpublished findings from our own research on climate change vulnerability and adaptation in relation to fisheries, agriculture, and municipal planning in Vestvågøy, Hammerfest, and Lebesby (Hovelsrud et al. 2010; Keskitalo et al. 2011; West and Hovelsrud 2010).

Vulnerability and Adaptation Narratives

Coastal Fishers

Across our case study sites, coastal fishers ($n = 44$) who were interviewed in both formal and informal interview contexts and took part in group discussions overwhelmingly emphasized nonclimatic issues such as regulations and socioeconomic aspects when describing their overall livelihood vulnerability (Hovelsrud et al. 2010; West and Hovelsrud 2010), supporting the view that climate change impacts must be considered in connection with other societal issues (Smit and Wandel 2006). Despite the lack of emphasis on climate change as being an important source of vulnerability for livelihoods, fisheries respondents are well aware of the climatic factors and changes to which their livelihoods are exposed, and they identify a number of ways in which these impinge on their livelihoods. For example, local informants described the northward shift in winter fisheries for spawning cod in Lofoten as a response to a shift in the distribution of the fish, which they related to a warming sea. This has had serious implications for the land-based fish industry that depend on deliveries from the fishers and received less catch on its quay. The need to make it affordable for fishers to travel the extra distance to deliver their catch represented an intolerable rise of cost for some fish buyers that ultimately contributed to their closure (see also Hovelsrud et al. 2010). Fishers in Lebesby reported a shift in the behavior and composition of fish species due to warmer ocean temperatures in recent years, including an increased sighting of greater quantities of southerly fish species, such as blue whiting (*Micromesistius poutassou*), monkfish (*Lophius piscatorius*), and mackerel (*Scomber scombrus*). Adapting to these changes requires that fishers obtain relevant quotas and employ fishing gear that might not be permitted under current fisheries regulations (West and Hovelsrud 2010).

Despite the observed changes in distribution of cod and other species of fish, climate variability and change are not currently perceived to be a challenge to fishing livelihoods because such changes fall within the existing and recognized high natural variability in local weather patterns and in the resource base (i.e., fish stocks). Nor is climate change perceived to be a future challenge despite research showing the projected implications of warming ocean temperatures for fish distribution, particularly for the cod stock (Sundby and Nakken 2008). When confronted with information about the possible consequences of climate change for their livelihoods, such as a warming sea or changes in the timing and spatial distribution of fishing seasons, fishers cited weather events that happened up to a hundred years ago and interpreted climate change projections as a part of the high natural and seasonal variability with which they are familiar (Hovelsrud et al. 2010).

Although recognized by the fishers, climate change and the consequences for local fisheries were interpreted in the context of cyclical changes in fishing conditions that form part of fishers' collective memory and experience. The causes of past and current fluctuations, particularly in local fish stocks, were framed in terms of wider marine ecosystem dynamics and interactions that include the effects of other fish and marine mammal stocks on feeding conditions and overfishing (see also West and Hovelsrud 2010). Thus, for the same reasons that scientists have difficulty in identifying direct causal links between climate change and changes in particular fish stocks, local fishers find it difficult to understand that changes in fish stocks can be attributed to climate change.

Nonclimatic factors, however, feature strongly in the vulnerability narratives of fishers. They are generally able to catch their allotted quotas, so the main socioeconomic determinants of their livelihoods include the size and cost of landings, fish prices, and the size of their quotas. Norwegian fisheries are exposed to an ongoing restructuring and rationalization process, driven by political and market factors, which has resulted in fewer and larger vessels and an overall reduction in employment in the sector over time (Fiskerirådgivning AS 2006; Jentoft and Mikalsen 2004). Recruitment of younger fishers to the industry is another commonly identified challenge; many fishers are getting older and nearing retirement, and lament that there are few young men and women who are interested in the traditional coastal fishing lifestyle or who possess the financial means of acquiring fishing vessels and quotas to continue the tradition.

Although fishers perceive themselves to be vulnerable to the forces of nature, their narratives indicate that this vulnerability is usually overcome through a combination of individual creativity and ingenuity, heroic efforts, physical and mental toughness, and time-tested knowledge, skill, and experience (e.g., about when and where it is safe to fish). When it is not overcome (e.g., in the case where a fisherman dies at sea during a storm), it is explained as having to do with the economic pressures in the industry, which may induce fishers to go fishing alone in dangerous weather, or with the poor judgment of the individual fisher. Fishers see themselves as stewards of a coastal culture, tradition, and identity that are strongly rooted in the place where they live and deliver their catch. They, and other nonfishing community members,[2]

---

[2] Nonfishers stress the importance of a vital fishery industry for the local economy, stating for example that "it is what we get over the dock that we all are living from."

consider the fishing occupation as an important cornerstone of employment and income in their communities. In sum, fishers' narratives reflect occupational values of freedom and independence, mastery of a traditional and time-tested craft, maintenance of local traditions, knowledge, and identity, providing fresh, local food, creating local employment and income, and being resilient and adaptable to unpredictable conditions.

Farmers

Farmers in our study ($n = 18$) identified a number of potential vulnerabilities to a combination of climate- and weather-related factors, as well as socioeconomic factors. Being located on the coast, the case study sites receive considerable precipitation throughout the year, although it is often unevenly distributed in space and time. In Vestvågøy farmers ($n = 7$) complained that they had suffered several successive summers with drought, even in fields with so-called marsh soil, and that in 2008 the summer drought had for the first time created major challenges for farmers who wished to irrigate their fields. Farmers have in the past few years recently been faced with less water in the natural streams and rivers that are used for irrigation, and this has led to irrigation restrictions by the local water authorities. The drought in 2008 led to substantial local reductions in yield, with some farmers losing up to 30 percent of their harvest. Even though the farmers are compensated by the government, harvest failure leads to reduced income for the farmer, and losses below 30 percent of crop are not compensated.

In Lebesby, farmers ($n = 8$) note that a longer growing season due to warmer air temperatures has extended the tree line farther north and has led to invasion of southerly species and insects such as the autumnal moth. Variability of air temperatures around 0°C in combination with precipitation in the spring may result in a layer of ice developing on agricultural fields, killing emerging crops through a lack of oxygen. In Vestvågøy and Lebesby, farmers are similarly challenged when there is an abundance of rain in the harvest season, leading to problems with being able to harvest at the right time or the harvest resulting in soil compaction, which can negatively affect crop emergence and land productivity in subsequent seasons. Farmers may fit their tractors and equipment with wider tires to reduce the pressure on the ground; however, few have invested in this kind of equipment, because it is expensive and still not used on a regular basis. There were also stories of damaged and widespread loss of crops; in Vestvågøy some farmers lost nearly all of their potato yields one year due to rain, which would make them eligible for some compensation. Farmers also note a higher rate of regrowth on their pasture land in recent years, although they do not agree among themselves on the explanations for this phenomenon.

Although climatic fluctuations could seriously affect the profitability of a farm in a single year, and thus create vulnerabilities for farming livelihoods, such fluctuations were seen as falling within the expected range of variability. Similar to fishers, farmers across the cases emphasized nonclimatic factors, primarily agricultural policies, as being of prime importance for agricultural livelihoods (see also Kvalvik et al. 2011). One farmer in Vestvågøy, for instance, expressed that he was more concerned over the prospects of Norway becoming a member of the European Union because this would end the current national protection of Norwegian agriculture from competition from European agricultural products. Other farmers expressed worries over the low level of recruitment to the sector, explaining that farming is hard work and that limited profitability deterred people from taking over their parents' farms. When asked what it would take for them to adapt to the possible consequences of climate change, the farmers' replies pertained mainly to how they might benefit from higher average temperatures and longer growing seasons.

Farmers across the case sites see the potential for growing more or different crops if the climate continues to warm. They express optimism that a longer and warmer growing season could allow them to harvest more than one crop annually and to grow more nitrogen-fixing crops, which would improve the nutrient levels and structure of the soil. Their ability to profit from temperature changes will, however, depend on a number of factors, including (for farmers in Lebesby) the ability to make use of new land, which is currently constrained by limited financing and land-use conflicts; the ability to renovate or build larger storage facilities to increase the hay-storing capabilities if their dairy herds are increased; availability of new crop species and varieties that can tolerate the conditions to which they are currently vulnerable (e.g., icing); and continued national political and financial willingness and support to provide transfers and subsidies to the agricultural sector in northern regions. The continuity of a viable agriculture sector in Lebesby, for example, has been successful in part due to funds that were provided by the municipality to ensure a generational shift in agriculture within the municipality, whereby younger families from within or beyond the region took over older farms whose owners were nearing retirement.

Municipal Planners

For the departments and officials of the municipalities that deal with land-use planning, environmental protection, and primary industrial sectors, coping with variability in weather, climate, and natural resource availability is inherently a part of the institutional mandate and routines. Still, planning or adapting to climate change has until recently not been on the municipal planning agenda at all, and where it has started to be included, it is not always fully implemented in the municipal plans, planning routines, and operations (Amundsen et al. 2010; Dannevig et al. 2012).

Municipal planners in our study ($n = 8$) were concerned with the potential consequences of climate change in their communities, and with few exceptions they express the need to assess and respond to reduce the physical and social vulnerability of communities to such changes. During discussions and meetings with municipal officials, they raised a number of concerns about potential impacts of climate change on their communities, including an increase in the number and magnitude of avalanches and rockslides (Hammerfest, Lebesby), challenges with winter floods (Hammerfest, Lebesby), sea-level rise in combination with storm surges (all municipalities), and the effects of storms and extreme weather on physical infrastructure (all municipalities). Driven by their concern about such impacts, municipal officials in Hammerfest and Vestvågøy have initiated joint research and assessment projects. The municipality of Hammerfest has implemented changes in regulations for seafront property development to adapt to projected future sea-level rise. This was a result of increased information about sea-level rise in northern Norway and because of engaged municipal officials involved in research projects. Downscaled climate projections for precipitation and wind are being used when planning new housing areas and more recently with respect to the relocation of the airport, which currently is situated in a very turbulent area below a steep mountainside.

In Lebesby a fatal slide event in 1956 led the municipality to construct its first avalanche protection. The avalanche barriers have been upgraded and extended a number of times since then, and the risk of future events has been mapped by the Norwegian Geotechnical Institute (NGI). However, due to the high costs of constructing protection in all areas, the municipality has had to prioritize protection of the most exposed and high-risk areas. A number of areas thus remain at risk, including the local cultural tourism facility Foldal, where avalanche threats caused the evacuation of tourists as recently as 2009.

Municipal officials dealing with land-use planning and zoning of buildings expressed skepticism toward the inclusion of climate change adaptation in their work without it also being formalized in the job description. This reflects the main and common concern expressed by officials across municipalities about the lack of capacity to address climate change issues as an additional or separate concern within the limitations of staffing and funding flows and departmental priorities. The lack of capacity, both in terms of financial resources and people, is one explanation for not adding climate change adaptation work to the other tasks of protecting citizens and infrastructure. The link between impacts of climate change and the need for adaptation and other responsibilities to protect citizens and infrastructure is not readily made.

Discussion

The chapter thus far has described the collective vulnerability narratives of farmers, fishers, and municipal officials. In this section, we analyze the similarities and differences in these narratives by comparing them, as the descriptions of climate vulnerability and narratives regarding adaptation to climate variability and change differ considerably between the three occupational groups and across the municipalities.

This leads us to consider how the dominant values, perceptions, and priorities associated with these occupations might have a bearing within and beyond the community level.

While the climate vulnerabilities differ considerably between the groups, we find that the narratives of the nonclimatic vulnerabilities, such as demographics, national policies, and resource management and financial and human capacities, reflect a more concurring perception of the societal challenges faced by these groups. Our results also point to a shared value narrative expressed in the optimism and determination to meet future challenges – captured in the local northern Norwegian expression "*vi står han av*" (we always handle hardships). Fishers in northern Norway operate under highly variable and uncertain conditions, in weather, management, and markets. Such uncertainty requires an individualistic attitude and heavy reliance on the fishers' own knowledge and experiences. The fishers clearly state that what they treasure most about their job is freedom and independence. When discussing weather-related challenges, they express a high confidence in their own resilience and adaptability toward challenging weather conditions. Values that characterize fishers include maintaining traditional livelihoods; providing fresh, clean food; and keeping the houses lit along the coast. In addition, fishers' main interests are to ensure the profitability of their fishing activities while also ensuring safety and well-being. They act rationally within the frames of such an objective, in addition to being active participants within a broader national and international community.

Fishers do not readily accept scientific knowledge as authoritative in the same way as municipal officials. To the contrary, they find that their own experiences and knowledge are at odds with the natural science that informs Norwegian marine resource management. This skepticism to fisheries science is highly likely to be extended to climate science and their interpretation of scientific information about climate change. Within the cultural theory framework, northern Norwegian fishers fit quite well with within the *individualistic* way of life within the cultural theory of risks framework (Douglas and Wildavsky 1982; Thompson and Wildavsky 1990). To acknowledge climate change and adaptation appears to be perceived as a threat to their freedom, which also could explain why they deem climate change as having little relevance for fishery activities.

The farmers participating in our study clearly expressed the perception that their livelihoods are as much determined by national agricultural policies and international trade agreements as by climatic conditions or their own efforts. Central values expressed by the farmers pertain to their role as food producers, maintainers of traditions, and custodians of the land. Current national agricultural policies favor entrepreneurship and innovation and in many ways challenge the traditional farming occupation. With respect to cultural theory, farmers may be best represented as a mixture between the *individualist* and *fatalist* way of life – individualistic because the current political and economic situation favor an entrepreneurial attitude and fatalistic because the very same conditions determine the farmers' livelihoods to a greater degree than other factors. The farmers that were involved in our research projects have expressed an opportunistic attitude toward future climate change because of the projected increase in growing season and potential for expansion of existing, or introduction of new, crops. Our findings show that farmers are in general much less skeptical of climate science than fishers. This may be because they already rely on scientific information about, for example, crop development and improved agricultural practices and because agricultural science as currently institutionalized provides a basis of innovation in farming systems. In addition to actively utilizing scientific information, they rely heavily on their own experience-based knowledge for their livelihood.

The main role of municipalities is to ensure safe and well-functioning communities and maintain societal functions and services. Municipal planners' use of knowledge is institutionalized and specified for their roles and tasks. They rely on formal sources of knowledge in addition to local knowledge and experience of, for example, climate hazards. Municipal planners need a degree of predictability as the municipal investments are long term (e.g., coastal infrastructure and docks are built to last up to fifty years). They carry out these duties within a hierarchical organization, the municipal administration, which determines the sources of knowledge that are considered legitimate and how tasks and actions are to be carried out. This can be described as being heavily regulated with a high degree of formal group membership. Within the cultural theory framework, the municipal official can be associated with the *hierarchist* category, by virtue of accepting authoritative knowledge if the source is deemed legitimate, such as climate science.

Our findings show that the local participants in our studies are fully aware of existing climate risks. Indeed, adapting to weather variability has a strong place in local and regional identities and discourses. However, adaptive responses to change occur in a social context of competing values, identities, occupational mandates, and priorities where climate change, relative to other pressing social and economic challenges, is not perceived to pose a great threat. These findings help to explain why scientific information does not automatically translate into adaptation in local contexts and challenge the assertion that there is a simple disconnect between scientific and lay understandings of climate change risks. This suggests the need for not only improved communication of climate science, such as downscaled scenarios of locally relevant climate elements, but also communication that recognizes that both competing and collective values, identities, experiences, interests, and occupational roles and mandates shape adaptive responses in particular communities. Our observations of how occupations and *ways of life* influence how knowledge of climate change is taken into account corroborate with recent studies on the relationship between how people assess the risk of climate change and how this relates to way of life (Kahan et al. 2012).

In our findings we see that *individualists*, such as the Norwegian fishers, do not perceive high risks associated with climate change, while *hierarchists*, such as the Norwegian municipal planners, do.

One reason postulated for differences in scientific and lay perceptions and framings of climate risks is that communication of scientific knowledge is at odds with the situated, subjective, and normative conceptions of "environmental risk" held by individuals (Jasanoff 2010). Climate change science has constructed a narrative of climate as something manageable that humanity through downscaled projections of future climate can mitigate and/or adapt to in order to achieve a desired outcome. This narrative is based on highly abstract climate science and goes along with policy prescriptions based on macroeconomic calculations (Hulme 2008). We argue, based on our empirical data and results, that this does not readily resonate with local peoples' experiences.

Northern communities, and particularly those where natural resource-based activities are important, have historically adapted to highly fluctuating biophysical conditions (Nuttall 2005; Tyler et al. 2007). Climatic factors have always shaped the basis and conditions for settlement and livelihoods in the municipalities in this study, and communities' collective histories and identities are built on successful responses to climatic variability and change (West and Hovelsrud 2010; Amundsen 2012). In this sense, climate variability and change can be seen as a source of both societal vulnerability and resilience. Another rationale is that climate is also a cultural object that is tied to worldviews, meaning, belief systems, values and ways of life, which shape interpretations of the need to adapt, to what, by whom and in what ways (Hulme 2008; Jasanoff 2010; O'Brien and Wolf 2010; Amundsen 2013). Our study corroborates this rationale by illustrating the importance of occupational roles and mandates and connected experience, perceptions, values, and knowledge traditions for understanding why and how local vulnerability narratives may depart from scientific designations of particular regions and social groups as "highly vulnerable" to climate change.

The *"vi står han av"* discourse of adaptability and resilience (Amundsen 2012) does not necessarily mean a disconnect between scientific and local understanding of climate change but can be interpreted as an adaptation strategy that is motivated and shaped by the collective experience of living amid high natural climate variability. It expresses optimism and determination in the face of the unpredictable social, economic, and climatic conditions that are central features of life in northern, coastal communities. Despite differences in the climate vulnerability narratives of the three different occupational groups and regardless of respondents' differing views on climate change, all of the participants in our study described undertaking past and continuing adjustments (either as individual entrepreneurs or in a professional capacity) to deal with climate variability and change and other social and economic challenges.

The findings suggest that ensuring sustainable, adaptable communities in northern, peripheral contexts requires sustained efforts on a number of fronts. In the view of project participants, it requires jobs that can attract and retain skilled residents, a demographically varied and educated population base that includes young people who are willing to settle in, invest, or return to their communities as well as investments to secure the social, economic, and cultural attractiveness of the community to residents and outsiders (Amundsen 2013; Hovelsrud et al. 2010; West and Hovelsrud 2010). Residents across our case study sites recognize these collective goals and interests as being intertwined with their own occupational roles, priorities, and prospects. Observations and impressions gained from the research so far suggest that the discourse "*vi står han av*" is a positive force for motivating adaptive behavior and resilience toward societal challenges in the communities. This is in agreement with other findings from northern Norway (Amundsen 2012).

Conclusion

Our findings show that climate change adaptation will not take place on the basis of climate change information and projections about future change alone and that scientific information is interpreted within a context of competing values, perceptions, and priorities of different occupational groups within and across communities, which in turn is likely to shape adaptive responses locally. In particular, scientific claims may be contested by individuals and groups whose knowledge, experience, identity, and occupational roles and mandates are at odds with scientific framings of climate change as a major challenge (Kahan et al. 2012; O'Riordan and Jordan 1999). Insights from cultural theory help to explain the differing climate vulnerability narratives of fishers, farmers, and municipal planners in the municipalities studied and to nuance the perceived dichotomy between "objective scientists" and "subjective communities" as distinct, undifferentiated groups. This in turn opens up space for challenging and qualifying the notion that there is a simple disconnect between scientific and lay perceptions/framings of climate change risks.

Our research shows that while adaptation to climate change is undoubtedly needed in municipalities, it is not likely to be undertaken by communities as an isolated project but rather in connection with efforts to deal with pressing societal challenges and goals, such as the need to create local jobs and the need to protect communities from extreme events. Adaptation efforts undertaken in the communities we studied are likely to be shaped by individual and collective knowledge and experiences of dealing with past and ongoing changes, the agency and mandates of different occupational groups, and differences in perceptions of where the responsibility for adaptation lies (whether with the state, through effective policies, with the municipality, through effective planning, or with individuals, through persistence, creativity, innovation, and time-tested experience).

In addition to what is perceived to be relevant local and scientific information, our study suggests that local adaptation is likely to be influenced by persistent cultural discourses, such as that of "*vi står han av*," which are grounded in local realities, perceptions, and experiences and whose specific outcomes in terms of adaptation will require more detailed investigation. Whether local discourses and narratives such as those uncovered in our research discussed here have made communities and individual respondents more or less resilient to climate change is a central question that deserves further investigation, as it will indicate whether such discourses constitute a potential or a limitation to adaptation (Adger et al. 2009). On one hand, the discourse may constrain adaptation to climate change by overestimating community adaptive capacity and underestimating the need to adapt to changes that challenge the collective coping ability, including the cumulative knowledge and resources, of communities. On the other hand, it may motivate adaptation by engaging and mobilizing individuals, groups, and planners within the municipalities to improve the social and economic viability of their communities through new local initiatives, alliances, and projects, thereby enhancing their resilience and reducing their vulnerability to multiple challenges.

We conclude that the disconnect between scientific and lay perceptions of climate change observed in our study and that is exemplified in the narrative of high adaptability and resilience "*vi står han av*" creates challenges both for effective local adaptation and for the development of relevant and user-friendly climate science that can inform such adaptation. Climate science has so far failed to incorporate a proper understanding of the historical social and economic contexts and the differing values, priorities, and experiences that influence how occupational groups within and across communities interpret and use climate information. Understanding and integrating such differentiated local concerns and realities into the development of appropriate climate information constitutes a necessary first step for bridging the disconnect between climate scientists and communities. Applying cultural theory to our analysis of the vulnerability narratives of three occupational groups in three northern Norwegian municipalities helps to illuminate how and why local priorities may differ from those of scientists, and in turn may contribute to bridging this gap.

Acknowledgments

We gratefully acknowledge and thank our local partners in Hammerfest, Lebesby, and Vestvågøy, especially Tom Eirik Ness, Kåre Tormod Nilsen, Kjersti Isdahl, Nils Kaltenborn, and Are Johansen, for invaluable discussions and interaction throughout the project. Without their engagement and interest, the project would not have come to fruition. We also thank the administration and politicians in the three municipalities for their warm welcome. The research was funded by the PLAN project under the

Research Council of Norway's NORKLIMA program and by the CAVIAR project under the Research Council of Norway's IPY 2007–8 program.

References

Adger W. N., Dessai, S., Goulden, M., Hulme, M., Lorenzoni, I., Nelson, D. R., Naess, L. O., Wolf, J. and A. Wreford. 2009. "Are there social limits to adaptation to climate change?" *Climatic Change* 93 (3–4): 335–54.

Amundsen, H. 2012. "Illusions of resilience? An analysis of community responses to change in northern Norway." *Ecology and Society* 17 (4): 46.

Amundsen, H. 2013. "Place attachment as a driver of adaptation in coastal communities in Northern Norway." *Local Environment* 2013: 1–20.

Amundsen, H., Berglund, F. and H. Westskog. 2010. "Overcoming barriers to climate change adaptation- a question of multilevel governance?" *Environment and Planning C: Government and Policy* 28 (2): 276–89.

Berge, G. 1996. *Tørrfisk: thi handlet du red'lig, og tørket din fisk: En bok om tørrfiskkultur, Nord-Norge og Bergen* [Stockfish: Thus you behaved upright, and dried your fish: A book about stockfish culture]. Stamsund: Orkana.

Brox, O. 1966. *Hva Skjer i Nord-Norge? En Studie i Norsk Utkantpolitikk* [What happens in northern Norway? A study of Norwegian politics in the periphery]. Oslo: Pax Forlag.

Crate, S. A. 2008. "Gone the bull of winter? Grappling with the cultural implications of and anthropology's role(s) in global climate change." *Current Anthropology* 4 (49): 569–95.

Dannevig, H., Hovelsrud, G. K. and T. Rauken. 2012. "Implementation of adaptation at the local level." *Local Environment Special Issues* 17 (6–7): 597–612.

Douglas, M. 1982. "Cultural bias." In *In the Active Voice*, edited by Douglas, M. and P. Keagan, 183–254. London: Routledge.

Douglas, M. and A. Wildavsky. 1982. *Risk and Culture: An Essay on the Selection of Technical and Environmental Dangers*. Berkeley: University of California Press.

Drinkwater, K. F. 2005. "The response of Atlantic cod (*Gadus morhua*) to future climate change." *Ices Journal of Marine Science* 62 (7): 1327–37.

Fiskerirådgivning AS. 2006. "LU-fakta om fiskeri- og havbruksnæringen i Nord-Norge og Nord-Trøndelag. Landsdelsutvalget for Nord-Norge og Nord-Trøndelag" [Data on the fisheries and aquaculture in northern Norway and the County of Nord-Trøndelag]. F:\Fr\P 42-06\Rapport 60620. 2006: Fiskerirådgivning AS Tromsø, Norway.

Førland, E. J., Benestad, R. E., Flatøy, F., Hanssen-Bauer, I., Haugen, J. E., Isaksen, K., Sorteberg, A. and B. Ådlandsvik. 2009. "Climate development in North Norway and the Svalbard region during 1900–2100." Temarapport for NorACIA, Polarinstituttets rapportserie 128. Tromsø: Norwegian Polar Institute.

Hanssen-Bauer, I., Drange, H., Førland, J. E., Roald, L. A., Børsheim, K. Y., Hisdal, H., Lawrence, et al. 2009. "Klima i Norge 2100. Bakgrunnsmateriale til NOU Klimatilpasning" [Climate in Norway 2100: Background material for the Norwegian Green Paper on Climate Adaptation]. Norsk Klimasenter, Oslo, Norway.

Heyd, T. 2008. "Cultural responses to natural changes such as climate change." *Espace populations sociétés* 2008/1: 83–88.

Hovelsrud, G. K. and B. Smit, eds. 2010. *Community Adaptation and Vulnerability in the Arctic Regions*. Dordrecht: Springer.

Hovelsrud, G. K., Dannevig, H., West, J. and H. Amundsen. 2010. "Adaptation in fisheries and municipalities: Three communities in northern Norway." In *Community Adaptation and Vulnerability in the Arctic Regions*, edited by Hovelsrud, G. K. and B. Smit, 23–62. Dordrecht: Springer.

Hulme, M. 2008. "The conquering of climate: Discourses of fear and their dissolution." *Geographical Journal* 174 (1): 5–16.

Hulme, M. 2009. *Why We Disagree about Climate Change: Understanding Controversy, Inaction and Opportunities*. Cambridge: Cambridge University Press.

Jasanoff, S. 2010. "A new climate for society." *Theory, Culture and Society* 27 (2–3): 233–53.

Jentoft, S. and K. H. Mikalsen. 2004. "A vicious circle? The dynamics of rule making in Norwegian fisheries." *Marine Policy* 28 (2): 127–35.

Kahan, D. M., Peters, E., Wittlin, M., Slovic, P., Ouellette, L. L., Braman, D. and G. Mandel. 2012. "The polarizing impact of science literacy and numeracy on perceived climate change risks." *Nature Climate Change* 2 (6): 1–4.

Keskitalo, C., Dannevig, H., Hovelsrud, G. K., West, J. J. and Å. G. Swartling. 2011. "Adaptive capacity determinants in developed states: Examples from the Nordic countries and Russia." *Regional Environmental Change* 111 (3): 579–92.

Kvalvik, I., Dannevig, H., Dalmannsdottir, S., Hovelsrud, G. K., Uleberg, E. and L. Rønning. 2011. "Climate change vulnerability and adaptive capacity in the agricultural sector in northern Norway." *Acta Agricultura Scandinavica* 61 (1): 27–37.

Ministry of Environment. 2011. "Oppdatert forvaltningsplan for Barentshavet og havområdene utenfor Lofoten. NOU 10 2010–2011" [Updated management plan for the Barents Sea and the seas off the coast of Lofoten]. Green paper. Oslo, Norway.

Næss, L. O., Bang, G., Eriksen, S. and J. Vevatne. 2005. "Institutional adaptation to climate change: Flood responses at the municipal level in Norway." *Global Environmental Change Part A* 15 (2): 125–38.

Norgaard, K. M. 2011. *Living in Denial: Climate Change, Emotions, and Every Day Life*. Cambridge, MA: MIT Press.

Nuttall, M. 2005. "Hunting, herding, fishing and gathering: Indigenous peoples and renewable resource use in the arctic." In *Arctic Climate Impact Assessment: Scientific Report*, 649–90. Cambridge: Cambridge University Press.

O'Brien, K. and G. Hochachka. 2010. "Integral adaptation to climate change." *Journal of Integral Theory and Practice* 5 (1): 89–102.

O'Brien, K. and J. Wolf. 2010. "A values-based approach to vulnerability and adaptation to climate change." *Wiley Interdisciplinary Reviews: Climate Change* 1 (2): 232–42.

O'Riordan, T. and A. Jordan. 1999. "Institutions, climate change and cultural theory: Towards a common analytical framework." *Global Environmental Change* 9 (2): 81–93.

Øseth, E. 2010. "Klimaendringer i norsk Arktis. Konsekvenser for livet i nord" [Climate change in the Norwegian Arctic: Consequences for life in the North]. Norsk Polarinstitutt Rapportserie 136.

Szerszynski, B. and Urry, J. 2010. "Changing climates: Introduction." *Theory, Culture, and Society* 27: 1–8.

Sabates-Wheeler, R., Mitchell, T. and F. Ellis. 2008. "Avoiding repetition: Time for CBA to engage with the livelihoods literature?" *IDS Bulletin* 39 (4): 53–59.

Smit, B. and Wandel, J. 2006. "Adaptation, adaptive capacity and vulnerability." *Global Environmental Change* 16: 282–92.

Sundby, S. and O. Nakken. 2008. "Spatial shifts in spawning habitats of Arcto-Norwegian cod related 828 to multi-decadal climate oscillations and climate change." *ICES Journal of Marine Science* 65: 953–62.

Thompson, M., Ellis, R. and A. Wildavsky. 1990. *Cultural Theory*. Oxford: Westview Press.

Tyler, N. J. C., Turi, J. M., Sundset, M. A., Strøm Bull, K., Sara, M. N., Reinert, E., Oskal, N., et al. 2007. "Saami reindeer pastoralism under climate change: Applying a generalized framework for vulnerability studies to a sub-arctic social-ecological system." *Global Environmental Change* 17 (2): 191–206.

Watten, J. 2004. "Natur som kilde til innovasjon" [Nature as a source of innovation]. In *Nord-Norge møter framtiden*, edited by Nilsson, J.-E. and A. Rydningen, 116–34. Tromso: Norut.

West, J. and G. K. Hovelsrud. 2008. "Climate change in northern Norway: Toward an understanding of socio-economic vulnerability of natural resource-dependent sectors and communities." Report 2008:04. Oslo: CICERO.

West, J. J. and G. K. Hovelsrud. 2010. "Cross-scale adaptation challenges in the coastal fisheries: Findings from Lebesby, northern Norway." *Arctic* 63 (3): 338–54.

12

Changes in Organizational Culture, Changes in Adaptive Capacity?

Examples from the Norwegian and Swedish Electricity Sectors

TOR HÅKON INDERBERG

Although adaptation to climate change is done by individuals, organizations are the primary socioeconomic units within which processes of adaptation will take place (Berkhout et al. 2006, 136). Organizations can achieve more than the sum of individuals, at the same time as they or their environments can be barriers for effective adaptation (Pelling 2011). Therefore, improving organizational adaptive capacity is important for securing necessary adaptations. This accentuates the need for looking at adaptive capacity within the organizational realm.

Organizational activity can be observed not only for individual organizations or private companies but also on the sector level, where public ministries, agencies, and public and private companies are the main actors (DiMaggio and Powell 1983; Scott 2008). In fact, many of the factors influencing organizational adaptive capacity are to be found on the sector level, where public regulations (explicitly) and shared norms and values (implicitly) put limits on or facilitate specific behaviors (Inderberg 2011).

While there is a growing focus on company and organizational adaptation to climate change (Furrer et al. 2011; Haigh and Griffiths 2011; Hoffmann et al. 2009; Inderberg and Arntzen Løchen 2012; Linnenluecke and Griffiths 2010; Linnenluecke et al. 2012; Winn et al. 2011) and within public organizations and multilevel governance (Glaas et al. 2010; Keskitalo 2009; Næss et al. 2005), organizational behavior at sector level in the nexus between public regulations and private enterprises has received less attention. While taking care of vital public interests, quasi-public network services like the postal and rail services and road and electricity sectors are all influenced by climate change and represent excellent opportunities for studying sector-level organizational adaptive capacity in such a setting.

Most analyses of adaptive capacity to climate change emphasize factors related to formal structures (see Eakin et al. 2010; Glaas et al. 2010), which include the written regulations, rules, and command lines that decide who can do what either within companies or through public regulations (Christensen and Peters 1999). The formal structure thus regulates and defines roles incentives and sanctions, both internally and between organizations. Although such a focus illuminates important aspects of adaptive capacity, this approach can overlook important cultural factors that promote or constrain adaptive capacity (Pelling 2011). An analysis of organizational culture, conversely, directs attention toward the dominant norms and values within an organizational field (Christensen and Peters 1999). Norms and values are often shared between organizational actors on the sector level and determine what types of behavior and decisions are regarded as appropriate (Thornton and Ocasio 2008) and therefore also adaptive capacity.

This chapter argues that organizational culture is a vital yet often overlooked dimension of adaptive capacity and shows that including it in an integrated analytical framework can modify or even change conclusions about adaptive capacity based on analyses of formal organizational factors. Drawing on organizational theory, this chapter presents a comparative analysis of how organizational changes in two quasi-public network services, the Norwegian and Swedish electricity grid sectors, influenced adaptive capacity to extreme weather events that can be associated with climate change. Two perspectives are used to underline this important message: an instrumental perspective and an institutional-cultural perspective.

A comparative analysis illuminates different and important aspects of how changes in formal structure and organizational culture influence adaptive capacity. During the 1990s the Norwegian and Swedish electricity sectors both experienced what can be considered revolutionary change (Bye and Hope 2006; Högselius and Kaijser 2007). They went from consisting of vertically integrated companies where both generation and transport of electricity were performed by the same companies to an unbundled structure where the two functions then had to be legally separated. This required transformation of the regulatory frameworks and led to changes in incentive structures for investments and the building out of electricity grids. The production parts of the sector were exposed to markets, while the electricity grid activities were placed under revenue regulation. In addition, cultural changes took place within both countries' electricity industries, characterized as a shift from an engineering to an economic institutional logic. The basis on which decisions were legitimized was thus altered: whereas the engineers focused on building robust infrastructure (often disregarding cost), the new economic paradigm valued cost efficiency. Such extensive organizational transformation can be expected to influence the electricity sectors' capacity to adapt to climate change (Berkhout et al. 2006, 136).

Comparisons made between the Norwegian and the Swedish electricity grid sectors show that formal structure and organizational culture influence adaptive capacity differently in the two contexts. The study is based on a comparison between two phases: before and after the respective energy reforms of 1991 and 1996, covering a total time span from 1985 to 2005. The research involved thirty-nine semistructured interviews as well as an analysis of formal documents such as laws and regulations, green and white papers, and secondary literature. The interviews were conducted with representatives from the industry, regulators, ministries, and researchers in Norway and Sweden in the period 2008–11. The interviewees held positions with senior or executive responsibility in the regulatory agencies or company they represented: the Norwegian Watercourse and Energy Directorate (NVE) and the Swedish Energy Market Inspectorate (EI) from the public agencies, and from the private sector, large companies such as Vattenfall, Hafslund, E.ON Sweden, and Agder Energi, as well as smaller ones like Karmark Energi and Stange energi. The interviewees had a varied number of years' experience, had the opportunity to provide free-form contributions, and were also asked specific questions about regulatory frameworks, organizational practices, legitimacy of behavior, and changes in the sector during the period of analysis. Information gathered about cultural change in particular provided valuable insights due to the limited availability of such data through other sources. Information about official regulations is freely available for both countries, and the interviews provided critical detail about how the grid companies related to these regulatory frameworks.

The first section of this chapter briefly outlines the theoretical framework for analyzing adaptive capacity within an organizational field such as the electricity sector. Then the state of the sectors in Norway and Sweden and the changes they have experienced through the reforms are described to provide a basis for the analysis. These changes are subsequently discussed based on the expectations from two theoretical perspectives, before addressing the question of how changes in formal organizational structure and culture influence a sector's adaptive capacity to climate change and extreme weather experience.

Two Perspectives on Adaptive Capacity

In terms of climate change, adaptive capacity is understood as the ability to change by reducing "vulnerability or enhancing resilience in response to observed or expected changes in climate or associated extreme events" (Adger et al. 2007, 720). Nonetheless, most discussions in the literature tend to be devoted to what influences adaptive capacity, with a particular focus on two broad directions (Adger et al. 2007; Oppermann 2011). One direction focuses on technical-scientific aspects, which tend to emphasize access to technology, resources, knowledge, the structure of institutions, risk spreading, and human and social capital (see Yohe and Tol 2002). Often these factors can be easily quantified, and although social factors such as human capital are included in this direction, they are often presented in a simplistic manner, with little attention to the changing social, economic, and cultural contexts (O'Brien et al. 2006). An alternative direction emphasizes the norms, values, and cultural contexts (Adger et al. 2007). In contrast to the technical-scientific direction, the institutional-cultural focuses on social factors as facilitators for or barriers to adaptive action and considers them an important component of adaptive capacity (Adger et al. 2009).

The two directions are represented in the chapter as the instrumental perspective and the institutional-cultural perspective.

In the *instrumental perspective*, a key assumption is that formal structures, understood as written rules and procedures, influence and channel attitudes and actions (Christensen et al. 2007, 144). Therefore organizational structures (inter- and intraorganizational) are tools or instruments for goal achievement. Adaptive capacity is limited by the availability of resources and information (Brunsson 2003, 168), and the rational instruments are embedded within the formal organizational structure through directives, regulations, incentives, and sanctions. "Formal organizational structure" here includes the whole sector, from grid companies via the administrative level and the regulators to the ministry level. By changing and balancing the formal organizational structure to fit the various goals set, the actions will be influenced differently, leading to a change in organizational outcome and capacity. A high adaptive capacity from the instrumental perspective includes a clear distribution of responsibility for robustness, resilience, and adaptation and clear expectations of and incentives for the grid companies on issues, investments, mapping of vulnerabilities, and maintenance that are relevant for adaptation.

The *institutional-cultural perspective* represents a social dimension of adaptive capacity. Organizational actors are seen as constrained by *institutional factors*: "more-or-less taken-for-granted repetitive social behaviour that is underpinned by normative systems and cognitive understandings that... enable self-reproducing social order" (Greenwood et al. 2008, 4–5). Institutional factors then include relatively stable routines, norms, and values that both constrain and empower action. Individuals and organizations fulfill or enact identities by following informal rules and procedures that they imagine as appropriate to the situation they are facing (March 1994). Actors act out what they believe is expected of them, bound by norms and values. In other words, action is what is deemed appropriate by an actor in a process whereby situation and roles are matched (Christensen and Røvik 1999; March and Olsen 1989). Also, these normatively founded behaviors can manifest in *institutional logics* within the same institutional setting and are systematic sets of normative systems with a central logic that provides actors with vocabularies of motive and a sense of identity (Thornton and Ocasio 2008). They can be coupled to professional background (Reay and Hinings 2009) and may compete, coexist, or get defined out through change in organizational demography or location (Lounsbury 2007; Reay and Hinings 2009).

Adaptive capacity depends on a normative and legitimate basis for avoiding or overcoming resistance to changes in practices (Næss et al. 2005). Consequently, changes in institutional logics can be important influences for adaptive capacity. Barriers to adaptation will exist where organizational culture in the form of institutional logics does not provide a legitimate basis for the implementation of these measures. From this perspective, a high adaptive capacity in the electricity sector is evident when prevailing institutional logics, as mapped empirically, provide a legitimate basis for maintenance of infrastructure, robustness of the system and security of supply, and an awareness of climate change. The institutional logics legitimize or delegitimize decisions for taking adaptive measures with the goals of reducing vulnerability to climate change and securing delivery of electricity.

Changes in the Norwegian and Swedish Electricity Sectors

Norway

Historically, the Norwegian energy system developed under a mutual understanding between the sector and the government. An abundant power supply was provided for industrial development and economic growth in return for resources and a legal basis that enabled massive development of large-scale hydropower (Inderberg and Eikeland 2009). Prior to reform in 1991, the electricity price did not distinguish between the cost of the electricity itself and the cost of transporting it, reflecting the fact that publicly owned companies performed both production and distribution of electricity within the same unit. The price was politically decided by the parliament and often took the form of long-term contracts (Midttun and Summerton 1998).

In the 1980s the organization of the electricity sector came under increasing pressure. When the regulator, the NVE, was divided in 1986, the utility parts of the directorate (grid and production) were separated from the regulatory functions. The new company, Statkraft, publicly owned and vertically integrated (like most of the sector), owned 30 percent of national electricity production and most of the national grid (Olsen 2000, 184). It was still under the direct authority of the Ministry of Petroleum and Energy (OED).

Interviewees from the NVE and industry representatives indicated that the dominant institutional logic at the time was an engineer-based thinking, developed through the dominant position of this profession. Belonging to a sector "building the country," these engineers were seen as "midwives of the welfare society" (Skjold 2009, 120).

There are indications of economic thinking prior to 1990, even if bringing electrical power to the nation was still the imperative. Indeed, Norway had a working market-based trade in excess power since 1971, even if this was not based on profit motivation (Skjold 2009, 19). All investments in production and capacity during this period were subject to cost reimbursements (Bye and Hope 2006, 22). In this way the sector contributed to the industrial modernization of Norway, where cheap electricity played an important part in developing a strong industry (Olsen 2000, 123). The engineering approach and the inflated investment budgets also meant that the physical structures were often overdimensioned to compensate for uncertainties about weather patterns (Nilsen and Thue 2006, 180). This, however, changed with the Energy Act of 1990.

On January 1, 1991, the formal structure of the electricity sector was completely changed. Most interviewees point to the fact that although the reform was facilitated by several important factors, economic inefficiency in the sector itself was an important driver. At the root of the Norwegian reform was the economic efficiency paradigm, increasing the need for economists in the sector. As a consequence, the Norwegian sector underwent large demographic changes. Even before the Energy Act, more economists were starting to appear in the sector, reflecting a cultural trend that accelerated after 1991. As one interviewee from a large Norwegian grid company expressed it, "yes, the 'economism' is seeping through the whole sector," and "I am an engineer but I have also taken two years of economics classes." No precise demographical data on professional backgrounds within the sector exist, but all interviewees described a similar change. The discourse changed from a technical reasoning with security of supply as a pillar to a social economic rationale. A clear difference is that all decisions from then on increasingly found legitimacy in arguments of social economic efficiency (Inderberg 2011). Arguments that were not based on economic efficiency became harder to defend. Such decisions are usually not taken now, and this has been a major cultural change.

The grid companies were regulated accordingly by economic incentive regulation, with the main purpose of increasing economic efficiency (Langset 2007; Olsen 2000). After some years of light-handed regulation, from 1997, the new model was used for the estimation of *revenue caps* for the network utilities (Langset et al. 2001). The NVE decides each company's maximum revenue, based on performance relative to a modeled reference company. Since the maximum revenue is set, the grid companies can only increase their profits by reducing costs. Previously overdimensioned grids contributed by allowing the grid companies to reduce reinvestment rates and focus on economic efficiency (Inderberg and Eikeland 2009). This is reflected in the investment rates in the sector as well as in the aggregate number of employees – both gradually going down since the reform (Statistics Norway 2010).[1]

The new set of goals reflected the New Public Management (NPM) reforms of the time, which were also a source of inspiration (Bye and Hope 2006; Inderberg et al. 2014). These reforms presented solutions for increased efficiency by market orientation, devolution, managerialism, and the use of contracts (Christensen and Lægreid 2001). In 2001 the NVE added "qualitative dependent revenue caps" to the model (called KILE), where the costs of failure were added in the model to influence the caps on revenue (Langset et al. 2001). The objective was thus to make it unprofitable for companies to reduce system reliability to unacceptable levels.

In a report by the European sector association Eurelectric, the Norwegian regulatory scheme scored an achievable rate of return "significantly below regulatory

---

[1] The trend has gradually changed, and since 2005 the general number of employees is again rising (Statistics Norway 2010).

rate of return," indicating a difficult investment climate for distribution companies (Eurelectric 2011, 20). Interviews with company representatives confirm this. They often felt it is not economically feasible to make necessary investments in the grid. The NVE claims that the companies have too short of a time perspective on the investments. At the same time, the quality of supply has actually been improved with fewer blackouts. One interviewee from the NVE also compared the Norwegian and the Swedish regulatory schemes and claimed that the Swedish reform led to "a much 'looser' regulation than the one in Norway, to the degree that the Swedish companies called for a change to get a clearer and more transparent system" (interview with the author, September 26, 2008).

The Norwegian energy sector, with its spatially dispersed electrical grid, has always been exposed to hard weather. Companies on the south coast, Agder Energi in particular, have had incidents of large amounts of wet snow breaking pylons and grid and leading to power failures. Interviews indicated a tendency that larger companies exposed to weather induced incidents have a more structured approach to anticipatory adaptation than smaller companies with little experience of this. However, the interviews also suggest that grid companies have difficulties in learning from other companies.

Sweden

Whereas Norway is almost fully dependent on hydropower, about 90 percent of Sweden's electricity is generated by approximately equal shares of hydro and nuclear power (IEA 2008). Contrary to the distributed production of hydropower in Norway, the majority of the Swedish hydropower is in the north, with relatively high transmission capacity to the south, where the bulk of consumption occurs. Therefore the main grid between the north and south of Sweden is relatively strong. The nuclear facilities are located in the south.

By the 1980s a self-regulatory system had been developed, in effect regulating the electricity price without real political interference. The state-owned energy company Vattenfall decided its own price, based on self-cost. Since Vattenfall had the majority of the "market," the other companies followed the Vattenfall price, including a profit margin. Privately owned companies, like Sydkraft[2] and Gullspång, also related to this price, contributing to the implicit self-regulation of the system. There were no formal agreements about prices, but an institutional safety valve in the form of a price council functioned as an arena for negotiations if any customers felt the price was unreasonably high (Söderberg 2008, 43).

As in Norway, the high pace of building grid and generating capacity that the Swedish sector experienced up to the mid-1980s flattened out along with consumption.

---

[2] Later to be acquired by German E.ON.

It was then natural to question whether a change in the system was desirable. While there was no large efficiency crisis in the Swedish sector, the arguments were directed at improving the system through unbundling and liberalization (Högselius and Kaijser 2007). The general structure of the reform and its NPM aspects are similar to those of Norway.

The Swedish sector is today fully unbundled, meaning that electricity transport and production are separated (IEA 2008). Distribution companies of highly varying size (and often owned by the production companies) largely own and run the distribution grid. These natural monopolies are subject to regulation by the EI, even though the regulatory scheme after the deregulation has been unstable.

Inefficiency was considered less of a general problem in the Swedish system than in the Norwegian, at least in the discourse in the period before the reform (Högselius and Kaijser 2007). This seemed to generate less of an undermining of the engineer's standing. The division of the company Vattenfall (Statkraft's equivalent in Sweden) happened in 1992 and was a prelude to the reform in 1996.

Interviewees claim that the Swedish regulatory scheme has been more volatile than the Norwegian one, and so far no model has lasted more than a few years. One of the more senior interviewees from the EI says that in such a system, "the utilities have been forced to guess what the 'reasonable revenue' would be by the end of the year. We have had some cases where the companies have guessed wrongly, and this has created some problems" (interview with the author, June 16, 2009). The Norm Model Regulation, which lasted from 2003 to 2007, was particularly controversial because of its lack of transparency. The reform led to low legitimacy of the regulatory scheme and resulted in several court proceedings between grid companies and the regulator. Furthermore, interviews from the EI and industry claim that regulations are different to Norway in that emphasis on efficiency is lower and that the cost norms give more room for investment. This is further supported by a Eurelectric report that indicates that the Swedish regulatory system provided a high rate of return (Eurelectric 2011, 20).

Another difference between Norway and Sweden that has resulted in changes is the impact of extreme weather. The 2005 storm Gudrun was attributed to climate change and raised awareness about system vulnerability. The storm caused large-scale damage to the southern Swedish distribution grid. The resulting outage left more than 500,000 customers without electricity, about 68,000 for a week or more (Palm 2008). Salient after the storm was the changed emphasis on security of delivery and a change from overland distribution grid to underground cabling. Many of the Swedish grid companies started programs to put extensive parts of the grid below the surface after *Gudrun*. The energy company E.ON's program Kraftag is one example; it has undergrounded 17,000 kilometers of its distribution grid (E.ON 2010). While Norway has had some experiences with rough weather over the years, albeit less extreme than Gudrun – most interviewees refer to the New Year storm of 1992 – these seem not to have modified sector behavior to the same degree.

The Implications of Organizational Change for Adaptive Capacity

The main changes of the Norwegian and Swedish grid sectors over time are summed up in Table 12.1. With such a transformation of both the formal structure and the dominant institutional logic in the field, a changed capacity to adapt to climate change within the organizational field should be expected.

The pre-1991-reform formal structure of the Norwegian sector was heavily directed toward building and maintaining a robust infrastructure, facilitating a high adaptive capacity as a baseline. The availability of funding for infrastructure investments, mapping of vulnerabilities, and maintenance necessary to cope with extreme weather contributed to a high adaptive capacity (Yohe and Tol 2002). Although there was little awareness of vulnerability to climate change per se, the system was expected to be resilient. Similar arguments can be made for Sweden. Worth noting, however, is that the efficiency crisis dominating the Norwegian field prior to the reform was much less present in the Swedish political discourse. Since the price was decided within the industry itself, the awareness of marginal cost principles prevailed, even within an engineer's logic. Sweden did seem to find a balance between considerations of robustness and economic efficiency; the sector was efficient while at the same time able to build robustness and adaptive capacity. The formal structure was clear: the grid companies were responsible for securing supply of electricity within a loosely self-regulating system. The availability of financial resources combined with clear responsibilities contributed to a high adaptive capacity in prereform Sweden, consistent with the instrumental perspective.

After the reforms in both countries, unbundling introduced a larger variety of specialized actors. The NPM reforms in both Norway and Sweden can, as elsewhere, be criticized for hollowing out capacity in general (Painter 2001, 209). The reforms contributed to a loss of political control due to fragmentation through increased horizontal and vertical specialization (Christensen and Lægreid 2001). Furthermore, they led to accountability issues and the establishment of a responsibility gap that undermines adaptive capacity to climate change (Palm 2008). The change in regulations means more freedom to the individual companies, and the question remains who is responsible if adaptations do not occur. The grid companies will have to make the adaptations, but the regulators are at the same time ultimately responsible for creating framework conditions for them to do so. This challenge is common to both countries and reduces adaptive capacity, in line with expectations following the instrumental perspective.

The Norwegian regulatory regime weakened the incentives to invest in the grid by punishing such decisions, effectively undermining adaptive capacity. The projects were no longer subject to cost reimbursements, and the main criterion for investments was now social economic feasibility. This change was paramount, and the rate of investments in the grid sank dramatically. This was also made possible due to the overdimensioned grid developed prior to the reform; the new regulatory regime

Table 12.1. Overview of organizational changes in Norway and Sweden

	1986	1991	1996	2001	2006
			Changes in formal structure		
Norway	Direct ownership, hierarchical	Reform: Unbundling, grid under loose regulation	Incentive regulation	Incentive regulation, KILE add-on	Gradual developments of one regulatory model
Sweden	Direct ownership, hierarchical	Direct ownership, hierarchical	Reform: Unbundling, grid under loose or no regulation	Loose regulation	Often changing regulatory models
			Changes in organizational culture/institutional logic		
Norway	Radical but delegitimized engineer's logic	Institutional vacuum filled by economist's logic	Economist's logic strong	Economist's logic dominant	Economist's logic dominant, but flaws appear
Sweden	Balanced engineer's logic	Balanced engineer's logic	Co-existing engineer's and economist's logic	Co-existing engineer's and economist's logic	Balanced economist's logic

seemed to take security of supply for granted and focused on economic efficiency. This reinvestment lag has further reduced adaptive capacity in Norway – and this is less the case for Sweden.

The quality of supply and the security of supply are two different entities. Whereas quality of supply reflects the general ability of the grid companies to constantly deliver electricity at the right frequency, security of supply reflects robustness and resilience of the system if it is exposed to stress. Whereas the former can be expressed with statistics, the latter is more difficult to quantify. Interestingly, the quality of supply has been improved in both countries since the reform. Based on investment rates, the Norwegian model seems to have been more successful at achieving efficiency than at prioritizing maintenance, but the frequency of power failures has gone down, both for planned and unplanned outages. This is, according to the interviewees, due to more efficient running of the system rather than security of supply.

The formal structure of the incentive-based regulatory framework in Norway did not encourage robustness. Even with the quality-dependent add-on in 2001, the incentives can be considered weak. However, even if the thinking was dominated by a shorter time horizon and the investment lag grew, companies did seem to develop an awareness of the risks of building up a maintenance and investment lag too large to catch up. Furthermore, institutional logics do not change overnight, and the remainder of the engineer's institutional logic legitimized investments "otherwise not made," as interviewees phrased it. Norwegian interviewees call this social responsibility, indicating that organizational culture is fragmented. It also indicates that the earlier paradigm coexists with the newer economic institutional logic and has influence on investments made.

The Swedish regulatory model – or models – shows a balance between other considerations. The Swedish regulatory scheme has changed radically several times since the reform, but the different models have tended to be geared less toward efficiency than the Norwegian scheme (Eurelectric 2011). The Swedish models have been more flexible, at least in practice, in the sense that they allow more financial leeway for the companies, grid investments add to the real value of the grid, which in turn generates a financial benefit for the company. The Swedish companies have had more access to resources for adaptation, as shown by the large investments made to underground the distribution grid after Gudrun. Such investments are more difficult in Norway due to a comparative lack of flexibility in the scheme and suggest a higher adaptive capacity to climate change in Sweden.

Decisions made in the sector have to be grounded by institutional logic, and when the logic changes, the legitimization of decisions changes accordingly. The institutional logics have shifted in the two countries, although differently. Norway has experienced the most radical change, and the move from a radical robustness-legitimizing thinking to an economic efficiency paradigm yields significant impact on adaptive capacity. The ability to make decisions that reduce vulnerability to climate change otherwise not grounded in economic arguments has clearly been undermined by the change in culture. With adaptations, especially under uncertainty, it can be difficult to argue social economic feasibility.

This does not mean that the prereform sector was institutionalized unilaterally toward security and quality of delivery. Particularly toward the end of the period, there are signs of this paradigm being undermined and elements of alternative arrangements and approaches were observed, such as the market-based power exchange in Norway (Schneiberg 2007). There seems to be a threshold, however, for the alternative cultures to yield much influence on the decisions about costs and projects undertaken (Pierson 2004). Because these alternatives did not really break through until the economist logic took over, the contributions of the engineer culture to a high adaptive capacity in Norway were evident throughout the prereform period (Inderberg 2011).

The lack of an efficiency crisis in Sweden contributed to less extreme cultural transitions. The Swedish trend runs parallel to, but was less radical than, the Norwegian, and the engineer's institutional logic exists alongside the economic paradigm. In addition, the rapid change of Swedish regulatory models has reduced the formal structure's ability to steer the companies' behavior and enabled cultural factors to exert more influence on company behavior (Christensen and Peters 1999, 9). The Swedish sector, more able to balance the cultural basis between an engineer's institutional logic and economist thinking, still provides a legitimate basis for investing in security of delivery. The formal structure makes resources available in combination with a willingness to invest, legitimized and encouraged by the cultural basis of the Swedish sector. This demonstrates a higher adaptive capacity than the Norwegian electricity sector based on the institutional-cultural perspective.

One reason for this difference between Norway and Sweden is that Gudrun thoroughly showed the intense vulnerability of the Swedish electricity system to extreme weather events. While Norway has had some examples of extreme weather, these have not yet led to a change in adaptive behavior in the sector. The cultural contexts are different, such that the Norwegian sector legitimizes economically efficient decisions more than vulnerability reduction. In other words, because of the prevailing institutional logic, learning from extreme weather events is on more fertile ground in Sweden than in Norway.

In Sweden we find less of an economist's institutional logic, partly influenced by the Swedish experience with extreme weather events. The product of this is that investments neither feasible nor possible in Norway are being made in Sweden. The undergrounding of large parts of the distribution grid is evidence of this, pointing toward a higher adaptive capacity stemming from the organizational culture. The large investments after Gudrun indicate either a shift in culture and a relegitimization of the engineer's institutional logic or that there never was as large a transformation of organizational culture as there was in Norway. Either way, this leads to a higher adaptive capacity resulting from the cultural dimension compared to Norway.

Because institutions both constrain action and provide sources for agency and change (Thornton and Ocasio 2008), the formal structure and institutional logics, at least over time, tend to reflect each other. While the causal relationship within this interdependency is difficult to separate, from the perspective of the Norwegian electricity sector's adaptive capacity, it is evident that shortcomings in the formal structure are compensated by older elements of institutional culture. This is visible in the grid companies' behavior, where investments were made which will not pay off. The Norwegian prereform structure justified spending on the grid, and the engineer-based thinking was in accordance with this. A robust grid was the result, capable of taking measures if vulnerabilities grew evident. Paradoxically, with a changing climate, the Norwegian grid seems to be getting weaker and is today in urgent need of reinvestment.

Conclusions: How Do Changes in Organizational Field Influence Adaptive Capacity?

This chapter has analyzed two directions of adaptive capacity to climate change in the Norwegian and Swedish electricity sectors. The instrumental and social directions are represented by formal structure and organizational culture in the electricity sector. Comparisons have been made between the Norwegian and the Swedish systems to show how the dimensions of adaptive capacity differ.

The findings show that both the formal structure and organizational culture highly influence adaptive capacity to climate change. These influences can be positive or negative, depending on context. The change in formal structure and organizational culture in the Norwegian electricity sector effectively reduced adaptive capacity to any challenge not encouraged by the regulatory regime and/or legitimized by economic efficiency arguments. Equivalent influences from formal and cultural factors are found for Sweden, although the changes in the sector have been less transformative and the considerations between efficiency and security of supply have been more balanced. On the basis of this research, we conclude that the capacity to adapt to climate change and extreme weather events is greater within the Swedish than within the Norwegian electricity grid sector.

Sometimes a change in the formal structure and institutional culture mutually reinforces influence on adaptive capacity. However, we have also seen that social context can compensate for reduced capacity resulting from formal regulatory factors. For example, cultural factors that encourage adaptive behavior can compensate for a formal structure that undermines it. Instrumental and cultural factors are therefore both necessary dimensions of adaptive capacity and are conditionally substitutes for each other. This finding does not support earlier claims that factors of adaptive capacity cannot be substituted (Tol and Yohe 2007). It may, however, indicate a hierarchy of necessary and contributing factors for adaptive capacity, which warrants further research.

The analysis has some important policy implications. The chapter shows that regulations should not be developed solely based on the assumption that organizational behavior is guided by instrumental preferences. Organizational adaptive capacity is influenced by the interplay between prevailing institutional logics and the formal regulatory structure. While the long-term influence between these two dimensions of institutional behavior should be investigated further (Scott 2008), the findings here show that not acknowledging the influence of culture will reduce important understandings about the ability to adapt to climate change.

There is a need for integrative perspectives that represent both the instrumental and the social dimensions of adaptive capacity to climate change. Although acknowledged in organization theory, this also seems to be an important message for the adaptation literature. The findings here provide useful lessons for the governance of organizations and have the potential to be applied across a wide array of sectors and over varying contexts. Not considering these dimensions could mean that important parts of adaptive capacity are ignored, potentially leading to policy recommendations that are difficult to implement in reality. Additionally, it is evident that organizational change in itself has influence on adaptive capacity to climate change and should be monitored.

Acknowledgments

I thank Tom Christensen, Karen O'Brien, Elin Selboe, Per Ove Eikeland, Jon Birger Skjærseth, Elin Lerum Boasson, and Jørgen Wettestad for valuable comments on earlier drafts and Susan Høyvik for language editing. This research has been part of the project "The Potentials of and Limits to Adaptation in Norway" (PLAN), funded by the Research Council of Norway.

References

Adger, W. N., Agrawala, S., Mirza, M. M. Q., Conde, C., O'Brien, K., Pulhin, J., Pulwarty, R., Smit, B. and K. Takahashi. 2007. "Assessment of adaptation practices, options, constraints and capacity." In *Climate Change 2007: Impacts, Adaptation and Vulnerability. Contribution of Working Group II to the Fourth Assessment Report of the Intergovernmental Panel on Climate Change*, edited by Parry, M. L., Canziani, O. F., Palutikof, J. P., van der Linden, P. J. and C. E. Hanson, 717–44. Cambridge: Cambridge University Press.

Adger, W. N., Lorenzoni, I. and K. O'Brien. 2009. "Adaptation now." In *Adapting to Climate Change. Thresholds, Values, Governance*, edited by Adger, W. N., Lorenzoni, I. and K. O'Brien, 1–22. Cambridge: Cambridge University Press.

Berkhout, F., Hertin, J. and D. M. Gann. 2006. "Learning to adapt: Organizational adaptation to climate change impacts." *Climatic Change* 78 (1): 135–56.

Brunsson, N. 2003. *The Organization of Hypocrisy: Talk, Decisions and Action in Organizations*. 2nd ed. Copenhagen: Copenhagen Business School Press.

Bye, T. and E. Hope. 2006. "Electricity market reform – the Norwegian experience." In *Competition and Welfare: The Norwegian Experience*, edited by L. Sørgard, 21–50. Bergen: Norwegian Competition Authority.

Christensen, T. and P. Lægreid. 2001. "A transformative perspective on administrative reforms." In *New Public Management: The Transformation of Ideas and Practice*, edited by Christensen, T. and P. Lægreid, 13–42. Aldershot, UK: Ashgate.

Christensen, T. and B. G. Peters. 1999. *Structure, Culture, and Governance: A Comparison of Norway and the United States*. Lanham, MD: Rowman and Littlefield.

Christensen, T. and K. A. Røvik. 1999. "The ambiguity of appropriateness." In *Organizing Political Institutions*, edited by Egeberg, M. and P. Legreid, 159–80. Oslo: Scandinavian University Press.

Christensen, T., Lægreid, P., Roness, P. G. and K. A. Røvik. 2007. *Organization Theory and the Public Sector: Instrument, Culture and Myth*. London: Routledge.

DiMaggio, P. J. and W. W. Powell. 1983. "The iron cage revisited: Institutional isomorphism and collective rationality in organizational fields." *American Sociological Review* 48 (2): 147–60.

Eakin, H., Eriksen, S., Eikeland, P. O. and C. Øyen. 2010. "Public sector reform and governance for adaptation: Implications of new public management for adaptive capacity in Mexico and Norway." *Environmental Management* 47 (3): 338–51.

E.ON. 2010. "Stora effekter av E.ONs investeringsprojekt Krafttag" [Large effects of E.ONs investment project Krafttag]. http://www.eon.se/templates/Eon2TextPage.aspx?id=60316&epslanguage=SV.

Eurelectric. 2011. *Regulation for Smart Grids: An Eurelectric Report*. Brussels: Eurelectric.

Furrer, B., Hamprecht, J. and V. H. Hoffmann. 2011. "Much ado about nothing? How banks respond to climate change." *Business and Society* 51 (1): 62–88.

Glaas, E., Jonsson, A., Hjerpe, M. and Y. Andersson-Skjöld. 2010. "Managing climate change vulnerabilities: Formal institutions and knowledge use as determinants of adaptive capacity at the local level in Sweden." *Local Environment* 15 (6): 525–39.

Greenwood, R., Oliver, C., Sahlin, K. and R. Suddaby. 2008. Introduction to *The SAGE Handbook of Organizational Institutionalism*, edited by Greenwood, R., Oliver, C., Sahlin, K. and R. Suddaby, 1–46. London: Sage.

Haigh, N. and A. Griffiths. 2011. "Surprise as a catalyst for including climatic change in the strategic environment." *Business and Society* 51 (1): 89–120.

Hoffmann, V. H., Sprengel, D. C., Ziegler, A., Kolb, M. and B. Abegg. 2009. "Determinants of corporate adaptation to climate change in winter tourism: An econometric analysis." *Global Environmental Change* 19 (2): 256–64.

Högselius, P. and A. Kaijser. 2007. *När folkhemselen blev internationell. Elavgleringen i historiskt perspektiv*. Stockholm: SNS Förlag.

IEA. 2008. "Energy policies of IEA countries: Sweden, 2008 review." Paris: OECD/International Energy Agency.

Inderberg, T. H. 2011. "Institutional constraints to adaptive capacity: Adaptability to climate change in the Norwegian electricity sector." *Local Environment* 16 (4): 303–17.

Inderberg, T. H. and P. O. Eikeland. 2009. "Limits to adaptation: Analysing institutional constraints." In *Adapting to Climate Change: Thresholds, Values, Governance*, edited by Adger, W. N., Lorenzoni, I. and K. O'Brien, 433–47. Cambridge: Cambridge University Press.

Inderberg, T. H. and L. Arntzen Løchen. 2012. "Adaptation to climate change among electricity distribution companies in Norway and Sweden: Lessons from the field." *Local Environment* 17 (6–7): 663–78.

Inderberg, T. H., Stokke K. B. and M. Windsvold. 2014. "The effect of new public management reforms on climate change adaptive capacity: A Comparison of urban planning and the electricity sector." In *Handbook of Climate Change Adaptation*, edited by W. Leal, chapter 35. New York: Springer.

Keskitalo, E. C. H. 2009. "Governance in vulnerability assessment: The role of globalising decision-making networks in determining local vulnerability and adaptive capacity." *Mitigation and Adaptation Strategies for Global Change* 14 (2): 185–201.

Langset, T. 2007. "KILE-ordningen – en stimulans til å opprettholde sikker levering av elektrisk kraft." In *Et kraftmarked blir til. Et tilbakeblikk på den norske kraftmarkedreformen*, edited by Moen, J. and S. Sivertsen, 84–87. Oslo: NVE.

Langset, T., Trengereid, F., Samdal, K. and J. Heggeset. 2001. "Quality dependent revenue caps – A model for quality of supply regulation." In *CIRED: 16th International Conference and Exhibition*. IEE Conference Publication 482. London: IEE.

Linnenluecke, M. and A. Griffiths. 2010. "Beyond adaptation: Resilience for business in light of climate change and weather extremes." *Business and Society* 49 (3): 477–511.

Linnenluecke, M. K., Griffiths, A. and M. I. Winn. 2012. "Extreme weather events and the critical importance of anticipatory adaptation and organizational resilience in responding to impacts." *Business Strategy and the Environment* 21 (1): 17–32.

Lounsbury, M. 2007. "A tale of two cities: Competing logics and practice variation in the professionalizating of mutual funds." *Academy of Management Journal* 50 (2): 289–307.

March, J. G. 1994. *A Primer on Decision-Making*. New York: Free Press.

March, J. G. and J. P. Olsen. 1989. *Rediscovering Institutions: The Organizational Basis of Politics*. New York: Free Press.

Midttun, A. and J. Summerton. 1998. "Loyalty or competition? A comparative analysis of Norwegian and Swedish electricity distributors' adaptation to market reform." *Energy Policy* 26 (2): 143–58.

Næss, L. O., Bang, G., Eriksen, S. and J. Vevatne. 2005. "Institutional adaptation to climate change: Flood responses at the municipal level in Norway." *Global Environmental Change* 15 (2): 125–38.

Nilsen, Y. and L. Thue. 2006. *Statens kraft 1965–2006. Miljø og marked*. Vol. 3. Oslo: Universitetsforlaget AS.

O'Brien, K., Eriksen, S., Sygna, L. and L. O. Naess. 2006. "Questioning complacency: Climate change impacts, vulnerability, and adaptation in Norway." *Ambio* 35 (2): 50–56.

Olsen, P. I. 2000. *Transforming Economies: The Case of the Norwegian Electricity Market Reform*. Oslo: Norwegian School of Management BI.

Oppermann, E. 2011. "The discourse of adaptation to climate change and the UK Climate Impacts Programme: De-scribing the problematization of adaptation." *Climate and Development* 3: 71–85.

Painter, M. 2001. "Policy capacity and the effects of new public management." In *New Public Management: The Transformation of Ideas and Practice*, edited by Christensen, T. and P. Lægreid, 209–30. Hampshire, UK: Ashgate.

Palm, J. 2008. "Emergency management in the Swedish electricity market: The need to challenge the responsibility gap." *Energy Policy* 36 (2): 843–49.

Pelling, M. 2011. *Adaptation to Climate Change: From Resilience to Transformation*. London: Routledge.

Pierson, P. 2004. *Politics in Time*. Princeton, NJ: Princeton University Press.

Reay, T. and C. R. Hinings. 2009. "Managing the rivalry of competing institutional logics." *Organization Studies* 30 (6): 629–52.

Schneiberg, M. 2007. "What's on the path? Path dependence, organizational diversity and the problem of institutional change in the US economy, 1900–1950." *Socio-Economic Review* 5 (1): 47–80.

Scott, W. R. 2008. *Institutions and Organizations: Ideas and Interests.* 3rd ed. Thousand Oaks, CA: Sage.
Skjold, D. O. 2009. *Power for Generations: Statkraft and the Role of the State in Norwegian Electrification.* Oslo: Universitetsforlaget.
Söderberg, M. 2008. "Four Essays on Efficiency in Swedish Electricity Distribution." Gothenburg: School of Business, Economics and Law, University of Gothenburg.
Statistics Norway. 2010. "Electricity statistic 2010." http://www.ssb.no/elektrisitetaar/tab-2010-05-26-01.html.
Thornton, P. H. and W. Ocasio. 2008. "Institutional logics." In *The SAGE Handbook of Organizational Institutionalism*, edited by Greenwood, R., Oliver, C., Sahlin, K. and R. Suddaby, 99–129. London: Sage.
Tol, R. S. J. and G. Yohe. 2007. "The weakest link hypothesis for adaptive capacity: An empirical test." *Global Environmental Change* 17 (2): 218–27.
Winn, M. I., Kirchgeorg, M., Griffiths, A., Linnenluecke, M. K. and E. Günther. 2011. "Impacts from climate change on organizations: A conceptual foundation." *Business Strategy and the Environment* 20 (3): 157–73.
Yohe, G. and R. S. J. Tol. 2002. "Indicators for social and economic coping capacity – moving towards a working definition of adaptive capacity." *Global Environmental Change* 12 (1): 25–40.

13

From Informant to Actor to Leader

Social-Ecological Inventories as a Catalyst for Leadership Development in Participatory Community Climate Change Adaptation

BRADLEY MAY

As communities face the challenges of adapting to climate change and pursuing sustainable trajectories, new methods of engaging actors and incorporating a diversity of perspectives are necessary. In Canada, a number of different community responses are emerging to address climate change. They range from formalized, strategic planning exercises within existing bureaucracies to more inclusive, participatory processes involving multiple actors. Over time, the flexibility and success of these various approaches for responding to change will be revealed. The potential benefit of more participatory approaches, however, is that they allow researchers the opportunity to contextualize climate change across a variety of actors and groups. This creates the opportunity for sense making that is more responsive to the diversity of views, values, and social complexity not necessarily captured in more managerial approaches. In addition, it also provides a structured method to delve into the question of how informants transition to become actors and eventually leaders in championing climate change adaptation.

In the Niagara Region, Ontario, Canada, a multiactor, collaborative process was developed and implemented to reflect the local needs and capacity of the community. The research study was designed as a collaborative effort from the outset and the ongoing process of adaptation observed. This chapter discusses findings related to the importance of social context and multiple actor views. It focuses on two of the key components examined in the initial phase of the research study. The first is the need for *social-ecological inventories* that capture the wealth of environmental knowledge and relationships that already exist within a study area. The second is how these inventories can be useful in developing observable, meaningful community *climate change adaptation leadership*. The components are described and the results of considering them early in the project are presented. As these components enhance an understanding of climate change in the broader context of sustainable trajectories, they also contribute to building community capacity for adaptation. The added perspective gained can serve to strengthen existing institutions, networks and emerging leaders.

Next, the chapter relates how these concepts are relevant to the Niagara Region study and some preliminary reflections are offered. After that, some conclusions are presented to address the question of how best to transfer these notions to other community climate change adaptation initiatives.

Conceptual Framework: Adaptive Collaborative Risk Management

In the search for meaningful approaches to community climate change adaptation, state- and federal-level authorities and other institutions are developing ways to use risk management as an integrative tool for making sustainable, transformative decisions. "Risk management approaches help decision-makers deal with the uncertainty of climate change" (Lemmen et al. 2008, 17). In Canada, a number of these tools have been developed as guidance for decision makers (e.g., Bizikova et al. 2008; Bruce et al. 2006; Chynoweth et al. 2011; Fenech and MacLellan 2007) and concepts developed have been or are already integrated into government policy (e.g., British Columbia 2007; Dawson Adaptation Project Team 2009; Government of Ontario 2009). In particular, these provincial- and territorial-level frameworks have been successful in prioritizing and integrating climate change into overall strategic decision making. The effective operationalization of these strategies is often left up to some combination of individual government departments, interdepartmental agencies and collaboratives, local municipalities, and groups of community actors. A key to success of these initiatives is to structure collaborative and participatory approaches that are integrated yet serve the mandates and interests of particular actors and actor groups.

The challenge faced is how to take these structured approaches and create more meaningful local collaborative initiatives that explore the range of perspectives within communities. Recently, the concept of adaptive collaborative risk management (ACRM) has been proposed to attempt to address some shortcomings identified in traditional risk management (May and Plummer 2011). In essence the familiar roles of risk management – identification, analysis, and control – are counterbalanced with the ideas of *collaboration* (as opposed to general treatments of *communication* as part of a stakeholder engagement process) and *adaptation and learning* (as opposed to *feedback and monitoring* of risk control or risk treatment options selected). By considering these ideas in an integrated, ongoing, deliberative fashion, the process becomes one of developing mutual understanding and social learning among communities, actor groups, and relevant outside interests, such as state or federal agencies. Figure 13.1 portrays this idea of ACRM, with collaboration, adaptation, and learning added to the conventional risk management framework.

One key shortcoming of existing risk management frameworks is that they are explicitly focused on decision making and often neglect broader issues of power sharing and active, collaborative engagement. ACRM attempts to address this shortcoming in a unified, understandable, iterative process.

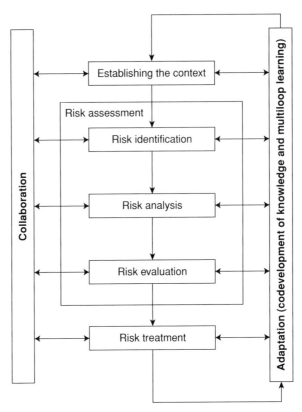

Figure 13.1. Adaptive collaborative risk management.
Source: May and Plummer (2011).

Linkages to Social-Ecological Inventories

Recently, the concept of ecological inventories from conservation biology has been elaborated to include notions of the important social aspects of social-ecological systems (SES) thinking (Schultz et al. 2007). The idea of social-ecological inventories (SEI) is a means for incorporating the important role of actors with detailed environmental knowledge of a particular SES in question in the overall makeup of stewardship and governance regimes. This idea has been applied in both biosphere reserves as well as community climate change adaptation initiatives (Mitchell 2011). Schultz et al. (2011) have gone further to incorporate this into the *Resilience Assessment Workbook* of the Resilience Alliance.[1] The five phases of the approach include *preliminary identification* of key actors and actor groups, *identifying key informants* through a process of preinterview screening, *interviewing key informants* to try and understand values and motive for participation, as well as gaining an insight into other actors that may not have been identified as part of the initial identification, *enriching the picture*

---

[1] http://www.resalliance.org/index.php/resilience_assessment.

through pausing to reflect as a research team on potential leaders in the process, power relationships, and existing knowledge bases, and finally *engagement*, using the most appropriate mix of facilitation tools (workshops, meetings, backgrounders, etc.). The enrichment of traditional stakeholder engagement through infusion with important local context ensures that issues that might otherwise be overlooked are identified. In addition, perspectives and local voices that might not otherwise be heard are explored in a structured fashion. The use of an *enriching the picture* step is key in developing a comprehensive SEI. It is envisioned that specific attention to the identification of such actors as bridging actors, gatekeepers, isolates, shadow networks, and local stewards (Schultz et al. 2011, 9) at the outset will increase the likelihood of success as the process unfolds.

For community climate change adaptation, SEIs play an important role in the *establishing the context* phase of the ACRM process (see Figure 13.1) when detailed profiles of potential stakeholders are developed and their bases of power, knowledge, and interest assessed to ensure that the participatory process is inclusive and ultimately more effective. This is extremely important as ACRM, as previously presented, is very much focused on collaboration, coproduction of knowledge, and learning throughout the process, from *risk assessment* through to *risk treatment*. Another aspect of a thorough and rigorous SEI is that it can increase the likelihood of developing a strong leadership core of actors and groups for climate change adaptation. The existing base of informants is utilized to develop actors for the process, which allows this leadership to emerge. The next section examines this key area of adaptive capacity.

Linkages to Climate Change Adaptation Leadership and Adaptive Capacity

Another aspect of the concept of ACRM that has relevance to participatory community climate change adaptation processes is how to create an environment for dialogue that facilitates sustainable outcomes. Only once an SEI process has been initiated and the relevant actors, bridging organizations and seats of environmental knowledge, have been identified can trust building and collaborative processes be designed and executed. Attention to the idea of leadership and how it develops becomes important. Facilitated workshops, a key aspect of participatory integrated assessments, can provide the platform for this dialogue (Bizikova et al. 2009). Thinking about the place of leadership and its important attributes can assist in creating lasting value for these integrated assessments. Environmental and natural resource managers have long understood the importance of the *power of personalities* in these assessments (Hanna 2007, 124) and the role of trust and relationship building. This is also true of the area of sustainability where certain attributes, discussed in more detail later, are important (Westley et al. 2011, 771).

At its most basic level, leadership is a "social influence process, operating within constraints" (Pfeffer 2000, 211). In this process of social influence, leaders play a role in *meaning making*, that is, defining meaningful action in a particular situation, and *value creation*, that is, operationalization of meaningful action and performance (Podolny et al. 2010). In addition, there is the recognition that social influence occurs at multiple levels within organizations. Kouzes and Posner (2007) express this influence through five methods: modeling the way, inspiring a shared vision, challenging the process, enabling others to act, and encouraging the heart (26). Coupled with this is the recognition that in any social interaction, followership, or being able to understand, acknowledge, and cede a certain degree of personal power to leaders in any given situation, can be as important as leadership itself (Collinson 2006). To lead, there must also be recognition that there are those individuals who are influenced and need to be supportive of actions contemplated or taken.

As with the distinction between adaptation and maladaptation, there is the need to distinguish between constructive and destructive leadership. As part of this, there is a moral obligation on the part of those assuming a leadership role and the need to be cognizant and avoid tendencies that the literature describes as destructive – "the systematic and repeated behavior by a leader, supervisor or manager that violates the legitimate interest of the organization by undermining and/or sabotaging the organization's goals, tasks, resources, and effectiveness and/or motivation, well-being or job satisfaction of subordinates" (Einarsen et al. 2007, 208). Nye (2010, 327) would attribute this to a leader's maladapted exercise of "contextual intelligence."

Recently, current models of leadership have been challenged based on the complexity faced by organizations and the speed at which adaptive challenges occur. Heifetz (1998) and Heifetz et al. (2009) make the distinctions between *technical challenges* – problems that can be clearly defined and addressed with known solutions or ones that can be developed by a few technical experts, as opposed to *adaptive challenges* – those that require significant shifts in people's habits, status, role, identity, and way of thinking, as one review puts it, "adaptive challenges are about people changing" (Cook 2009). Further along these lines, in arguing that current models of leadership are the products of traditional, top-down bureaucratic thinking, ill-suited to current realities, Uhl-Bien et al. (2007) propose a framework – complexity leadership theory (CLT), which combines aspects of adaptive, administrative, and enabling leadership (305), enmeshed in an existing "bureaucratic superstructure of planning, organizing, and missions" (302). They also go on to describe an "emergence dynamic," which uses contexts and mechanisms, driven by adaptability, learning, and creativity (see their Figure 1, 308) to respond to change. Theories such as these demonstrate the ongoing discourse that is occurring on what leadership is, or should be, in times of rapid change and transformation.

For the focus of this chapter, the constraints referred to previously are created by the challenge of being strategic about adapting to an uncertain climate change (Hallegatte 2009), as part of a broader set of integrated, organizing principles that include the reconsideration of vulnerability (Ribot 2011), the importance of sustainability

(Swart et al. 2003; Westley et al. 2011), and the need to examine adaptation as part of a more explicit treatment of integrated sustainable adaptation and mitigation (SAM) (Bizikova et al. 2008). Leadership is exerted by informants, actors, individuals, and organizations that can have a direct impact on the process and eventual success of adaptation. The literature variously refers to this assumption of a leadership role as policy entrepreneurship (Huitema and Meijerink 2010), scale-crossing brokerage (Ernstson et al. 2010), opinion leadership (Crona and Bodin 2010), and boundary crossing (Galaz et al. 2011). Adaptation is the end product of incremental decisions that can be monitored (performance), are contextual (result from clear visioning and meaning making), and are based on active, genuine engagement.

Looking more broadly at adaptive capacity and adaptive environmental governance, Armitage and Plummer (2010a) develop a number of thematic areas within which certain aspects shed light on what constitutes effective climate change leadership qualities. They include institutional diversity and flexibility, multiparty collaborations that can facilitate transformation, a recognition of the interconnectedness of governance and biophysical systems, systematic multiloop learning processes mutually developed with others, and development of sustainable frameworks to cope with stresses, maintain existing assets, and provide livelihood opportunities (10–13). Furthermore, leadership, in both individual and group contexts, is a core component in responsiveness and adaptive capacity for successful climate change adaptation. As part of an adaptive capacity wheel, Gupta et al. (2010) discuss the importance of leadership that is visionary (room for long-term visions and reformist leaders), entrepreneurial (room for leaders that stimulate actions and undertakings; leadership by example), and collaborative (room for leaders who encourage collaboration between different actors; adaptive co-management) and that supports learning (461). These ideas, in conjunction with the formulation of SEIs, provide further guidance as to what constitutes sound collaborative research process design.

What else can be said about leadership as it relates to climate change adaptation? When examining the notion of barriers to adaptation, Moser and Ekstrom (2010, Table SI-1) approach this idea of leadership by asking three basic questions:

1. Who is leading the process?
2. Do leaders have formal authority and/or the necessary skill and ability to facilitate the process?
3. Do leaders and others involved have the ability and willingness to develop a set of criteria to judge options?

In the context of participatory processes, which often develop emerging structures over time, it is necessary to ask these questions on an ongoing basis as the community engagement is undertaken. Leadership of the process can shift over time and also take on formal as well as informal characteristics. Their second question is also important in that it focuses on not only power and authority but also having participants with the requisite skills and abilities to engage in the process. With their third question, the focus is on value creation and a willingness to come together on issues related to evaluation and performance.

In examining the role of formal authority in question 2, the idea of power as an important component of social influence and leadership is also a key concept. Dengler (2007) expands on the idea of individual exercise of power to consider "spaces of power for action" (423) in the realms of science, policy, and local knowledge. Super-agents are "knowledge-brokers who serve in leadership roles of organizations" (430). It is important to note as well that the exercise of power can take a number of different forms (see Raik et al. 2008 for their discussion of four types – coercion, constraint, consent production, and real, i.e., "agents exercise power within preconditioned, structured social relations" (732)). Power can also be exercised in a number of specific ways (see Nye 2010 and his examination of how best to utilize hard, soft, and smart power, using "contextual intelligence, or the ability to understand context so that hard and soft power can be successfully combined into a smart power strategy" (327)). These various expressions of power, and how they are manifest, are key in understanding the dynamics of participatory processes for climate change adaptation.

In addition, related to question 2's reference to necessary skill and ability, one of the few articles that attempts to characterize climate change adaptation leadership suggests that there are in fact four types of relevance to climate change adaptation, policy leadership, connectivity leadership, complexity leadership, and sustainability leadership, and "each function requires the execution of specific leadership tasks which can be performed by different types of leaders, such as positional leaders, ideational leaders, sponsors, boundary workers, policy entrepreneurs or champions" (Meijerink and Stiller 2013, 240).

In terms of question 3, important information on climate, both historical and future scenarios, required for adaptation decision making, is usually the domain of scientists within government or academia. Detailed studies on impacts may be the domain of other agencies, such as state, local, or nongovernmental. Policy makers are most often tasked with using some form of structured approach, such as risk management or other analytical techniques to make decisions (Wilson and McDaniels 2007). Holders of local knowledge are in possession of the necessary ingredients and experience that make decisions relevant (Cave et al. 2011; Cohen et al. 2006). This makes the concept of ACRM with its focus on power sharing all the more useful in navigating the intersection between these various spaces.

In conclusion, thinking further on the role of CCAL in the context of overall approaches to enhancing adaptation efforts and adaptive capacity can be useful in building on questions originally posed by Smit et al. (2009) in their examination of the gross anatomy of adaptation. In their analysis, they posed three questions (with a fourth implied in their concluding section):

1. Adapt to what?
2. Who or what adapts?
3. How does adaptation occur?
4. Is adaptation effective?

If we ask these questions slightly differently, it is possible to envision an anatomy of CCAL that extends those of Moser and Ekstrom (2010):

1. Lead adaptation to what ends?
2. Who (individual) or what (institution) leads adaptation (and at what scale)?
3. How does adaptation leadership occur?
4. What constitutes effective adaptation leadership?

Table 13.1 summarizes these questions, alongside some key leadership concepts applicable to each question that have been previously discussed in this chapter.

From the discussion in this section, the concept of SEIs has been presented, as well as how they relate to the overall notion of risk management that is on one hand technical but also adaptive and collaborative (ACRM). SEIs are intended to enhance the potential effectiveness of participatory climate change adaptation. In addition, concepts surrounding the notion of impact, influence, and power were explored in developing ideas surrounding CCAL. The examination of both these concepts is currently under way in a study of climate change in the Niagara Region. The background of this study is presented next.

Niagara Region

Rationale for a Niagara Region Community Climate Change Network

The use of community-based, participatory research for climate change adaptation has a history in Canada that is continuing to enhance understanding of adaptation processes (see, e.g., Berkes and Jolly 2001; Cohen et al. 2006; Ford et al. 2007; Armitage and Plummer 2010b; Shepherd et al. 2006; Vasseur 2010). As such, this experience has led to the development of generalized approaches on how to design and execute participatory action research. Adaptation science activities in Canada have been undertaken federally to ensure "that Canadians understand the impacts of a changing atmosphere in order to reduce the adaptation deficit and take advantage of new opportunities that may arise" (Environment Canada 2009, 3). As more communities are involved in the integration of climate change adaptation into overall strategic planning and operational decision making, and are approaching this integration through a variety of techniques that range from the highly bureaucratic to more collaborative and participatory, understanding adaptation processes through modeling and accumulating case study evidence becomes increasingly important (Burton et al. 2007, 374).

Table 13.1. *Aspects of climate change adaptation leadership discussed*

Key climate change adaptation leadership questions (adapted from Smit et al. 2009)	Key references from chapter
Lead adaptation to what ends?	Vision creation and meaning making (Podolny et al. 2010) Creation of adaptive capacity (Gupta et al. 2010; Armitage and Plummer 2010a) Strategic decision making (Hallegatte 2009) Integration of adaptation, mitigation, and sustainability (Bizikova et al. 2008) Removing barriers (Moser and Ekstrom 2010) Vulnerability reduction (Ribot 2011)
Who or what leads adaptation (and at what scale)?	Informants and actors (via social-ecological inventories) (Schultz et al. 2007; Schultz et al. 2011; Mitchell 2011) Super-agents (Dengler 2007) Scale-crossing brokers (Galaz et al. 2011; Ernstson et al. 2010) Boundary workers (Lynch et al. 2008) Opinion leaders (Crona and Bodin 2010) Individuals, teams, and organizations (Kouzes and Posner 2007)
How does adaptation leadership occur?	Innovation and entrepreneurialism (Huitema and Meijerink 2010) Knowledge power spaces (science, policy, local) (Dengler 2007) Contextual intelligence (Nye 2010) Followership (Collinson 2006) Participatory integrated assessments (Bizikova et al. 2009) Collaboration and knowledge brokering (Kløcker Larsen et al. 2012) Structured decision making (Wilson and McDaniels 2007) Learning and sharing of best practices (Cohen et al. 2006; Niagara Climate Change Network 2011; Penney 2012)
What constitutes effective adaptation leadership?	Value creation and performance (Kouzes and Posner 2007) Complexity leadership theory (Uhl-Bien et al. 2007) Policy, connectivity, complexity, and sustainability leadership (Meijerink and Stiller 2013) Adaptive collaborative risk management (May and Plummer 2011)

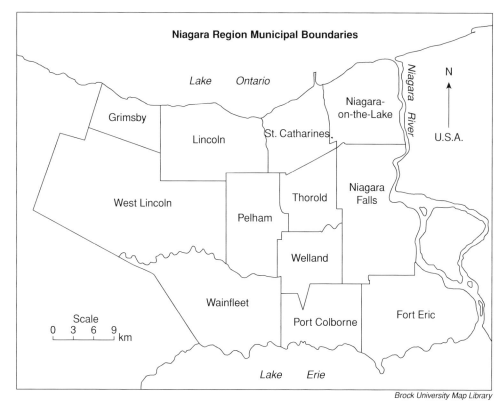

Figure 13.2. Niagara region, Ontario, Canada, administrative boundaries. From Brock University, Map Library (n.d.).

The Niagara Region was selected as a climate change adaptation research project for a number of different reasons. It is a regional municipality in the southwestern part of the Province of Ontario and bordered on two sides by the Great Lakes, Lakes Ontario and Erie (see Figure 13.2). As a result of a previous series of reports constituting the Toronto-Niagara Study (Bellissario and Etkin 2001; Craig 1999; Mills and Craig 1999; Moraru et al. 1999; Smoyer et al. 1999), there was already important background material and heightened interest within the community related to the challenges of climate change and sustainability. Specifically, these reports examined such topics as public health needs, municipal water systems and drought, heat-related mortality, and extreme weather events. The work under the present research program was intended to extend and update these baseline studies with the most up-to-date climate information, tools, and techniques available, using a participatory-based approach. In addition, there was an opportunity to examine these issues with a more holistic exploratory framework, in this instance ACRM.

The region is also undergoing structural economic change as it transitions from traditional manufacturing into other areas, such as green energy development and tourism. At the same time there is a community planning focus to strengthen the existing economic base in wineries, tender fruit, and agriculture (Niagara Economic Development Corporation 2009). It is also an area with unique biophysical characteristics, including the iconic Niagara Falls, as well as being part of the Niagara Escarpment Biosphere Reserve (Dempster 2004).

In terms of climate change, the region is expected to experience general warming in all seasons into the 2050s, as well as increased precipitation, particularly in the winter months (May and Gafarova 2010). Research done as part of the project also indicates that into the 2050s, there are expected to be more extreme hot days during the summer months, more precipitation falling as rain instead of snow during the winter season, and increased risk for vector-borne diseases such as Lyme disease (Fenech and Shaw 2010).

The region had also initiated a comprehensive review of its sustainability plan and provided a window of opportunity to examine the impacts of climate change as part of this larger effort. As such, there was potential in the study to examine the usefulness of SAM concepts, as set out by Bizikova et al. (2008) within the ACRM framework. SAM was envisioned as one of the key research areas, along with social learning and adaptive capacity indicator development. ACRM provided a useful way to advance the community-based research initiative mentioned above as well as inform region decision making, using already familiar risk management techniques.

The region had also completed its first carbon audit (Niagara Region 2010a), and planning staff were tasked with developing a broader climate change action plan where the

Project will provide much of the "planning framework" in what would normally be required in a plan for climate change adaptation. In October 2008 Regional Council endorsed the Climate Change Action Work Program, which among other things, included a component on assessing vulnerabilities resulting from potential climate change. The Brock-EC Project will go beyond assessing vulnerabilities, as it intends to engage those key actors in the community that might be affected. In this regard, the Brock-EC Project fits nicely with the Council endorsed work program and broader Business Plan Objectives (Niagara Region 2010b, 1).

In response to this interest, a science plan was developed to guide research activities that focused on four subelements:

1. SEI, social learning and their role in climate change adaptation
2. evaluation of the effectiveness of ACRM as an overarching process
3. the usefulness of SAM as an integrative tool for sustainability planning
4. CCAL and adaptive capacity indicators (Gafarova et al. 2010)

In November 2009, an experts group was convened by Brock University and Environment Canada to brainstorm various ideas surrounding a climate change adaptation project for the region. Representatives from the University of British Columbia,

International Institute for Sustainable Development, York University, the University of Waterloo, Wilfrid Laurier University, the Stockholm Environment Institute, and Six Nations First Nation Eco-Centre met to explore and help define the research plan. Prior to engagement of actors in a participatory process, an academic white paper was prepared (Armitage and Plummer 2010b), a socioeconomic profile for the region and compendium of adaptive capacity indicators developed (Khan 2009), and a survey of local government activities conducted (Crawford-Boettcher 2009). In addition, and germane to this chapter, a detailed SEI (Velaniškis 2009) provided the main insight into how best to structure the proposed participatory process from the perspective of CCAL.

As a result of the SEI, thirty-eight key informants from thirty-three organizations across nine sectors were interviewed. A series of workshops were held to explore climate change and its relevance to both the Niagara Region as the study area in general and their particular institutional spheres of influence more specifically. These actors represented regional and municipal government, education, media, emergency management, ecological management, health care, business, nongovernment organizations, and agriculture (Pickering et al. 2011). The top five key practice areas of stakeholders involved policy development and sustainability planning, education, water protection, carbon management, and environmental regeneration (Velaniškis 2010a). Water was identified as a key cross-cutting resource management challenge, as was responding to emergencies and public health (Velaniškis 2010a).

The first meeting of the group was a meet-and-greet style of gathering where there was an overall presentation of the project by the principal investigators and the research team was introduced. This was followed by a second workshop where the SEI results were presented and climate scientists presented their preliminary findings on historical trends and future scenarios using a standardized technique – the Rapid Assessment to the Impacts of Climate Change (RAICC) (Fenech and MacLellan 2007). Researchers from Vrije Universiteit, Amsterdam, the Netherlands, explained and administered a mind mapping exercise to participants (Haug et al. 2011; Velaniškis 2010b). In addition, a detailed exercise was administered to develop a baseline from which to measure social learning and affirm the observations in the initial survey by Crawford-Boettcher (2009).

By the third meeting, participants were presented with the results of the second survey and given a chance to share their experiences with climate change in the region. This meeting signified a transition to a less formal, more interactive process. The informants participated in an exercise designed to explore potential directions for the Niagara study after a guest speaker presented a number of similar climate change adaptation initiatives in other municipalities in Ontario. One of the key results was a desire to share information about community adaptation through an electronic portal. This led to the creation of the Niagara Climate Change Network (NCCN) (Pickering et al. 2011).

Subsequent to this, a core group of eighteen actors continued to meet and provide a focus for the broader network. A number of specific outcomes were achieved, including a decision to formalize a group called the NCCN, a recognition of the need for a charter that other potential members within the region could read and sign on to, and a white paper to further examine vulnerabilities and adaptive strategies within the region, all ultimately leading to a community climate change action plan that contained both adaptation and mitigation components (Pickering et al. 2011). These initiatives can be considered the boundary objects of the process, after Lynch et al. (2008). In September 2012, the steering committee made a decision to continue working as an informal network, supporting ongoing initiatives, such as the regional climate action plan.

Reflections on the Importance of the Linkages between Social-Ecological Inventories and Climate Change Adaptation Leadership

One of the key observations of the SEI relevant to this discussion relates to informants' perception of adaptive capacity for climate change. As a result of the interviews and surveys, it was found that "conditions necessary for achieving full capacity included [the need for] more collaborative approaches to planning for adaptation, adequate information about potential impacts, political will, and strong leadership" (Velaniškis 2010a, 20). Forty percent of interviewees identified that there was conditional capacity to adapt to climate change. This relates directly to all three of the barriers (understanding, planning, and managing) identified by Moser and Ekstrom (2010). From Dengler's (2007) perspective, this would suggest a signal from both local and policy power spaces for not only information on scientific knowledge but also a path forward to navigate the intersection of all three power spaces for action.

A second element from the SEI was the perception that, while two organizations were identified as playing a leadership role for adaptation in Niagara (the regional government and local Conservation Authority), 25 percent of interviewees cited that leadership was "absent" (Velaniškis 2010a). As the NCCN has developed, as described in the previous section, a core group of actors has taken up the challenge and moved on to develop a charter and took initial information provided by Environment Canada and Brock University (the science knowledge space) to further specify areas of risk for development of adaptation options and engage other sources of information to inform their climate change action plan. In this, the region played a key role. The SEI was useful in that it allowed for a process to develop where key informants assumed roles as actors and facilitated their own leadership development opportunities (Baird et al. 2014).

If we return to the leadership questions posed earlier, the overall goal of the Niagara Region study was to define, through the use of SEIs, the answers to the following questions.

Lead Adaptation to What Ends?

First, initial work by the research team in the first series of workshops identified a number of potential areas for adaptation focus. These included considerations of how extreme weather events would change under a changing climate, how public health agencies would be impacted, what adaptation efforts were possible in terms of protected areas and biodiversity, adaptation solutions for agriculture and viticulture, and adaption options for the tourism sector (Fenech and Shaw 2010; Chynoweth et al. 2011).

Second, the wording in the charter developed emphasized the need for overarching commitments related to reducing greenhouse gas emissions, building and enhancing capacity for adaptation, creating a web of solid relationships to implement best practices, and improving knowledge about climate change and its impact on the region through education and outreach (Niagara Climate Change Network, 2011). In addition, there was a desire to be more inclusive and reach out to groups not initially part of the process – who was missing?

Finally, this was examined via a sponsored white paper that recommended substantive actions to address identified gaps: leadership in climate change adaptation planning; coordination and sharing of weather data, trends, and climate projections; awareness and engagement activities; vulnerability assessments; infrastructure risk assessments; an inventory of existing adaptation actions and gaps; implementing adaptation; and monitoring impacts and adaptation (Penney 2012). While some of the white paper issues identified were the same as in the original scoping research, it does indicate a transition from the *realism-universalism* to the *realism-contextualism* quadrant of Kløcker Larsen et al.'s (2012) analytical framework.

Who (Individual) or What (Institution) Leads Adaptation (and at What Scale)?

The original goals related to the Niagara project were identified by the research team at a regional scale and supported by the regional government. The science plan was intended to contribute to an understanding of adaptation processes and was developed as a result of specific questioning in the areas of social learning, risk management, indicator development, and sustainability (Kløcker Larsen et al. 2012). The strategic direction given to region integrated planning staff by the regional council was also a driver (Niagara Region 2010a). The members of the NCCN developed as a group to move forward on adaptation. They identified a need for an electronic portal to share ideas. A smaller steering committee of committed informant-turned-actors composed a core individual leadership group driving change. They did so at a regional scale, but not in isolation. They actively sought out input from a broader network of contacts, for example, provincial agencies and other early adopters, federal atmospheric scientists, academics, and international organizations involved in community adaptation and social learning.

In addition, one of the SEI informant nongovernmental organizations emerged as an important bridging organization. The Niagara Sustainability Initiative (NSI) had already been formed through the provision of provincial-level funding to a local postsecondary institution, Niagara College. Their purpose was to engage the business community in sustainability, carbon emissions auditing, and management best practices. Their role on the NCCN Steering Committee aided in the development of the NCCN's charter and ensured representation of the local power space (Dengler 2007).

The focus of action on specific themes and adaptation projects among the group was uncertain, other than a desire to approach community climate change adaptation on the basis of social learning. The network of key actors explored initiatives that further defined the ends of the exercise, focused on a concrete climate action plan, an outreach component for those actors in the broader initial network identified through the SEI, as well as any others not a part of the network.

How Does Adaptation Leadership Occur?

Adaptation leadership evolved through the series of meetings and workshops involving the NCCN. Initially, participants wanted the research team to maintain the leadership role. While there was some frustration expressed over the process not occurring fast enough for some members, there was no clear assumption of leadership by the group. This was for a number of reasons, for example, time commitment involved, eventual outcome of the process, and clear authority for decision making. Eventually, certain actors stepped forward to assume responsibility for the process from the research facilitation team. The research team identified a project manager to provide a conduit and secretariat for support to the steering committee and broader group. As scientific knowledge needs arose, those requirements were relayed from the network to the research team via the secretariat. A key component of this was the ability of actors to develop a view of the problem as being broader than individual organizational interests.

Creating a vision where participant organizations can see where they fit into the process was important. A great deal of time was spent at the beginning of the study explaining the role of the research team as a source of knowledge and expertise for local decision making. The process was for the participants to control and lead (direct) the process, not the researchers. One member of the research team served as a project manager for the network, providing logistical support and being a conduit to identify any knowledge gaps that the researchers could help with. The charter that was developed by the NCCN clearly focused the effort of the group on a commitment to take action to reduce emissions, build capacity for adaptation, create a web of relationships, and improve knowledge through education and outreach (Niagara Climate Change Network 2011).

Another important result of the group's success in managing the process was a degree of embeddedness that occurred as a result of the regional government's ongoing sustainability activities. One pillar of the research plan was to explore the ideas developed for SAM. In their second phase of the Sustainable Niagara initiative, the region identified a number of indicators of progress, one of which was action on greenhouse gas emissions; the NCCN was specifically identified in the summary document as a possible champion group within the region (Niagara Region 2011, 19). It was intended that the group would go on to take ownership (further clarifying and monitoring) of the specific sustainability indicator. Within the report opportunities exist to contribute in some of the other sustainability areas, for example, indicator species abundance, economic diversification, participation in community activities, and volunteerism (Niagara Region 2011). Going forward, the NCCN has an emerging credibility that is reflected in official government policy documents.

What Constitutes Effective Adaptation Leadership?

Ongoing research will be able to shed light on evaluative criteria such as the degree to which social learning has occurred, the strength of developing social networks through social network analysis, the degree of inclusiveness of the network, and the number of collaborative efforts that result from the project. By far the most challenging aspect faced by the group was the time spent in the creation of a vision and sense of what adaptation means to the community and how it should be integrated into other initiatives. A sense of urgency for decision making, commonly identified as a key ingredient for transformative action (Anderies et al. 2006), was created more through process than an external driver. There was some direction from local politicians through council resolution to develop a climate action plan and the intersection of interests with university and government researchers. There was no presence of an actual, imminent threat, recent iconic extreme, or similar climate-related loss to galvanize action.

Effective adaptation leadership will be ultimately reflected in the NCCN's ability to influence not only adaptation policy making but also concrete action on the part of members in line with the developed charter and its recognition as an example of adaptation best practice for other communities. This will continue to occur through the now informal commitment of "organizations, businesses, community groups, and local residents working together" (Niagara Climate Change Network 2011).

Conclusion

Community climate change adaptation processes evolve over time. As conventional risk management, as envisioned by ACRM, suggests, *establishing the context* is extremely important (May and Plummer 2011). SEIs are an important aspect of identifying key informants and bridging organizations that can influence adaptation outcomes. While realizing that there are specific barriers to adaptation (Moser and Ekstrom 2010) and that actors operate within definite spheres of power for action

(Dengler 2007), involving the right mix of stakeholders and interests from the beginning is an important ingredient for success. The technique is flexible enough to revisit group composition over time. In addition, the presence of suitable boundary objects is necessary to create an environment for common understanding (Lynch et al. 2008).

Having individuals – super-agents, scale-crossing brokers, opinion leaders or policy entrepreneurs who can navigate the varied demands and expectations of knowledge, policy, and local knowledge spheres – is key to defining CCAL. Furthermore, these activities also have to acknowledge that informants, actors, and agents need to either possess leadership skills required to engage in collaborative decision making or allow for enough time that these skills develop during the participatory process. As these types of action research activities develop over time, different actors and agents can assume leadership at any given time, to meet the demands of the group. In this sense, reflecting on CCAL is an important ingredient to the adaptation process. Asking questions such as those identified in this chapter will contribute to the robustness of research activities and sound research design.

In the Niagara Region study, an SEI approach was used to frame a case study that had multiple dimensions (Baird et al. 2014). This entry point was then used to work with communities to strategize adaptation options and assess the development of ongoing, emergent CCAL. Whereas the initial SEI results indicated the perceptions of a lack of leadership for climate change adaptation in the region, the resultant process was able to indicate that this was not necessarily so. "When case studies acquire political momentum and become conduits for local actors to access decision makers, it is because the collaboration between researchers and local communities has facilitated the creation of such a platform" (Kløcker Larsen et al. 2011, 29). Acknowledging, examining, and mapping this collaboration over time is an important aspect of understanding how to more effectively design and facilitate approaches for communities faced with responding to the challenges of climate change.

Acknowledgments

Ryan Plummer and Derek Armitage have been strong supporters in this project from the beginning and have provided invaluable guidance. The Niagara Climate Change Network is the result of the dedicated contributions of all stakeholders that have participated and are participating in the process. Thanks also to the research team whose work is highlighted here, in particular Jonas Velaniškis and Kerrie Pickering. Don MacIver, Adam Fenech, and Neil Comer, formerly of Environment Canada, have assisted in supporting research into how communities can effectively adapt to a changing climate. Thanks also to colleagues at the Environmental Change and Governance Group at the University of Waterloo and Wilfrid Laurier University, Waterloo, Canada, for their insights.

References

Anderies, J. M., Walker, B. H. and A. P. Kinzig. 2006. "Fifteen weddings and a funeral: case studies and resilience-based management." *Ecology and Society* 11 (1): 21. http://www.ecologyandsociety.org/vol11/iss1/art21/.

Armitage, D. and R. Plummer. 2010a. "Integrating perspectives on adaptive capacity and environmental governance." In *Adaptive Capacity and Environmental Governance*, edited by Armitage, D. and R. Plummer, 1–19. Berlin: Springer.

Armitage, D. and R. Plummer. 2010b. *Adaptation, Learning and Transformation in Theory and Practice: A State-of-the-Art Literature Review and Strategy for Application in Niagara*. Final report, Contract KM 170-09-1257, submitted to Adaptation and Impacts Research Section – Environment Canada, Toronto.

Baird, J., Plummer, R. and K. Pickering. 2014. "Priming the governance system for climate change adaptation: The Application of a social-ecological inventory to engage actors in Niagara, Canada." *Ecology and Society* 19 (1): 3. doi:10.5751/ES-06152-190103.

Bellissario, L. M. and D. Etkin. 2001. *Climate Change and Extreme Weather Events in the Toronto-Niagara Region*. Report and Working Paper Series. Report 2001-2. Toronto-Niagara Study on Atmospheric Change. Environment Canada, Toronto, 34 pp.

Berkes, F. and D. Jolly. 2001. "Adapting to climate change: social-ecological resilience in a Canadian western Arctic community." *Conservation Ecology* 5 (2): 18. http://www.consecol.org/vol5/iss2/art18/.

Bizikova, L. 2009. *Challenges and Lessons Learned from Integrated Landscape Management (ILM) Projects*. International Institute for Sustainable Development. http://www.iisd.org/publications/pub.aspx?id=1109.

Bizikova, L., Burch, S., Cohen, S. and J. Robinson. 2009. "A participatory integrated assessment approach to local climate change responses: Linking sustainable development with climate change adaptation and mitigation." In *Climate Change, Ethics, and Human Security*, edited by O'Brien, K. L., St. Clair, A. L. and B. Kristoffersen, 157–79. Cambridge: Cambridge University Press.

Bizikova, L., Neale, T. and I. Burton. 2008. *Canadian Communities' Guidebook for Adaptation to Climate Change. Including an Approach to Generate Mitigation Co-benefits in the Context of Sustainable Development*. 1st ed. In association with and published by Environment Canada and University of British Columbia, Vancouver.

British Columbia. 2007. *Climate Action Plan*. Province of British Columbia, Canada, Victoria.

Brock University, Map Library. n.d. *Niagara Region Municipal Boundaries*. http://www.brocku.ca/maplibrary/maps/outline/local/stcathDT.jpg.

Bruce, J. P., Egener, I. D. M. and D. Noble. 2006. *Adapting to Climate Change: A Risk-Based Guide for Ontario Municipalities*. Final report submitted to Climate Change Impacts and Adaptation Programme – Natural Resources Canada, Ottawa. http://adaptation.nrcan.gc.ca/projdb/pdf/176a_e.pdf.

Burton, I., Bizikova, L., Dickinson, T. and Y. Howard. 2007. "Integrating adaptation into policy: upscaling evidence from local to global." *Climate Policy* 7 (4): 371–76.

Cave, K., General, P., Johnston, J., May, B., McGregor, D., Plummer, R. and P. Wilson. 2011. "The power of participatory dialogue: Why talking about climate change matters." *Indigenous Policy Journal* 22 (13): 13. http://indigenouspolicy.org/Articles/VolXXIINo1/ThePowerofParticipatoryDialogue/tabid/225/Default.aspx.

Chynoweth, A., May, B., Fenech, A., MacLellan, J. I., Comer, N. and D. MacIver. 2011. "Tools for integrating climate change and community disaster management in Canada." In *Practical Solutions for a Warming World: American Meteorological Society Conference on Climate Adaptation*. http://ams.confex.com/ams/19Applied/webprogram/Paper190059.html.

Cohen, S., Neilsen, D., Smith, S., Neale, T., Taylor, B., Barton, M., Merritt, W., et al. 2006. "Learning with local help: Expanding the dialogue on climate change and water management in the Okanagan Region, British Columbia, Canada." *Climatic Change* 75 (3): 331–58.

Collinson, D. 2006. "Rethinking followership: A post-structuralist analysis of follower identities." *The Leadership Quarterly* 17 (2): 179–89.

Cook, I. 2009. *Book Reviews for Managers. Leadership Development and Teambuilding. The Practice of Adaptive Leadership: Tools and Tactics for Changing Your Organization and the World.* Fulcrum Associates. http://www.888fulcrum.com/.

Craig, L. 1999. *Atmospheric Change in the Toronto-Niagara Region – An Assessment of Public Health Department Activities and Needs.* Report and Working Paper Series 99-2. Toronto: Toronto-Niagara Study on Atmospheric Change, Environment Canada.

Crawford-Boettcher, E. 2009. *Bridging Science and Policy for Community Climate Change Adaptation: Is Climate Science "Useable" for Local Practitioners?* Unpublished report submitted to Adaptation and Impacts Research Division – Environment Canada, Toronto.

Crona, B. and Ö. Bodin. 2010. "Power asymmetries in small-scale fisheries: A barrier to governance transformability?" *Ecology and Society* 15 (4): 32. http://www.ecologyandsociety.org/vol15/iss4/art32/.

Dawson Adaptation Project Team. 2009. *Dawson Climate Change Adaptation Plan.* City of Dawson. Northern Climate ExChange, Whitehorse.

Dempster, B. 2004. "Canadian Biosphere Reserves: Idealizations and realizations." *Environments* 32 (3): 95–101.

Dengler, M. 2007. "Spaces of power for action: Governance of the Everglades Restudy process (1992–2000)." *Political Geography* 26 (4): 423–54.

Einarsen, S., Aasland, M. S. and A. Skogstad. 2007. "Destructive leadership behaviour: A definition and conceptual model." *The Leadership Quarterly* 18 (3): 207–16.

Environment Canada. 2009. *Environment Canada's Adaptation and Impacts Science Plan.* Adaptation and Impacts Research Section – Environment Canada, Toronto, 27 pp.

Ernstson, H., Barthel, S., Andersson, E. and S. T. Borgström. 2010. "Scale-crossing brokers and network governance of urban ecosystem services: The case of Stockholm." *Ecology and Society* 15 (4): 28. http://www.ecologyandsociety.org/vol15/iss4/art28/.

Fenech, A. and J. MacLellan. 2007. *Rapid Assessment of the Impacts of Climate Change: Building Past and Future Histories of Climate Extremes: Linking Climate Models to Policy and Decision-Making.* Final report submitted to Adaptation and Impacts Research Section – Environment Canada, Toronto.

Fenech, A. and T. Shaw. 2010. *Niagara Region's Changing Climate: Preliminary Results.* Presentation to the Niagara Climate Change Network, St. Catharines, Ontario.

Ford, J., Pearce, B., Smit, B., Wandel, J., Allurut, M., Shappa, K., Ittusujurat, H. and K. Qrunnut. 2007. "Reducing vulnerability to climate change in the Arctic: The case of Nunavut, Canada." *Arctic* 60 (2): 150–66.

Gafarova, S., May, B. and R. Plummer. 2010. *Adaptive Collaborative Risk Management and Climate Change in the Niagara Region: A Participatory Integrated Assessment Approach for Sustainable Solutions and Transformative Change.* Adaptation and Impacts Research Section – Environment Canada, Toronto.

Galaz, V., Moberg, F., Olsson, E.-K., Paglia, E. and C. Parker. 2011. "Institutional and political leadership dimensions of cascading ecological crises." *Public Administration* 89 (2): 361–80.

Government of Ontario. 2009. *Adapting to Climate Change in Ontario: Toward the Design and Implementation of a Strategy and Action Plan.* Final report, by the Expert Panel on Climate Change Adaptation, submitted to the Minister of the Environment, Toronto.

Gupta, J., Termeer, C., Klostermann, J., Meijerink, S., van den Brink, M., Jong, P., Nooteboom, S. and E. Bergsma. 2010. "The Adaptive Capacity Wheel: A method to assess the inherent characteristics of institutions to enable the adaptive capacity of society." *Environmental Science and Policy* 13 (6): 459–71.

Hallegatte, S. 2009. "Strategies to adapt to an uncertain climate change." *Global Environmental Change* 19 (2): 240–47.

Hanna, K. S. 2007. "Implementation in a complex setting: Integrated environmental planning in the Fraser River Estuary." In *Integrated Resource and Environmental Management*, edited by Hanna, S. and D. S. Slocombe, 119–36. Oxford: Oxford University Press.

Haug, C., Hultema, D. and I. Wenzler. 2011. "Learning through games? Evaluating the learning effect of a policy exercise on European climate policy." *Technological Forecasting and Social Change* 78 (6): 968–81.

Heifetz, R. 1998. *Leadership without Easy Answers*. Cambridge, MA: Harvard University Press.

Heifetz, R., Grashow, A. and M. Linsky. 2009. *The Practice of Adaptive Leadership: Tools and Tactics for Changing Your Organization and the World*. Boston, MA: Harvard Business Press.

Huitema, D. and S. Meijerink. 2010. "Realizing water transitions: The role of policy entrepreneurs in water policy change." *Ecology and Society* 15 (2): 26. http://www.ecologyandsociety.org/vol15/iss2/art26/.

Khan, Z. A. 2009. *The Socio-economic Profile of the Niagara Region*. Final report submitted to Adaptation and Impacts Research Section – Environment Canada, Toronto.

Kløcker Larsen, R., Swartling, Å. G., Powell, N., May, B., Plummer, R., Simonsson, L. and M. Osbeck. 2012. "A framework for facilitating dialogue between policy planners and local climate change adaptation professionals: Cases from Sweden, Canada and Indonesia." *Environmental Science and Policy* 23: 12–23.

Kløcker Larsen, R., Swartling, Å. G., Powell, N., Simonsson, L. and M. Osbeck. 2011. *A Framework for Dialogue Between Local Climate Adaptation Professionals and Policy Makers Lessons from Case Studies in Sweden, Canada and Indonesia*. Research report submitted to Stockholm Environment Institute, Stockholm. http://www.sei-international.org/publications?pid=1882.

Kouzes, J. M. and B. Posner. 2007. *The Leadership Challenge*. 4th ed. San Francisco: John Wiley.

Lemmen, D. S., Warren, F. J. and J. Lacroix. 2008. "Synthesis." In *From Impacts to Adaptation: Canada in a Changing Climate 2007*, edited by Lemmen, D. S., Warren, F. J., Lacroix, J. and E. Bush, 1–20. Ottawa: Government of Canada.

Lynch, A. H., Tryhorn, L. and R. Abramson. 2008. "Working at the boundary: Facilitating interdisciplinarity in climate change adaptation research." *Bulletin of the American Meteorological Society* 89 (2): 169–79.

May, B. and S. Gafarova. 2010. *Climate Change Scenarios for the Niagara Region, Using Information from the Canadian Climate Change Scenarios Network*. Edited by N. Comer. Ensemble Scenarios for Canada. Adaptation and Impacts Research Section – Environment Canada, Toronto. http://www.cccsn.ca/.

May, B. and R. Plummer. 2011. "Accommodating the challenges of climate change adaptation and governance in conventional risk management: Adaptive collaborative risk management (ACRM)." *Ecology and Society* 16 (1): 47. http://www.ecologyandsociety.org/vol16/iss1/art47/.

Meijerink, S. and S. Stiller. 2013. "What kind of leadership do we need for climate adaptation? A framework for analyzing leadership functions and tasks in climate change adaptation." *Environment and Planning C* 32 (2): 240.

Mills, B. and L. Craig, eds. 1999. *Atmospheric Change in the Toronto-Niagara Region: Towards an Integrated Understanding of Science, Impacts and Response*. Proceedings of a workshop. University of Toronto, May 27–28.

Mitchell, R. 2011. *Social-Ecological Inventories: Building Resilience to Environmental Change within Biosphere Reserves*. Resilience Alliance. http://www.resalliance.org/cdirs/raprojects/index.php/11.

Moraru, L., Kreutzwiser, R. and R. de Loë. 1999. *Sensitivity of Municipal Water Supply and WastewaterTreatment Systems to Drought*. Report and Working Paper Series 99-3. Toronto-Niagara Study on Atmospheric Change, Environment Canada, Toronto.

Moser, S. C. and J. A. Ekstrom. 2010. "A framework to diagnose barriers to climate change adaptation." *PNAS* 107 (51): 22026–31. http://www.pnas.org/cgi/doi/10.1073/pnas.1007887107.

Niagara Climate Change Network. 2011. *Niagara's Climate Change Charter*. Ontario: Niagara Region.

Niagara Economic Development Corporation. 2009. *Annual Report 2009*. Ontario: Niagara Region.

Niagara Region. 2010a. *Niagara Region 2006 Corporate GHG Emissions Inventory*. Final report submitted to Niagara Regional Council, Thorold, Ontario.

Niagara Region. 2010b. *Adapting to Climate Change in the Niagara Region: An Update on a Developing Partnership between the Region, Brock University, and Environment Canada*. Final report submitted to Council – Integrated Community Planning, Thorold, Ontario.

Niagara Region. 2011. *Sustainable Niagara Phase 2: Measuring Progress*. Sustainable Niagara – Niagara Region, Thorold, Ontario.

Nye, J. S., Jr. 2010. "Power and leadership." In *Handbook of Leadership Theory and Practice: A Harvard Business School Centennial Colloquium*, edited by Nohria, N. and R. Khurana, 305–32. Boston, MA: Harvard Business Press.

Penney, J. 2012. *Adapting to Climate Change: Challenges for Niagara*. WaterSmart Niagara. Ontario: Niagara Region.

Pfeffer, J. 2000. "The ambiguity of leadership." In *ASHE Reader on Organization and Governance in Higher Education*, edited by C. M. Brown, 205–13. Upper Saddle River, NJ: Prentice Hall.

Pickering, K., Plummer, R. and B. May. 2011. "Niagara Climate Change Project: A collaborative participatory approach to climate change adaptation." In *Practical Solutions for a Warming World: American Meteorological Society Conference on Climate Adaptation*. http://ams.confex.com/ams/19Applied/webprogram/Paper190532.html.

Plummer, R., General, P., Cave, K. and B. May. 2009. *A Transboundary Dialogue on Climate Change and Water in the Great Lakes Basin: Exploring the Vulnerability and Adaptive Capacity of Aboriginal Peoples*. Indigenous Adaptation Network. http://www.indigenousadaptationnetwork.com/.

Podolny, J. M., Khurana, R. and M. L. Besharov. 2010. "Revisiting the meaning of leadership." In *Handbook of Leadership Theory and Practice: A Harvard Business School Centennial Colloquium*, edited by Nohira, N. and R. Khurana, 65–106. Boston, MA: Harvard Business Press.

Raik, D. B., Wilson, A. L. and D. J. Decker. 2008. "Power in natural resources management: An application of theory." *Society and Natural Resources* 21 (8): 729–39.

Ribot, J. 2011. "Vulnerability before adaptation: Toward transformative climate action." *Global Environmental Change* 21 (4): 1160–62.

Schultz, L., Folke, C. and P. Olsson. 2007. "Enhancing ecosystem management through social-ecological inventories: Lessons learned from Kristianstads Vattenrike, Sweden." *Environmental Conservation* 34 (2): 140–52.

Schultz, L., Plummer, R. and S. Purdy. 2011. *Applying a Social-Ecological Inventory: A Workbook for Finding the Key Actors and Engaging Them*. Final report submitted to the Resilience Alliance – Stockholm. http://www.resalliance.org/index.php/resilience_assessment.

Shepherd, P., Tansey, J. and H. Dowlatabadi. 2006. "Context matters: What shapes adaptation to water stress in the Okanagan?" *Climatic Change* 78 (1): 31–62.

Smit, B., Burton, I., Klein, R. J. T. and J. Wandel. 2009. "An anatomy of adaptation to climate change and variability." In *The Earthscan Reader on Adaptation to Climate Change*, edited by Schipper, E. L. F. and I. Burton, 63–88. London: Earthscan.

Smoyer, K. E., Rainham, D. G. C. and J. N. Hewkoet. 1999. *Integrated Analysis of Heat-Related Mortality in the Toronto-Windsor Corridor*. Report and Working Paper Series 99-1. Toronto-Niagara Study on Atmospheric Change. Toronto: Environment Canada.

Swart, R., Robinson, J. and S. Cohen. 2003. "Climate change and sustainable development: Expanding the options." *Climate Policy* 3 (S1): S19–40.

Uhl-Bien, M., Marion, R. and B. McElvey. 2007. "Complexity leadership theory: Shifting leadership from the industrial age to the knowledge era." *The Leadership Quarterly* 18 (4): 298–318.

Vasseur, L. 2010. "Championing climate change adaptation at the community level by using an ecosystem approach: An example from Greater Sudbury in Ontario, Canada." In *Building Resilience to Climate Change: Ecosystem-Based Adaptation and Lessons from the Field*, edited by Pérez, Á. A., Fernández, B. H. and R. C. Gatti, 150–59. Switzerland: International Union for Conservation of Nature (IUCN).

Velaniškis, J. 2009. *Proposed Strategies for Social-Ecological Boundary Identification and Stakeholder Engagement for Climate Change Adaptation and Collaborative Risk Management in the Niagara Region*. Draft report submitted to Adaptation and Impacts Research Section – Environment Canada, Toronto.

Velaniškis, J. 2010a. *Report on the Social-Ecological Network Characteristics and Stakeholder Perceptions about Climate Change in the Niagara Region*. Final report submitted to Adaptation and Impacts Research Section – Environment Canada, Toronto.

Velaniškis, J. 2010b. *Report on the Methodology for Coding and Analyzing Concept Maps*. Final report submitted to Adaptation and Impacts Research Section – Environment Canada, Toronto.

Westley, F., Olsson, P., Folke, C., Homer-Dixon, T., Vredenburg, H., Loorbach, D., Thompson, J., et al. 2011. "Tipping toward sustainability: Emerging pathways of transformation." *AMBIO* 40 (7): 762–80.

Wilson, C. and T. McDaniels. 2007. "Structured decision-making to link climate change and sustainable development." *Climate Policy* 7 (4): 353–70.

14

Participation and Learning for Climate Change Adaptation

A Case Study of the Swedish Forestry Sector

ÅSA GERGER SWARTLING, OSKAR WALLGREN, RICHARD J. T. KLEIN,
JOHANNA ULMANEN, AND MAJA DAHLIN

Researchers and decision makers have begun to recognize the need to build capacity for adaptation to climate change at all relevant levels in society. Over the last decade notions of adaptation and adaptive capacity have increasingly been a core theme in a diverse body of environment and resource literature as well as applied research domains such as risks and hazards, climate change, political ecology, social-ecological systems, and resilience (Plummer and Armitage 2010). The term *adaptive capacity* has been introduced to capture the ability or potential of a system to respond successfully to climate variability and change (Adger et al. 2007) and can thus be defined as the ability of stakeholders to plan, prepare for, facilitate, and implement adaptation options. Adaptation could include adjustments in behavior as well as changes in the management of resources and the use of technologies. Factors that determine adaptive capacity to climate change include economic wealth, technology and infrastructure, information, knowledge and skills, institutions, equity, and social capital (Smit et al. 2001).

In addition to highlighting the usefulness of the concept for understanding the adaptation process, scholarly work has also identified (1) gaps between the theoretical existence of adaptive capacity and the actual response to climate variability and change (O'Brien et al. 2006); (2) uncertainties in assessing adaptive capacity, in particular across different scales of analysis (Vincent 2007); and (3) the uneven distribution of adaptive capacity among and within countries, societies, and localities, as well as across age, class, gender, health, and social status (Adger et al. 2007).

A growing body of literature investigates how adaptive capacity manifests in specific localities and resource systems (Armitage 2005; Brooks et al. 2005; Keskitalo et al. 2010, Nadine et al. 2013), and despite recent analyses of the links between learning, adaptive capacity, and governance (Armitage et al. 2010, Berkhout et al. 2006; Diduck 2010; Pelling et al. 2008; Tschakert and Dietrich 2010), there has been surprisingly little research on how learning could benefit adaptation decision making and shape resilient livelihoods (Tschakert and Dietrich 2010).

This chapter focuses on the role of learning in advancing climate change adaptation. The main research questions are, How can participatory processes contribute to learning on climate change and adaptation? What characterizes this learning process and how, if at all, does it contribute to a self-perceived increase in the capacity to adapt to climate change? The chapter uses a case study of the Swedish forestry sector to address these questions. It presents an empirical analysis of data obtained from twelve focus groups, a joint stakeholder workshop, and follow-up interviews. The analysis then provides the basis for a discussion of the potential for participatory processes and learning to build adaptive capacity to climate change.

Perspectives on Participation and Learning

Several determinants of adaptive capacity identified in the wider literature, such as social capital, human capital, and knowledge exchange, are largely a product of the outcomes of interactions and relationships between stakeholders (Mendis et al. 2003). Social capital has been defined as "connections among individuals – social networks and the norms of reciprocity and trustworthiness that arise from them" (Putnam 2000, 19) and relates to dimensions of culture, community cooperation, equity, institutions, and institutional linkages (Mendis et al. 2003). When trying to understand learning on climate change and impacts, we can thus expect insights to be gained from observing deliberative processes on climate change adaptation.

Stakeholder participation in environmental policy and management is often advocated for normative, substantive, and instrumental reasons (Fiorino 1990). The *normative* argument refers to the democratic right of interested and affected parties to be involved in decision making that concerns their lives. Participation is *substantive* because it draws on the diverse knowledge of participating groups that could improve the quality of decision making and produce socially robust science and policy. For example, local knowledge has been shown to be important for ensuring sustainability outcomes in ecosystem management (Berkes and Folke 1998; Olsson and Folke 2001), and local people may bring knowledge and experience relevant to the decision-making process to which scientists and policy makers typically do not have access (Irwin 1995; Wynne 1996). Stakeholder involvement in policymaking is *instrumental* because it is likely to foster trust, learning, engagement, and compliance (see also Pahl-Wostl 2006). Deliberative processes encourage observation, reflection, and opportunities for communication and persuasion among social groups in contexts of high uncertainty (Lee 1993, 1999).

There are also scientific arguments for engaging stakeholders in the research process. Science-based stakeholder dialogues can play an important role in (1) providing a reality check for the research, (2) identifying socially relevant and scientifically challenging research questions, (3) improving considerations of ethical issues in research, and (4) providing access to data and knowledge that otherwise would remain unknown or difficult to access (Welp et al. 2006).

Over the last decades, the learning dimension of resource management and environmental governance has attracted researchers from different perspectives and disciplines. The goal of learning may be justified for various reasons. In a comanagement process, participants may strive to learn for normative reasons; it may provide an opportunity to alter management practices that may prove more economically beneficial to resource users, and learning may also be a reaction to social-ecological change, where new management strategies, rules, or procedures are needed in the face of resource depletion and livelihood disruptions (Armitage et al. 2009).

Much scholarly work has focused on understanding the levels of learning in policy and management, ranging from simple learning (so-called single-loop learning), in which new knowledge is absorbed through error correction, through double-loop learning, in which underlying objectives, values, and norms are modified as new knowledge becomes available, through to more advanced, reflexive "learning to learn," triple-loop learning, which seeks to correct errors by addressing or designing governance protocols and norms (Argyris and Schön 1978, 1996; Armitage 2008; Fiorino 2001; Flood and Romm 1996; Keen et al. 2005; May and Plummer 2011). Related distinctions are technical, conceptual, and paradigmatic learning (Van de Kerkhof and Wieczorek 2005) and first-order, second-order, and third-order learning (Bennett and Howlett 1992).

Since the 1990s, research on learning has proliferated and expanded into diverse fields such as policy learning (Bennett and Howlett 1992; Fiorino 2001; Haas 2000; Sabatier 1993), organizational learning (Berkhout et al. 2006; Fitzpatrick 2006; Levitt and March 1988), social learning (Pahl-Wostl et al. 2007; Parson and Clark 1995; Steyaert and Jiggins 2007; Tabara et al. 2009; Wals 2007), collaborative learning (Allen et al 2001; Keen and Mahanty 2005), and transformative learning (Percy 2004; Tarnoczy 2011; Vulturius and Gerger Swartling 2015). In recent years, social learning has attracted particular attention, generating rich models of concerted action, collaborative inquiry, and learning for sustainability (Diduck 2010), and has become a normative goal and process in the environmental governance sphere. Social learning has been highlighted as particularly relevant for addressing the complexity of adaptation processes involving multiple actors on different scales and with different perspectives (Hinkel et al. 2010). In the context of environmental governance, there is growing interest in the role of social learning as a policy alternative for dealing with issues that are "wicked" in nature, meaning that there are no simple one-off solutions and that many different perspectives need to be considered (Blackmore et al. 2007; Rittel and Webber 1973). There is evidence that stakeholder participation facilitates and stimulates social and other forms of learning among stakeholders (Cundill 2010; Diduck 2004; Johannessen and Hahn 2013; Wibeck 2014).

Despite this widespread support for learning as a goal and process toward sustainability, concepts, perspectives, and approaches to learning remain vague and have sometimes been uncritically applied, without due attention given to the role of power and multilevel learning (Armitage et al. 2008). The recent literature on social learning calls for greater conceptual precision over its meaning and criteria involved (Gerger Swartling et al. 2011; Reed et al. 2010; Wals 2007). Reed et al. (2010) identify three key problems with the contemporary conceptualization of social learning. First, social learning as a concept is frequently confused with the conditions or methods necessary to facilitate social learning, such as stakeholder participation. Second, there is often confusion between the concept itself and the potential outcomes it may have. Third, there is little distinction made between individual and wider social learning. The authors suggest that, to be classified as social learning, a process must demonstrate a change in understanding in the individuals involved; that the change exceeds the individual "and becomes situated within wider social units of communities of practice and occur through social interactions and processes between actors within a social network" (Reed et al. 2010, 6).

In summary, recent research suggests that, when studying learning on climate change and adaptation, it is important not only to assess whether learning has taken place but also to identify the type of learning, and how it manifests in relevant processes and outcomes. Drawing on these theoretical observations, we apply the research questions to the case of climate change adaptation in the Swedish forestry sector.

Participation and Learning in Practice: The Case of Climate Change Adaptation in the Swedish Forestry Sector

Case Study Context and Method

Sweden is one of the world's largest exporters of pulp, paper, and sawn timber, and more than 50 percent of Sweden is covered by forests. The forest industry accounts for 10–12 percent of total employment, turnover, and added value within Swedish industry and for 11 percent of Sweden's exports (Swedish Forest Agency 2014). Swedish forest management is governed by the Forestry Act, which has two objectives with equal weight: sustainable production of timber and wood, and biodiversity conservation. Recently, recognition of the importance of forests for recreation, leisure, and other social purposes has increased. In addition, attention has been drawn to the global importance of forest management for climate change mitigation through the production of biofuels and the role of forest soils, standing forests, and forest products as carbon sinks.

Although Swedish forests are often seen as a common national resource, no less than 50 percent of the productive forest land is managed by 330,000 small-scale owners, typically individuals or family-owned farming operations with forest holdings. The remaining landownership is evenly split between major forest companies and the state. Management strategies and the underlying reasons for owning forest land can differ significantly from the major industrial and public forest owners. Within the research community and the forestry sector, there is a growing recognition that knowledge and

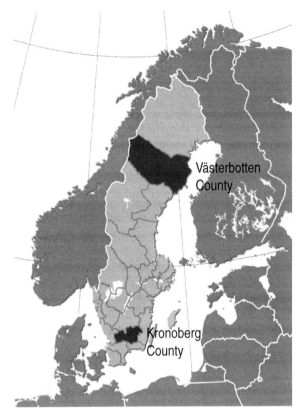

Figure 14.1. The case study regions of Västerbotten County and Kronoberg County. *Source*: SEI. *Credit*: Hugo Ahlenius, Nordpil, 2011.

expectations of climate change should influence decisions on forest management and reforestation.

Data and results reported here have been generated within a case study of the first phase of the research program Mistra-SWECIA (2008–2011; http://www.mistra-swecia.se). The study focuses on small-scale forest owners and forestry professionals and how they are affected by, deal with, or intend to manage extreme events that affect forests. The Swedish forestry specific data on climate change and effects and have been generated using the regional climate model RCA3_ECHAM5 and the ecosystem model LPJ-Guess and downscaled to the two case study regions (see Figure 14.1). The regional scenarios indicate higher temperatures and a longer growing season with likely increased growth in Swedish forests but also modified risk of damage caused by pathogens and pests, in particular spruce bark beetles (Jönsson et al. 2012; Jönsson and Bärring 2011). Climate change is also expected to affect the water balance and a number of other climate and weather-related conditions in Swedish forests. Indirectly, through reduced deep freezing and wetter winters, the ecosystem model results indicate increased vulnerability to storms (Lagergren et al.

Table 14.1. *Focus group themes, content, and participatory exercises*

	Meeting theme	Issues explored	Exercises
1.	"Orientation and exploration of climate change risks"	Perceived risks and challenges today and in the future General understanding of, and perspectives on climate change	Brainstorming Reflections on media clippings
2.	"Further exploration of climate change impacts and adaptation"	How adaptation is framed and what it means to adapt What information is required, how scientific knowledge is understood, how it should be communicated, and to whom	Open discussion and collage production/presentation Interactive presentations of climate change and impacts scenarios, feedback to scientists
3.	"Ways forward"	What are the barriers to and opportunities for adaptation within forestry Who is important for facilitation and implementation of climate adaptation and why, how	Ranking of, discussion on critical factors Actor analysis and dependencies
4.	"Joint stakeholder workshop"	What is required to get from today's situation to a desirable future General reflections, feedback	Time-line ranking with content discussion Break-out group session Presentation and discussion on outcomes of break-out group session in plenary

2012; Lagergren and Jönsson 2010). The empirical analysis draws on twelve focus group sessions that took place over a six-week period in the first half of 2010, a subsequent workshop, and a series of semistructured follow-up interviews held eight months after the workshop with the aim of assessing any longer-term learning. The discussion topics of the stakeholder meetings are summarized in Table 14.1. For a more detailed description, see André (2013).

The process involved 27 forest officials and forest owners, recruited from two geographical regions: Västerbotten County in the north and Kronoberg County in south (Figure 14.1) of Sweden. These regions reflect the diversity of biophysical conditions and variations in socio-economic characteristics relevant to Swedish forestry. The selection of participants was informed by a mapping and analysis of relevant stakeholders (André et al. 2012).

As shown in Table 14.2, thirteen participants were forest professionals operating in the two counties. Their professional interests included silviculture and management,

Table 14.2. *Summary of stakeholders participating in the study*

Stakeholders	Kronoberg	Västerbotten	Total
Private forest owners – small-scale private forest owners	8	6	14
Forest professionals – authorities, forest owners associations, forest and saw mill companies, etc.	7	6	13
Total	15	12	27

consultancy for private forest owners, and purchasing of timber. The remaining participants were private forest owners from the same regions, the majority of whom resided on or nearby their properties. Almost all of them were members of forest owner associations.

As illustrated in Table 14.1, the first meeting explored participants' perceptions of risk to which they may be exposed, without introducing specific information about climate change. In the second meeting, participants were provided with scientific information (scenarios of climate change, impacts on land use and hydrology) by invited experts and discussed the implications for forestry in general and their forest land in particular. The third session then focused on participants' expectations of future adaptation needs and activities within Swedish forestry. In the subsequent workshop, participants from all four groups were brought together to share experiences and perspectives on climate adaptation efforts and to propose future adaptation action within Swedish forestry.

The case study has generated data in the form of transcripts, questionnaire responses, and stakeholder-produced material (see further André 2013; André et al. 2012). The data analysis included the following steps: (1) preliminary exploratory analysis of transcripts and questionnaire responses, development of initial themes for coding data; (2) refinement of themes from data, creation of coding scheme including themes and underlying subcategories that address key research questions; (3) in-depth review of transcripts using the coding scheme, interpretation of findings; and (4) comparison across transcripts from various stakeholder events.

Results

Our analysis points to a fair degree of change in perceptions and views among stakeholders over the course of the process, in terms of framing of climate change and impacts and new insights into complexity of factors affecting adaptation needs and options within forestry. One general observation is that the perceived risk of climate change and the possible need to adapt increased during the process, but it varied considerably across individuals.

In the first session, the greatest risks and challenges identified were ones not primarily associated with climate change. Rather, climate change was perceived to be an issue for national and global policy and markets to deal with. Discussions centered on concerns about maintenance of or increase in forest production, carrying capacity of forest land, forest ditching, forest sensitivity to game damage, and conflicts over forest land due to emerging nature conservation policy requirements. Many stakeholders felt that they would not need to take extensive adaptation actions to protect their forest land from the impacts of climate change: "it's nothing that I have to bother about today" (forest owner, Västerbotten). "I don't think I'll have to change my [forest] management practices that much" (forest owner, Kronoberg).

Comments also reflected the uncertainty involved in assessing the impacts of climate change as well as a more relaxed "wait and see" position and in some cases skepticism. The perceived uncertainty was not only based on judgment on scientific evidence but also on a general reluctance to accept new scientific findings, because such findings have historically been used to justify changes in forestry policy opposed by the forest owners. Another recurring theme was the possible benefits of climate change for forestry: "Long ago when I took over the property I received leaflets from the Swedish Forest Agency ordering me to take away deciduous trees. – Clear away deciduous trees! Today you should leave it be. You never know when such tricks appear" (forest owner, Västerbotten); "I'm probably more positive about this. I don't believe climate change to be as widespread as they claim" (forest professional, Västerbotten).

Among forest owners in Kronoberg, the discussions were more focused on climate-related risks in relation to forest management practices and adaptation was more widely accepted as one plausible, albeit not urgent, strategy for the future: "If we get higher temperatures in the longer term it might force us to change our tree species composition"; "It might require new practices in forest management, trenching might not be enough." This difference compared to the Västerbotten groups may well be linked to two recent severe storms that hit the region in December 2005 (Gudrun) and in January 2007 (Per). The more severe of them, Gudrun, hit Kronoberg hard. Nationally, it resulted in 75 million cubic meters of felled wood: the equivalent of the total amount of harvested wood in all of Sweden in one full year (Svensson et al. 2011). Most of the forest owners shared their experience with storm Gudrun, and some noted that extreme weather events will probably become more frequent in the future: "If we get significant change in the climate in the future with stronger precipitation then I believe that we will face problems."

Yet, as noted, also in this group, most participants did not seem to feel any real urgency to plan and prepare to adapt their forestry to climate change but were primarily concerned about forest productivity. Sentiments expressed built on the notion that it is impossible to prepare for an unknown climate future, and climate change was also associated with something beneficial: "To me it feels good to be forest owner from an economic perspective for various reasons, we have the future ahead of us" (forest owner, Västerbotten). Such sentiments reflect recent scientific findings, suggesting that forestry and agricultural sectors in Northern Europe may benefit from limited temperature rise in terms of increased crop yields and increased forest growth in the short term (EEA 2010; SOU 2007).

In the second focus group session, the stakeholders started with a discussion on the relevance of climate change adaptation in their forest activities and business. After this participants were presented with scientific model results on regional climate change and regional to local impacts on forestry. The information was presented by a climate scientist of the Rossby Centre, the climate modeling research unit at the Swedish Hydrological and Meteorological Institute, and an ecosystem scientist from the Department of Physical Geography and Ecosystem Science at Lund University. This session provided an opportunity to interact with experts, including asking questions, making comments and suggestions for further model improvements, and discussing freely with each other about implications on their forest land and the possible need to change forest management practices. The discussions focused on the landscape scale and enabled a lively exchange of views and experiential knowledge between participating stakeholders. In response to the scientific information, it was concluded that vital and healthy forests have a general capacity to withstand future risks of damage from bark beetle attacks, fungus, reduced levels of ground frost, storms, and droughts: "I've come to the conclusion, no matter what type of changes that will come, a vital forest is better prepared to resist it. That's why I'm trying to keep my forest vital through clearing and thinning and maintaining the trench systems" (forest owner, Kronoberg).

Overall the analysis of the second meeting demonstrates a wider appreciation of the opportunity to share both scientific and experiential knowledge in an environment conducive to learning: "This is a really good way to learn about what the science says about all this" (forest owner Kronoberg); "The great thing is that here we can ask questions and get things explained in simple words and bounce ideas back and forth with each other and the researchers" (forest professional, Västerbotten). Our observations also suggest that learning occurred with respect to the relevance of several forestry specific risks: fungus, infestations, "wet snow," amount of water, depth and duration of ground frost, tree felling, and timing and length of warm periods. Though a number of participants had previous knowledge about climate-related risks such as storms and bark beetle outbreaks and attacks on storm-felled trees, the outcome of the meetings indicates a diffusion of knowledge on climate change and adaptation across larger segments of stakeholders as well as deepened understanding of the complexity of climate change and impacts within Swedish forestry. Moreover, the expected negative, long-term risks, were more widely recognized alongside the more positive effects of faster growth and longer vegetation periods that had dominated the initial discussions: "we forest people tend to think it [climate change] only means positive things for us here in the North but we've got to realize it's far more complex than that, if we look at the longer-term perspective" (forest owner Västerbotten).

The third session focused on the views of forest stakeholders on adaptation needs, barriers, and possible pathways. Participants raised and discussed a wide range of possible adaptation measures that had not been raised or were not clearly articulated earlier. The solutions referred to local management strategies as well as changes in regulation and policies notably at the national level. While it was generally perceived to be difficult to choose "the right" adaptation strategies, there was some agreement that climate change will require a faster turnover time of forest: "We have to turn it [the forest] over faster and then we can pretty easily minimize the damages in terms of volume" (Forest owner, Västerbotten). To improve the carrying capacity of forests that was initially suggested in the first meeting, several suggestions for adaptive measures were made, for example, more intensified forestry; earlier harvesting; laxer national market policy including company taxation; and technology development.

Moreover, our observations from this meeting indicate a new or previously not displayed attention to and concern about the policy dimension of adaptation within Swedish forestry. Policies that were seen as problematic or were expected to become problematic in relation to climate change include the EU Water Framework Directive and national policies on the management of game, ditching, the choice of seedlings, and the rotation period: "It will also be necessary to take a new look at the existing policies that aren't clearly meeting the demands of the future climate change in the forests" (forest professional, Västerbotten).

Overall the analysis of the third round of focus groups demonstrates a continued, strong influence of the climate and forestry data presented on the articulation of plausible adaptive measures and insights into new issues such as level of risks of spruce bark beetle outbreaks and deteriorated carrying capacity of roads and transport damages. Most discussions referred directly to data and scenarios presented. Nonetheless also in this session a number of stakeholders expressed skepticism about a perceived lack of certainty in the scientific data presented earlier. This was particularly evident among forest professionals: "I believe this model uncertainty prevents you from taking a stand regarding a valid line of argument or approach" (forest professional, Kronoberg); "The knowledge base is still inept to make clear suggestions for clearing to prevent storm damages" (forest professional, Västerbotten). It was learned among a considerable proportion of participants, and a confirmed fact for others, that there is inevitably some scientific uncertainty that makes it difficult to determine desirable adaptation options: "There's still very little concrete advice we [forest agency advisers] can give to forest owners based on this. To some extent, we can tell them what they can do to prevent transport damages during frost free winters. Other than that, I feel a bit uncertain about what to advise people" (forest professional, Kronoberg).

While it was widely comprehended what climate and impact models can offer in terms of informing about the extent of risks and estimated time horizons, further model simulations were called for to serve as effective decision support for adaptive forest management. Gaps in existing data on climate change impacts identified included: fungus and insects, "wet snow," depth of ground frost, tree felling, amount of water in spring, and timing and length of warm periods. In light of the general perceptions and understanding of risk articulated in the first meetings, this deeper appraisal of contemporary scientific knowledge and identification of research gaps in forestry specific risks indicate that learning took place over the course of the focus group process.

The fourth and final time the stakeholders met was in a larger workshop in Stockholm in May 2010. In addition to sharing experiences among groups, participants highlighted the general lack of arenas for learning across forest actors and "learning environments." To facilitate local climate adaptation within forestry, participants agreed they need better access to research results on concrete problems related to climate adaptation in the forestry sector, such as damage from spruce bark beetle. They also called for improved collaboration and communication with national authorities and across local forest actors regarding road maintenance and game management, as well as between communities of science and practice. This wish for multistakeholder collaboration and interaction was "new" – no such opinions surfaced in the focus group sessions. Many participants also emphasized that their primary goal remains to maintain or increase forest productivity, irrespective of their interest in climate change adaptation.

The follow-up interviews eight months later were an attempt to capture stakeholders' reflections on the process as well as the possible long-term effects of learning beyond the participatory exercises. Table 14.3 outlines a selection of main arguments of the respondents. Overall, we observe a self-reported shift in mind-sets and learning as a result of their engagement in our study. Learning gains appears to have had limited effect on forest professionals and moderate impact on forest owners: already highly scientifically informed professionals typically argued that the scientific material and discussions largely consolidated and deepened their previous knowledge and the perceived value of the process itself was moderate, but mostly mildly positive. On the contrary, change in mind-sets and learning was frequently reported by the forest owners. Respondents' comments referred to the value of the participatory process as such, as well as to self-perceived changes in perceptions/attitudes, knowledge acquisition, and behavior change. The interviews also revealed certain limitations with regard to learning in participatory processes, including the amount of information presented to participants, the domination of certain actors in the discussions, and the difficulty in grasping the long time perspective of climate change and to separate it from other change processes.

Table 14.3. Selection of stakeholder views on long-term effects of participatory process and learning from follow-up interviews

General reflections, change in mind-set	Behavior change	Skepticism, no difference
It provided me a new perspective on relevant topics, the program has highlighted the issues.	I do read the related articles now.	There were difficult questions to have an opinion about.
It was nice, one gets a new fresh perspective.	It highlighted the issue, I bring up the issue more often with other forest owners and I've been reading more on the topic.	No it has not had any direct influence, we have got training in this before.
It was interesting, a different environment and approach compared to how one normally works within forestry.	I've been discussing it with the other forest owners that participated and I raise the issues in my advisory role.	It was interesting but I haven't been influenced the way I view these issues.
It was very interesting to get insight into the scenarios and that we could see maps that where pretty clear and informative.	The knowledge is sustained since there are several who I meet in Södra [the largest forest owner association in Sweden] etc., and I bring this up in our discussions.	I'm skeptical about the format, there was lack of forestry competence among researchers.
It opened my eyes a bit, I understand more about this now, it's okay for it to take some time though since it's a slow process to grow trees.	I now read articles on the topic.	I did not get anything concrete out of it.
I've got more knowledge now, a firmer ground to stand on than earlier.	I'm more confident in my assessments which allows me to raise it and argue for it when discussing forestry.	The experience still support the previous forest production objective.
Very stimulating, it highlighted the issues that concern my work.	I read more but I've had a passive forest year.	It has highlighted the subject, but there is no need for a rapid adaptation.
It has widened my thinking about forest management in change and over time.	I've revisited the scientific material when planning my forest management practices.	The general question in my mind is how will they get anything descent out of this?
I hadn't made the connection between climate change and forests if it hadn't been for this project, now I see it differently, I now see an adaptation need in the longer term.	I bring these ideas with me in discussions about forest planning.	I didn't gain that much personally from it since I'm already informed about climate change in the context of forests.
I have more knowledge now, I know better how to deal with climate change.	I talk about it in a totally different way than before.	
I think about climate change in a different way now, I haven't been thinking about the forests in this way before.	I bring it up in different situations, e.g., as chair person of the [forest association].	
I care much more now, you get influenced by this new philosophy and then you think about actions to be taken.	When I see an article, then I read it.	
The outcome is that I've got to use as many means as possible. For example I've changed my mind about *Contorta*. It has increased my awareness, I see an increased need for vital forests.	I reconnect to it when reading something related or when it comes up in the media.	
It has been brought up on my own agenda, I think about it more now, nowadays I think that things will change over time and I didn't do that before.		
I think it was very exciting project, I learned from it and it has triggered my interest in climate adaptation.		

Discussion and Conclusions

Overall we have observed a number of changes in perceptions, values, and understanding over the course of our participatory study. Against this background, the question arises whether the process has created a sustained increase in adaptive capacity, that is, the ability or potential of the "system" of forest stakeholders to respond successfully to climate variability and change (Adger et al. 2007). The conventional understanding is that factors that determine adaptive capacity include economic wealth, technology and infrastructure, information, knowledge and skills, institutions, equity, and social capital (Smit et al. 2001). Some of these determinants can meaningfully be discussed in the context of our study. The participatory process has clearly provided access to shared *information*, and the results indicate the level of *knowledge* among the stakeholders has increased to a varying degree along the way. Those outcomes were apparent during and after the second meeting when there were direct interactions and knowledge exchange between stakeholders and scientists. *Social capital* is likely to have increased as a result of social networks (Putnam 2000) evolving. Interestingly, stakeholders themselves highlighted the value of this dimension, through calls for multilevel, multistakeholder arenas for knowledge sharing and collaborations on climate change and adaptation within the Swedish forestry sector.

However, it is difficult to assess the level of change in adaptive capacity of participating stakeholders over the process as such. This difficulty partly reflects the qualitative nature and relative short-term scope of our case study, but it also points to a more general challenge. The term *adaptive capacity* describes the aggregate ability of actors to plan, prepare for, facilitate, and implement adaptation options. Some research tends to indicate that, by knowing the change in adaptive capacity determinants, we will automatically know the change in the capacity itself. However, our case study on Swedish forestry points to the challenges of understanding how the determinants interact and whether minimum requirements linked to certain determinants (e.g., for access to economic resources) have to be met for others to come into play (e.g., knowledge or social capital). Trying to understand the determinants of adaptive capacity in causal and quantitative terms remains a valid, but different, inquiry.

This chapter has explored how participatory processes can contribute to learning on climate change and adaptation needs and options. It has done so by identifying changes in perceptions and learning manifested over the course of a participatory process with Swedish forest stakeholders. The research has also aimed at identifying key aspects that characterize the learning process and assessing how, if at all, it has contributed to a self-perceived increase in capacity to adapt to climate change. The first finding is that the participatory process has yielded changes in (expressed) perceptions and learning about climate change effects, vulnerability, and potential adaptation measures over the course of the process. The opportunity to engage in the process has been widely appreciated by participants, suggesting that the research is beneficial for stakeholders themselves, in terms of bringing them together and creating a space for sharing knowledge and views on challenges and opportunities they face. However, from the participants' perspective, the researchers are just making a short visit into their everyday lives. This points to the shortcomings of short-lived participatory exercises, especially if adaptation does not feature prominently in participants' everyday decision making, and when there are other issues on their agendas and potential goal conflicts. The temporal dimension of learning and its outcomes must therefore be recognized in participatory exercises on climate change adaptation. Nonetheless there is some evidence of self-perceived, enhanced adaptive capacity on the part of a number of stakeholders as a result of our stakeholder engagement process.

Second, our findings highlight the scientific value of taking a participatory approach to studying local adaptation processes. Participatory approaches tend to provide insights and perspectives that remain hidden when using other research methods (Forrester and Gerger Swartling 2010). Moreover, much learning occurs through communication and knowledge sharing with others, so participatory processes are a natural way of both fostering learning and examining the conditions for learning and how learning takes place. In this regard, it is important to consider what role the setup and the very implementation of the participatory process can have for the outcomes and that there is no one-size-fits-all approach to learning.

Third, our study illustrates the importance of distinguishing between the various forms of learning that take place. We have identified ample evidence of change in perceptions, mind-sets, and self-perceived learning that points to increased individual learning over the somewhat longer term. However, following Reed et al.'s (2010) definition of social learning, our results do not indicate that wider social learning has occurred since the learning is a result of limited research interventions with a smaller community of forest stakeholders rather than learning situated within a wider social unit or community of practice. The empirical study also indicates a mix of single- and double-loop learning among participating stakeholders. Single-loop learning is manifested in terms of stakeholders' identification of alternative forest management strategies, although underlying goals (notably forest productivity) remain unchanged. The identification of new collaborative and organizational frameworks points to the occurrence of double-loop learning, characterized by collective reflection and the formulation of new goals (notably resilient forests), alongside the wish to maintain high forest productivity. There is, conversely, no indication of triple-loop learning in the study, manifested in terms of stakeholders learning about how to modify their actions, values, and underlying governance norms with respect to climate change adaptation.

Finally, our research points to the need to address and better understand how participatory approaches can be scaled up. Is it possible to formulate general advice on how to replicate and mainstream local success in participatory learning on adaptation into sector programs, funding schemes and formal training efforts? Drawing on experiences from natural resource management in India, Farrington and Boyd (1997, 389) argued that scaled-up approaches have to be "based on multi-agency partnerships in which each can play a role according to its comparative advantage." This perspective is applicable to our case study of the Swedish forestry sector in which stakeholders should seek joint approaches to knowledge sharing and collective learning and action. The Forestry Agency holds a key role here as a convener, but the challenge of outreaching 330,000 individual forest owners requires active engagement of forest owners' associations and actors purchasing forest products for the Swedish forestry sector to improve its ability to adapt to climate change. Future research will have to identify successful examples of mainstreaming participatory learning approaches for adaptation, beyond the local and experimental scale at which research reported here was carried out.

Acknowledgments

The authors would like to thank all stakeholders for their time and enthusiasm to participate in this study as well as Louise Simonsson (Swedish Defense Research Agency, formerly Linköping University) and Karin André (Stockholm Environment Institute, formerly Linköping University) for their contributions to the empirical study. We are also grateful to the two anonymous reviewers for their valuable comments on an earlier version of the text. Funding for this research was made available by the Swedish Foundation for Strategic Environmental Research (Mistra) through its program Mistra-SWECIA. This chapter also contributes to the Nordic Centre of Excellence for Strategic Adaptation Research (NORD-STAR), which is funded by the Norden Top-Level Research Initiative subprogram "Effect Studies and Adaptation to Climate Change."

References

Adger, W. N., Agrawala, S., Mirza, M. M. Q., Conde, C., O'Brien, K., Pulhin, J., Pulwarty, R., Smit, B. and K. Takahashi. 2007. "Assessment of adaptation practices, options, constraints and capacity." In *Climate Change 2007: Impacts, Adaptation and Vulnerability*, edited by Parry, M. L., Canziani, O. F., Palutikof, J. P., van der Linden, P. J. and C. E. Hanson, 717–43. Contribution of Working Group II to the Fourth Assessment Report of the Intergovernmental Panel on Climate Change. Cambridge, UK: Cambridge University Press.

Allen, W., Bosch, O., Kilvington, M., Oliver, J. and M. Gilbert. 2001. "Benefits of collaborative learning for environmental management: Applying the integrated systems for knowledge management approach to support animal pest control." *Environmental Management* 27 (2): 215–23.

André, K. 2013. *Climate Change Adaptation Processes: Regional and Sectoral Stakeholder Perspectives*. Doctoral thesis, Linköping Studies in Arts and Science.

André, K., Simonsson, L., Gerger Swartling, Å. and B. O. Linnér. 2012. "Method development for identifying and analysing stakeholders in climate change adaptation processes." *Journal of Environmental Policy and Planning* 14 (3): 243–61.

Argyris, C. and D. A. Schön. 1978. *Organizational Learning: A Theory of Action Perspective*. Reading, MA: Addison-Wesley.

Argyris, C. and D. A. Schön. 1996. *Organizational Learning II: Theory, Method and Practice*. Reading, MA: Addison-Wesley.

Armitage, D. 2005. "Adaptive capacity and community-based natural resource management." *Environmental Management* 35 (6): 703–15.

Armitage, D., Berkes, F. and N. Doubleday. 2010. *Adaptive Co-management: Collaboration, Learning and Multi-level Governance*. Vancouver: University of British Columbia Press.

Armitage, D., Marschke M. and R. Plummer. 2008. "Adaptive co-management and the paradox of learning." *Global Environmental Change* 18 (1): 86–98.

Armitage, D., Plummer, R., Berkes, F., Arthur, R. I., Charles, A. T., Davidson-Hunt, I. J., Diduck, A. P., et al. 2009. "Adaptive co-management for social–ecological complexity." *Front Ecol Environ* 7 (2): 95–102.

Bennett, C. J. and M. Howlett. 1992. "The lessons of learning: Reconciling theories of policy learning and policy change." *Policy Sciences* 25 (3): 275–94.

Berkes, F. and C. Folke, eds. 1998. *Linking Social and Ecological Systems: Management Practices and Social Mechanisms for Building Resilience*. Cambridge: Cambridge University Press.

Berkhout, F., Hertin, J. and D. M. Gann. 2006. "Learning to adapt: Organisational adaptation to climate change impacts." *Climatic Change* 78 (1): 135–56.

Blackmore, C., Ison, R. and J. Jiggens. 2007. "Social learning: An alternative policy instrument for managing in the context of Europe's water." *Environmental Science and Policy* 10 (6): 493–98.

Brooks, N., Adger, W. N. and M. Kelly. 2005. "The determinants of vulnerability and adaptive capacity at the national level and the implications for adaptation." *Global Environmental Change* 15 (2): 151–63.

Cundill, G. 2010. "Monitoring social learning processes in adaptive comanagement: Three case studies from South Africa." *Ecology and Society* 15 (3): 28.

Diduck, A. 2004. "Incorporating participatory approaches and social learning." In *Resource and Environmental Management in Canada*, edited by B. Mitchell, 497–527. Don Mills, Ontario: Oxford University Press.

Diduck, A. 2010. "The learning dimension of adaptive capacity: Untangling the multi-level connections." In *Adaptive Capacity and Environmental Governance*, edited by Armitage, D. and R. Plummer, 199–221. Berlin: Springer.

EEA. 2010. *European Environment State and Outlook*. Report. Copenhagen.

Farrington, J. and C. Boyd. 1997. "Scaling up the participatory management of common pool resources development." *Policy Review* 15 (4): 371–91.

Fiorino, D. 1990. "Citizen participation and environmental risk: A survey of institutional mechanisms." *Science, Technology and Human Values* 15 (2): 226–43.

Fiorino, D. 2001. "Environmental policy as learning: A new view of an old landscape." *Public Administration Review* 61 (3): 322–34.

Fitzpatrick, P. 2006. "In it together: Organizational learning through participation in environmental assessment." *Environmental Assessment and Policy Management* 8 (2): 157–82.

Flood, R. L. and N. R. A. Romm. 1996. *Diversity Management: Triple Loop Learning*. Chichester, UK: Wiley.

Forrester, J. and Å. Gerger Swartling, eds. 2010. "Overcoming the challenges of 'doing participation' in environment and development: Workshop summary of lessons learned and ways forward." *SEI Working Paper*, Stockholm Environment Institute.

Gerger Swartling, Å., Lundholm, C., Plummer, R. and D. Armitage. 2011. "Social learning and sustainability: Exploring critical issues in relation to environmental change and governance." *SEI Project Report*, Stockholm Environment Institute.

Haas, P. M. 2000. "International institutions and social learning in the management of global environmental risks." *Policy Studies Journal* 28 (3): 558–75.

Hinkel, J., Bisaro, S., Downing, T. E., Hofman, M. E., Lonsdale, K., McEvoy, D. and J. D. Tábara. 2010. "Learning to adapt: Reframing climate change adaptation." In *Making Climate Change Work for Us: European Perspectives on Adaptation and Mitigation Strategies*, edited by Hulme, M. and H. Neufeldt, 113–34. Cambridge: Cambridge University Press.

Irwin, A. 1995. *Citizen Science: A Study of People, Expertise and Sustainable Development*. London: Routledge.

Johannessen, Å. and T. Hahn. 2013. "Social learning towards a more adaptive paradigm? Reducing flood risk in Kristianstad municipality, Sweden." *Global Environmental Change* 23 (1): 372.

Jönsson, A. M. and L. Bärring. 2011. "Future climate impact on spruce bark beetle life-cycle in relation to uncertainties in regional climate model data ensembles." *Tellus A* 63 (1): 158–73.

Jönsson, A. M., Schroeder, L. M., Lagergren, F., Anderbrant, O. and B. Smith. 2012. "Guess the impact of Ips typographus – an ecosystem modelling approach for simulating spruce bark beetle outbreaks." *Agricultural and Forest Meteorology* 166–67: 188–200.

Keen, M., Brown, V. and R. Dybal. 2005. *Social Learning in Environmental Management*. London: Earthscan.

Keen, M. and S. Mahanty. 2005. "Collaborative learning: Bridging scales and interests." In *Social Learning in Environmental Management*, edited by Keen, M., Brown, V. A. and R. Dyball, 104–20. London: Earthscan.

Keskitalo, C. H., Dannevig, H., Hovelsrud, G., West, J. J. and Å. Gerger Swartling. 2010. "Adaptive capacity determinants in developed states: Examples from the Nordic countries and Russia." *Regional Environmental Change* 11 (3): 579–92.

Lagergren, F. and A. M. Jönsson. 2010. "Climate change and forests' sensitivity to storm and spruce bark beetle damage." *Mistra-Swecia Newsletter* 1: 10. http://www.mistra-swecia.se/.

Lagergren, F., Jönsson, A. M., Blennow, K. and B. Smith. 2012. "Implementing storm damage in a dynamic vegetation model for regional applications in Sweden." *Ecological Modelling* 247: 71–82.

Lee, K. N. 1993. *Compass and Gyroscope: Integrating Science and Politics for the Environment*. Washington, DC: Island Press.

Lee, K. N. 1999. "Appraising adaptive management." *Conservation Ecology* 3 (2): 3.

Levitt, B. and J. G. March. 1988. "Organizational learning." *Annual Review of Sociology* 14 (1): 319–40.

May, B. and R. Plummer. 2011. "Accommodating the challenges of climate change adaptation and governance in conventional risk management: Adaptive collaborative risk management (ACRM)." *Ecology and Society* 16 (1): 47.

Mendis, S., Mills, S. and J. Yantz. 2003. *Building Community Capacity to Adapt to Climate Change in Resource-Based Communities*. Prepared for the Prince Albert Model Forest.

Nadine, A., Marshall, R. C., Tobin, P. A., Marshall, M. G. and A. J. Hobday. 2013. "Social vulnerability of marine resource users to extreme weather events." *Ecosystems* 16 (5): 797–809.

O'Brien, K., Eriksen, S., Sygna, L. and L. O. Naess. 2006. "Questioning complacency: Climate change impacts, vulnerability, and adaptation in Norway." *Ambio* 35 (2): 50–56.

Olsson, P. and C. Folke. 2001. "Local ecological knowledge and institutional dynamics for ecosystem management: A study of Lake Racken Watershed, Sweden." *Ecosystems* 4 (2): 85–104.

Pahl-Wostl, C. 2006. "The importance of social learning in restoring the multi-functionality of rivers and floodplains." *Ecology and Society* 11 (1): 10.

Pahl-Wostl, C., Craps, M., Dewulf, A., Mostert, E., Tabara, D. and T. Taillieu. 2007. "Social learning and water resources management." *Ecology and Society* 12 (2): 5.

Parson, E. A. and W. C. Clark. 1995. "Sustainable development as social learning: Theoretical perspectives and practical challenges for the design of a research program." In *Barriers and Bridges to the Renewal of Ecosystems and Institutions*, edited by Gunderson, L. H., Holling, C. S. and S. S. Light, 428–61. New York: Columbia University Press.

Pelling, M., High, C., Dearing, J. and J. Smith. 2008. "Shadow spaces for social learning: A relational understanding of adaptive capacity to climate change within organisations." *Environment and Planning A* 40 (4): 867–84.

Percy, R. 2005. "The contribution of transformative learning theory to the practice of participatory research and extension: Theoretical reflections." *Agriculture and Human Values* 22 (2): 127–36.

Plummer, R. and D. Armitage. 2010. "Integrating perspectives on adaptive capacity and environmental governance." In *Adaptive Capacity and Environmental Governance*, edited by Plummer, R. and D. Armitage, 1–19. Berlin: Springer.

Putnam, R. 2000. *Bowling Alone: The Collapse and Revival of American Community*. New York: Simon and Schuster.

Reed, M., Evely, A. C., Cundill, G., Fazey, I. R. A., Glass, J., Laing, A., Newig, J., et al. 2010. "What is social learning?" *Ecology and Society* 15 (4): 1.

Rittel, H. and M. Webber. 1973. "Dilemmas in a general theory of planning." *Policy Sciences*, 4 (2): 155–69.

Sabatier, P. A. and H. Jenkins-Smith, eds. 1993. *Policy Change and Learning: An Advocacy Coalition Approach*. Boulder, CO: Westview Press.

Smit, B., Pilifosova, O., Burton, I., Challenger, B., Huq, S., Klein, R. J. T. and G. Yohe. 2001. "Adaptation to climates change in the context of sustainable development and equity." In *Climate Change 2001: Impacts, Adaptation and Vulnerability*, edited by McCarthy, J. J., Canziani, O. F. and N. Leary, 877–912. Contribution of Working Group II to the Third Assessment Report of the Intergovernmental Panel on Climate Change. Cambridge: Cambridge University Press.

SOU. 2007. *Sweden Facing Climate Change – Threats and Opportunities*. Swedish Commission on Climate and Vulnerability. Stockholm: The Ministry of the Environment. http://www.sweden.gov.se/sb/d/574/a/96002.

Steyaert, P. and J. Jiggins. 2007. "Governance of complex environmental situations through social learning: A synthesis of SLIM's lessons for research, policy and practice." *Environmental Science and Policy* 10 (6): 575–86.

Svensson, S., Bohlin, F., Bäcke, J.-O., Hultåker, O., Ingemarson, F., Karlsson, S. and J. Malmhäll. 2011. *Ekonomiska och sociala konsekvenser i skogsbruket av stormen Gudrun*. Swedish Forest Agency Report No. 12. Jönköping: The Swedish Forest Agency.

Swedish Forest Agency. 2014. *Swedish Statistical Yearbook of Forestry 2014*. Jönköping: Swedish Forest Agency.

Tabara, J. D., Cots, F., Dai, X., Falaleeva, M., Flachener, Z., McEvoy, D. and S. Werners. 2009. "Social learning on climate change among regional agents and institutions: Insights from China, Eastern Europe and Iberia." In *Interdisciplinary Aspects of Climate Change*, edited by Leal Filho, W. and F. Mannke, 121–50. Frankfurt: Peter Lang.

Tarnoczy, T. 2011. "Transformative learning and adaptation to climate change in the Canadian Prairie agro-ecosystem." *Mitigation and Adaptation Strategies for Global Change* 16 (4): 387–406.

Tschakert, P. and K. A. Dietrich. 2010. "Anticipatory learning for climate change adaptation and resilience." *Ecology and Society* 15 (2): 11.

Van de Kerkhof, M. and A. J. Wieczorek. 2005. "Learning and stakeholder participation in transition processes towards sustainability: Methodological considerations." *Technological Forecasting and Social Change* 72 (6): 733–47.

Vincent, K. 2007. "Uncertainty in adaptive capacity and the importance of scale." *Global Environmental Change* 17 (1): 12–24.

Vulturius, G. and Å. Gerger Swartling. 2015. "Overcoming social barriers to learning and engagement with climate change adaptation." *Scandinavian Journal of Forest Research* 30: 217–25.

Wals, A. E., ed. 2007. *Social Learning towards a Sustainable World*. Wageningen: Wageningen Academic.

Welp, M., de la Vega-Leinert, A., Stoll-Kleemann, S. and C. C. Jaeger. 2006. "Science-based stakeholder dialogues: Theories and tools." *Global Environmental Change* 16 (2): 170–81.

Wibeck, V. 2014. "Enhancing learning, communication and public engagement about climate change – some lessons from recent literature." *Environmental Education Research* 20 (3): 387–411.

Wynne, B. 1996. "May the sheep safely graze? A reflexive view of the expert-lay knowledge divide." In *Risk, Environment and Modernity: Towards a New Ecology*, edited by Lash, S., Szerszynski, B. and B. Wynne, 44–83. London: Sage.

15

Integral GIS

Widening the Frame of Reference for Adaptation Planning

LYNN D. ROSENTRATER

Adaptation to climate change moderates harm and exploits beneficial opportunities associated with observed and expected changes in climate. It is guided by various decision-making frameworks that represent idealized models of participatory planning, usually in the form of an iterative learning cycle that emphasizes action and reflection (Reason and Bradbury 2008). Frameworks such as the U.K. Climate Impacts Programme's Adaptation Wizard (UKCIP 2013) and the United Nations Environment Programme's (UNEP) Programme of Research on Climate Change Vulnerability, Impacts, and Adaptation (PROVIA 2013) articulate their step-by-step processes differently, but the workflow is generally the same: the starting point is some kind of scoping exercise to single out problems and define objectives, followed by information gathering aimed at identifying adaptation needs; subsequent steps focus on appraising adaptation options and, finally, monitoring and evaluating actions; here the cycle begins again with a new starting point for subsequent planning.

Identifying adaptation needs is a subjective process, making planning procedurally difficult: a diverse assembly of interested parties, each with its own values and agendas, deliberates over the harmful or unwanted consequences of climate change. These discussions are informed by vulnerability assessments that describe how climate is projected to change and how those changes may affect people and places. Assessing vulnerability also sheds light on the significance of climatic risks in relation to other risks managed by communities, which allows adaptation needs to be prioritized. However, a critical examination of vulnerability assessments suggests that they often fall short of expectations to inform adaptation policy. Assessment results tend to convey what is already known about climatic risks but without sufficient spatial or temporal detail to provide definitive information for planning optimum responses (Rosentrater 2010). Vulnerability assessments also do not typically account for the invisible losses and nonmaterial concerns (e.g., identity, beliefs, and values) that are important to stakeholders (Turner et al. 2008). When all is said and done, stakeholders seem reluctant to embrace representations of vulnerability that differ from their own evaluations of climatic risks, thus limiting planning.

Studies of environmental psychology show that tacit knowledge shapes how individuals perceive and manage risk, making subjective filters both significant and influential when responding to climate change (Rosentrater et al. 2013). To reach a deeper understanding of vulnerability that reveals multiple entry points for adaptation, assessments must consider the impact of climate change on the values and priorities of diverse social groups (O'Brien and Wolf 2010). In this chapter I describe a process for assessing vulnerability that accounts for the subjective judgments that influence decisions about adaptation needs. The purpose is to harmonize the myriad perspectives inherent to participatory planning through a deliberative process that explores and reconciles competing, or conflicting, points of view. The process is best suited to small-scale place-based assessments, although it can be applied to any multiactor, multicriteria decision context.

I begin by discussing how vulnerability is a relative condition best understood as a product of both objective and subjective evaluations. Next, I describe how geographic information systems (GIS) are used to address spatial problems like adaptation to climate change. I then introduce integral GIS as an approach for assessing vulnerability that combines evidence and values-based dimensions of climate change. The approach is designed to broaden the frame of reference when gathering information about specific planning decisions. I provide a brief example of how integral GIS can be used to assess vulnerability and its relative advantages over conventional assessments. The chapter concludes with a discussion of guiding principles for using integral GIS.

Understanding Vulnerability: An Integral Perspective

The premise of vulnerability is to understand the potential for harm (Adger 2006). Assessing vulnerability to climate change requires data produced at different spatial and temporal scales, thus in many adaptation processes GIS is used to ground analyses in spatially nuanced evaluations of risk (Preston et al. 2011). GIS produces graphical representations – usually in the form of maps – that communicate conditions, processes, and trends across the human landscape. Vulnerability maps are accessible and powerful communication tools that convey at a glance the location of vulnerable people and places. Methods for mapping vulnerability are dominated by indicator-based approaches that combine representations of potential impacts and potential adaptation into a composite measure (Metzger and Schroter 2006). Such indices are usually based on the Intergovernmental Panel on Climate Change (IPCC) definition of vulnerability as a function of exposure, sensitivity, and adaptive capacity (McCarthy et al. 2001). But these mappings are neither neutral nor unproblematic with regard to representing the multidimensional nature of vulnerability.

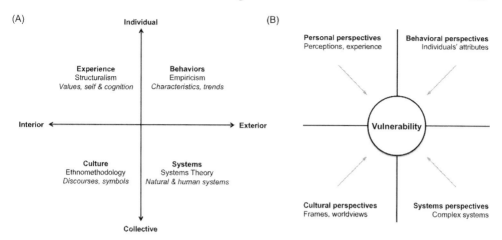

Figure 15.1. (a) The perspectives, methodological families, and information associated with the integral quadrants and (b) representative knowledge used to understand vulnerability to climate change from an integral perspective.

All knowledge about climate change – a complex problem that is often perceived as distant from day-to-day experience – is contingent on four basic aspects of inquiry: *who* is acquiring the knowledge (epistemology), *what* feature is being investigated (ontology), *where* and *when* (in space and time) the inquiry is situated, and *how* (methodologically) the inquiry is conducted (Carolan 2004; Eddy 2005; Esbjörn-Hargens 2005). When one aspect of the inquiry changes, the resulting knowledge also changes. This poses a challenge for policy making in that "the same phenomena can be looked at from multiple perspectives, at multiple scales... and be simultaneously valued differently according to differences in circumstance" (Eddy 2006, 13). When identifying adaptation needs, people see the world in different ways and will consequently prioritize different things worthy of safeguarding, which is to say that climate change means different things to different people. Some people look at a vulnerability map and see an issue of safety, others see a threat to their identity, and still others may see nothing at all. This multiplicity calls for a better understanding of why and to whom climate change matters and suggests that a reflexive, polycentric approach is needed to capture the full significance of climate change for adaptation purposes.

One way to account for the multiple perspectives that exist among adaptation actors is through a transdisciplinary framework based on integral theory – a field of study that conceptualizes the complex interaction between ontology, epistemology, and methodology (Eddy 2008; Esbjörn-Hargens 2009). Integral theory is conceived of as quadrants created at the intersection of two polarities that describe the interior and exterior and the individual and collective dimensions of a phenomenon (Figure 15.1a). The left-hand quadrants are characterized by subjectivity and correspond to first-person perspectives derived from structuralism – methods of interpretation and analysis of human cognition, behavior and experience – and ethnomethodology, which is used to understand how people make sense of the world through social orders and ideological systems of values. The right-hand quadrants consist of objective, third-person perspectives based on empiricism and systems theory that describe the behaviors, responses, and interactions between nature and society. The individual quadrants do not stand in isolation but rather integrate and shape one another. Integral theory posits that phenomena are construed through a "combination of data, knowledge and meaning, and their interrelations involve non-linear interaction of inductive and deductive reasoning processes" (Eddy 2006, 20). In other words, an object of interest is interpreted through information associated with each of the quadrants, which gives rise to a holistic understanding of the phenomenon.

Understanding vulnerability involves knowledge associated with objective, exterior dimensions of climate change that focus on negative material outcomes as well as subjective, interior dimensions that address the meaning and relevance of those outcomes for specific individuals and groups (O'Brien and Wolf 2010). Figure 15.1b illustrates how these objective and subjective dimensions integrate to produce a comprehensive view of vulnerability. The model builds on the IPCC definition of vulnerability to include the perceptions and beliefs of those who bear the consequences of climate change. These subjective dimensions affect not only vulnerability outcomes but also the priorities people set with regard to adaptation (O'Brien 2009).

The quadrants on the right side of the figure represent "objective" vulnerability, in the sense that the components are externally measurable. The upper-right quadrant pertains to whom or what may be vulnerable. Data here consist of empirical observations representing the exterior aspects of people or places that determine vulnerability, that is, age, gender, physical health, socioeconomic circumstances, and so on. The lower-right quadrant represents the complex systems that determine potential impacts and potential adaptation. Potential impacts are typically described by scenario-based data that quantify the nature and magnitude of climate variation (exposure to climate change) and the degree to which people or places could be affected (sensitivity). Examples include changes in climatic variables (i.e., temperature and precipitation), extreme events (i.e., drought and sea level rise), and demand for particular resources (i.e., water, food, energy). The human, technological, and financial resources that are available in a given location characterize potential adaptation (also known as adaptive capacity), which interacts with and shapes both exposure and sensitivity.

The quadrants on the left side of Figure 15.1b consist of interior dimensions, which cannot be easily or directly observed by others, that represent "subjective" vulnerability. These subjective, or interior, dimensions mediate the objective, or exterior, dimensions on the right. In the upper-left quadrant are personal perspectives based on individuals' perceptions of risk, their knowledge, and their experience. These include affective appraisals of impacts, consequences, and possible coping responses that initiate powerful rationalizing defenses for some and validation for others. On one hand, direct, personal experience of climate change can make people feel more exposed and at risk, especially when the impacts of climate change affect what individuals value in terms their cultural and/or spiritual identity, their sense of place, their way of life, and their visions of the future (O'Brien and Wolf 2010). On the other hand, the absence of close encounters with climate change can lead to overconfidence in one's coping ability and even an emotional distancing from the implications of climate variability and change (Norgaard 2011).

The lower-left quadrant represents cultural perspectives that structure the other three perspectives by reinforcing assumptions, mind-sets, and worldviews. These social constructions are called frames and are used to link concepts that give meaning to vulnerability. Frames are sense-making devices that enact a particular interpretation of a phenomenon by emphasizing select information (de Boer et al. 2010). They can be considered a product of the values and priorities of the observers. For example, natural scientists and those oriented toward scenarios of the future tend to emphasize exposure to climate change and frame vulnerability as an *outcome* of potential impacts, whereas social scientists and those interested in the broader set of stressors people face draw attention to adaptive capacity and express vulnerability in terms of the *context* that describes local circumstances (O'Brien et al. 2007). Frames are powerful communication devices and largely determine the kind of analysis that is done and the responses that are legitimate and prioritized (Hulme 2009).

Spatial representations of exposure, sensitivity, and adaptive capacity that locate objectively vulnerable people or places serve policy makers' needs for simple, generalized, and actionable information, but these exterior dimensions only partially describe the phenomenon that is vulnerability. The dominance of so-called objective measures in assessment activities downplays the role of subjectivity in explaining the causes of vulnerability and identifying appropriate responses (O'Brien 2009). Integral theory provides a useful framework for identifying adaptation needs because it rejects the notion that vulnerability – and climate change, more generally – is any one singular phenomenon. The four-quadrant model places subjective dimensions of vulnerability on equal footing with more familiar, objective dimensions. This multiple and systematic understanding of vulnerability brings transparency to the shared and private ideologies that act as filters through which adaptation actors interpret climatic risks and judge their implications. Moreover, an integral framework necessitates the use of an inclusive, participatory process that openly reflects the diverse values represented in communities, which is vital to successful adaptation planning.

GIS: Practices and Challenges

GIS is generally defined as a set of practices that act across various scales – geographic, temporal, and governance – to solve spatial problems (Schuurman 2004). Conventional GIS practice pertains to the measurement, analysis, and display of geographically referenced data that describe the natural and social worlds. GIS is commonly used for generic tasks in public administration, such as maintaining inventories of resources and assets, exchanging data across departments and agencies, and producing illustrations for communications. Through the techniques of spatial query (i.e., data processing), GIS also enables analysts to describe and predict spatial patterns that inform public policy. This practice assumes a particular understanding of GIS as essentially positivist, quantitative, and grounded in Cartesian geometry. It is a largely deterministic and instrumental practice based on deductive reasoning and objectivity. However, social theorists question the neutrality of GIS, especially with regard to how spatial problems are represented in GIS and thereby addressed through policy (Schuurman 2000).

GIS is both a storage system and a communication platform for geographic information: a structured set of spatially referenced data, which functions as a model of reality, is stored in a database; once located in a central repository, the data can be easily analyzed and used to visualize an approximation of what is known about the world through maps and other types of output. Every application of GIS requires attention to what should be represented and how. As with all representations of the world, GIS analysts must somehow limit the amount of detail that is captured when developing a geographic database and creating a visualization. Choices are made throughout the process, and all GIS representations are necessarily partial because they must ignore real-world features that are too complex or otherwise ambiguous. Maps and other kinds of GIS representations are therefore contingent on how the four dimensions of inquiry – ontology, epistemology, methodology, and space – are treated in the process of map making (Eddy 2006, 2008). Consequently, the practice of GIS simultaneously describes, masks, and distorts reality (Carolan 2009).

Theoretical discourses such as critical realism (Danermark et al. 2002) and postnormal science (Brown et al. 2010) suggest that objectivity as a completely detached and value-neutral perspective does not exist in practice. All observations are theory laden, and all actors are inherently biased by their cultural experiences and worldviews. Yet there is a widespread belief that externally measureable, quantitative data are somehow independent and therefore objective and free from bias. Kwan (2002) writes about the subjectivity of those who use GIS and the decisions involved in simplifying spatial models to build representations of the world. Different individuals see the world in different ways, and what is left out of a representation is often just as important as what is included. Today, after more than a decade of debate among GIS scholars and practitioners, it is commonly understood that GIS does not present a value-neutral view of the world; rather, it provides a formal framework for the reconciliation of different worldviews. This postpositivist ethic recognizes the contingency of mapped data, which is to say that geographic information is neither right nor wrong but contingent on who, what, where and when, and how it is mapped. As a result, GIS has evolved to accommodate subjective appraisals of physical and human realities

Table 15.1. *Comparisons between conventional GIS and qualitative GIS*

	Conventional GIS	Qualitative GIS
Approach	Top-down: Expert driven based on aggregated indicators	Bottom-up: Guided by the preferences and values of local-level actors
Data	Cartesian spatiality: Discrete objects, continuous fields, raster cells, and their attributes	Non-Cartesian spatiality: Complex connections, human experience, collective meanings
Methods	Deductive analysis: Cluster detection, density and distance, map algebra, regression	Inductive analysis: Surveys, interviews, focus groups, geocoding/geotagging, triangulation
Scale	Generalizes local variation: Autocorrelation coefficients, geographically weighted regression	Emphasizes interactions across scales: Multidimensional visualization; spatial, temporal, thematic navigation
Outcomes	Data visualization: Generalization, statistical representations	Knowledge production: Explanation, theoretical representations

by incorporating qualitative research methods and mixed data types. Many technical barriers to data integration have fallen, and qualitative GIS has emerged as a practice that acknowledges the positionality of GIS and the knowledge that can be produced with it (Wilson 2009).

The practice of qualitative GIS prioritizes stakeholder involvement to broaden the knowledge base when characterizing spatial problems. It is assumed that better decisions are implemented with less conflict and more success when stakeholders drive them (McCall and Dunn 2012). Table 15.1 compares conventional and qualitative forms of GIS practice. Both are associated with graphical representations (i.e., maps) that help decision makers understand and manage a host of human activities. However, the outcomes of the respective practices differ and are achieved through contrasting approaches, data inputs, methods, and representations of space. Much like conventional GIS practice, qualitative GIS pertains to the measurement, analysis, and display of data, but it seeks situated understandings of social and environmental problems rather than statistical descriptions or generalizable predictions. Qualitative GIS adapts existing geospatial techniques for the interpretative analysis of geographic information – which is often expressed as narratives, texts, photographs, drawings, videos, and animations – to represent people's "lived experiences" (Kwan 2002). Where conventional GIS relies on a mastery of spatial science techniques, quantitative methods, objective ways of knowing, and positivist methods for decision making, qualitative GIS attempts to model human reasoning, relying on qualitative research methods and promoting subjective ways of knowing that result in more inclusive policy interventions (Cope and Elwood 2009).

When it comes to assessing climate vulnerability, conventional GIS summarizes important demographic, environmental, and socioeconomic conditions (e.g., Preston et al. 2009), whereas qualitative GIS concentrates on the knowledge and experience of populations at risk (e.g., Tembo 2013). Although top-down approaches are necessary to ensure that vulnerability assessments are linked to policy, bottom-up approaches are also needed to reflect the values and beliefs in the community (Yuen et al. 2013). Because both forms of GIS inform adaptation in meaningful ways, a critical reinterpretation of GIS is needed that bridges conventional and qualitative GIS practices: a set of practices that facilitates dialogue in the contested space between ephemeral and measurable knowledge where several interpretations of vulnerability can be valid at the same time, a new understanding of GIS that eschews dualisms like qualitative–quantitative or subjective–objective, operating rather within a unit of both–and that mutually legitimates opposing perspectives by combining them into a common vision.

Linking Integral Theory and GIS

Integral GIS is a discursive strategy to merge conventional and qualitative GIS practices where different forms of knowledge and knowledge seeking are combined to generate more comprehensive models of reality. Integral GIS is enacted through a set of practices that are always being negotiated and open to experimentation. It is an iterative and reflexive process that facilitates social learning where adaptation actors – government representatives, outside experts, and various publics – work together to reconcile their different viewpoints. The purpose is to support collaborative processes in which diverse participants author flexible spatial narratives that explore different problem understandings as a means of generating a common, shared understanding.

To obtain a comprehensive mapping of vulnerability, highly contextual knowledge must augment systemic or rule-based knowledge. The premise is that the integration of both subjective and objective dimensions of vulnerability will yield insights by exposing the discrepancies between the stories told by different types of knowledge. Figure 15.2 outlines the methods and data for mapping vulnerability, thereby illustrating how integral GIS creates a framework to develop a multiple and systematic understanding of the mechanisms by which climate change interacts with the local context. Conventional GIS is used to analyze objective, exterior dimensions where traditional geospatial techniques are used to map aggregated indicators of vulnerability. The weighting of indicators is often contentious, but garnering input from stakeholders can increase the legitimacy of the resulting vulnerability index. This can be achieved through either direct engagement with stakeholders (e.g., Preston et al. 2009) or software that allows individuals to customize the weightings interactively (e.g., Carter et al. 2014). Mapping with conventional GIS yields information about objective vulnerability – negative outcomes and the context in which they occur – but generally provides too little detail about the causes or implications of vulnerability to discern specific adaptation interventions (Rosentrater 2010; Yuen et al. 2013).

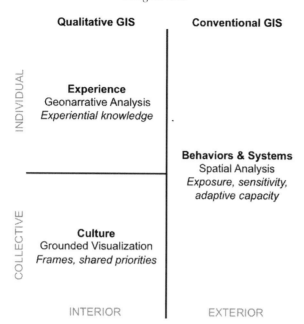

Figure 15.2. GIS methods and representative information for assessing vulnerability in an integral framework. This framework can be applied to any multiactor, multicriteria decision context.

Qualitative GIS, conversely, delivers interior dimensions that generate a deep understanding of vulnerability, revealing multiple entry points for adaptation. Geo-narrative analysis (Kwan and Ding 2008) is used to analyze the experiences of those who are objectively vulnerable through their tacit knowledge, personal beliefs, and emotions. Personal experience with local changes or extreme events generates information about the determinants of vulnerability that is not accounted for in externally measurable indicators. At the same time, stories of the people, places, and institutions that foster cohesion are useful for generating ideas about potential adaptations. These subjective mappings of vulnerability can also be collected through direct engagements with stakeholders (e.g., Tembo 2013) or software that allows various media types (i.e., text, photos, and sound) to be attached to specific geographic entities in the database (e.g., Rinner et al. 2008).

It is worth noting that there is no definite correspondence between data types and the various perspectives of vulnerability; interior dimensions are not represented exclusively by qualitative data, nor are exterior dimensions solely quantitative. Geo-narratives are generated through numbers, words, and multimedia so personal perspectives of vulnerability can be qualitative and quantitative as well as spatial and aspatial. Similarly, although exposure and sensitivity data are represented by direct variables that are always quantitative, adaptive capacity is often represented by qualitative data that provide insights into the interaction between people and their physical and social surroundings (e.g., O'Brien et al. 2004).

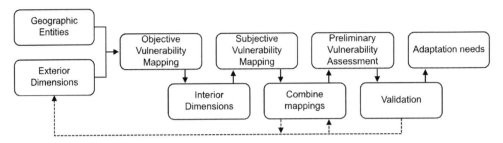

Figure 15.3. Idealized workflow for assessing vulnerability with integral GIS to identify adaptation needs. Solid lines denote the direction of key steps in the process; dashed lines indicate opportunities for iteration.

To combine the mappings of objective and subjective vulnerability, grounded visualization (Knigge and Cope 2006) is used to identify the relationships between the different perspectives, revealing the ways in which vulnerability is framed and, subsequently, how adaptation is conditioned. Careful design of the engagement process is important so that conflicts can be aired and meaningfully resolved. Neutral facilitators – who act as knowledge brokers, promoting the potential value of different types of knowledge to adaptation actors – are helpful in weaving together, to the extent possible, the various perspectives and negotiating a common viewpoint. The value of integral GIS lies not in the results of any single query or visualization but in the range of explorations and analyses that can be carried out by closing information gaps, enhancing learning, and reconciling competing demands. Integral GIS takes the most passionately held parts of competing points of view and combines them to identify what is important to adaptation actors. Taken as a whole, the methods associated with integral GIS represent recursive and reflexive strategies that facilitate the discovery of patterns and processes in a way that is impossible when only one set of data or methodology is used.

Applying Integral GIS

Vulnerability assessments are social arenas where ideas about adaptation needs are developed. Examining vulnerability from different vantage points shows how the concept of vulnerability carries a variety of meanings that imply quite different responses. Figure 15.3 illustrates an idealized workflow for identifying adaptation needs using integral GIS. Integral GIS is a platform for both data management and data analysis throughout the vulnerability assessment. But it should also be seen as process for knowledge construction where the focus is on the collaborative extraction of information from data and shared meaning from that information (Eddy 2006). The act of mapping objective vulnerability and subjective vulnerability creates boundary objects for adaptation actors to consult and negotiate. The visual tools embedded in GIS allow actors to access mapped locations to which arguments refer, attach a geographical reference to each new argument, and connect new arguments to a map object or region. By jointly interacting with the maps, actors can begin to understand where their conceptions of vulnerability coincide and where they differ. Differences in interpretations are reconciled by combining the mappings through moderated discussions in the manner of participatory GIS (Jankowski 2009). The objective is to work iteratively, toward a shared understanding of vulnerability, by juxtaposing disparate representations to explicitly demonstrate the partiality of each point of view (Ramsey 2008). This involves social learning that exposes the assumptions, limitations, and distortions contained within different understandings of vulnerability, including the partiality of one's own interpretation. Mapping vulnerability can thereby become more grounded, more systematic, and more relevant than the results from conventional assessments.

Data collected in Norway for a regional assessment of health impacts in the Nordic countries (Carter et al. 2014) can be used to illustrate how integral GIS leads to a broader and deeper view of vulnerability than conventional vulnerability assessments and reveals multiple entry points for adaptation. We assessed the vulnerability of people aged sixty-seven and older to extreme weather events such as high summer temperatures, low winter temperatures, and freeze–thaw conditions. Because of their diminished ability to regulate body temperature, the elderly are more sensitive to extreme temperatures than younger adults. They are also more likely to live alone, have reduced social contacts, and experience poor health, all of which contribute to their general vulnerability. Moreover, the elderly are a growing population cohort in Norway and thus represent an emerging issue for public health planners. The project was concerned with developing alternative approaches for characterizing objective vulnerability rather than informing a specific adaptation process, but informal conversations with stakeholders provided information post hoc to sketch out some subjective perspectives of vulnerability and a range of adaptation needs indicated by these perceptions.

We began by mapping objective vulnerability through an online interface that allowed vulnerability indicators to be weighted interactively.[1] Sociodemographic characteristics at the municipal level were combined with potential impacts (indicators of exposure and sensitivity) and potential adaptation (indicators of adaptive capacity) to reveal differential vulnerability across Norway. The pattern of vulnerability changed depending on the particular combination of indicators used and their ratios to one another. However, generally speaking, rural areas – which are gradually losing inhabitants, thus reducing local services and creating an aging demographic profile – show the highest levels of vulnerability to extreme events.

---

[1] Complete details about the data and interactive tool are presented in Carter et al. (2014).

Exploratory interviews based on snowball sampling were conducted prior to a stakeholder workshop to evaluate the mapping tool. We contacted two types of stakeholders – representatives of national and regional organizations who have responsibility for care of the elderly, and individuals over the age of sixty-seven – to ask about the effects of temperature and other weather-related challenges for the elderly. Interestingly, the interviews revealed that the issue of climate change vulnerability among the elderly was not relevant to the activities or interests of any of the stakeholders. However, from these discussions, a picture of subjective vulnerability (or lack thereof) begins to emerge. The health sector representatives expressed skepticism that vulnerability of the elderly was an issue worthy of examination, believing instead that climate-sensitive diseases like Lyme disease pose a bigger risk to Norwegian public health. They acknowledged the threat of climate change but felt that adaptive capacity at the national level would facilitate autonomous adaptation as necessary. The elderly individuals we contacted perceived themselves, by and large, as neither vulnerable nor as being elderly, even if, according to external measures, they could be regarded as both. They focused on their resilience and ability to avoid the worst parts of winter by traveling to warmer places like Spain for months at a time (Ruud 2010). Instead, it was the implications of climate change on future generations that resonated most with this group, especially on a political level. Many of the individuals with whom we spoke were involved with Besteforeldres klimaaksjon (Grandparents' Climate Campaign) and were active in lobbying the Norwegian government for domestic cuts in carbon emissions.

Most vulnerability assessments stop at the point of mapping objective vulnerability. If they do engage with stakeholders, the purpose is often to elicit how the mapping might be refined to enhance its usefulness (Yuen et al. 2013). Our project on the health impacts of climate change in the Nordic countries was no different. The mapping of objective vulnerability we developed reveals the increasing risks of extreme weather events and their impacts on the elderly, especially in rural areas. The mapping frames climate change as a technical problem that can be managed by applying established know-how and procedures. Given the emphasis on the contextual conditions that determined vulnerability in the objective mapping, the adaptation needs indicated here might include increases in economic, human, and technological capital so that communities can better respond to extreme events. The stakeholder feedback we received indicated that municipality-scale information is only useful on national to regional scales and that local decision makers require more spatial detail to inform their adaptation processes (Carter et al. 2014). However, had we taken an integral approach to assessing vulnerability, we could have revealed a broader suite of potential adaptations that are applicable across a range of scales.

On the basis of the information derived from the exploratory interviews, the subjective mapping in this case would likely have included expressions of uncertainty, perhaps linked to a new round of objective mapping assessing climate-sensitive disease risks, to address the concerns expressed by the caregivers. Additional data might have included spatial links to areas in Europe elderly Norwegians are known to travel and the health risks that exist there, or more explicit consideration of the temporal dimension and the effect of climate change on future generations. These new features suggest several potential adaptations not revealed by the objective mapping, among them research on how best to safeguard vulnerable populations given that there is uncertainty about specific impacts; awareness raising about early warning systems in popular travel destinations and translation services for making use of them; and mitigation of anthropogenic climate change to reduce the exposure of future generations. The advantage of integral GIS is the ability to reveal what cannot be measured directly (i.e., beliefs and emotions) and to provide input to adaptation processes that is more sensitive to contextual dynamics than statistical measures alone. Without this level of nuance and detail, vulnerability assessments risk providing limiting information about potential adaptation needs.

Conclusion

Vulnerability maps are simple and powerful tools that communicate the risks associated with climate change. The information they provide is intended to help identify adaptation needs, but they rarely succeed in representing the multidimensional nature of vulnerability. A subtle reductionism occurs when subjectivity is overlooked, and this chapter has discussed the importance of augmenting rule-based knowledge about vulnerability with contextual knowledge that situates the experiences and concerns of all adaptation actors. Subjective, interior dimensions of vulnerability affect not only outcomes but also the priorities people set with regard to adaptation. To identify multiple entry points for adaptation, vulnerability assessments must therefore ask why and to whom climate impacts matter.

Integral GIS is a facilitated multistakeholder approach that supports public participation, discussion, and analysis in spatial decision making. The approach incorporates multiple modes of inquiry and data types by fusing qualitative research methods and conventional GIS practice. Using standard tools embedded in GIS, adaptation actors can exchange information, compare perspectives, and build a collective model of vulnerability that incorporates the interests and values of diverse social groups. To integrate these views, vulnerability must be problematized in a way that is understandable, relevant, and personally important. The process must be true to the underlying science of vulnerability while negotiating between possible, probable, and preferable outcomes. By using the deliberative aspects of spatial decision making, data and sense making can be brought together to answer questions about how projected outcomes of climate change may affect the lives and values of those who can be considered objectively vulnerable. Making the interior dimensions of vulnerability more transparent and the process of mapping vulnerability more open to question and debate is one way to reduce the objective veneer currently given to adaptation.

References

Adger, W. N. 2006. "Vulnerability." *Global Environmental Change-Human and Policy Dimensions* 16 (3): 268–81.
Brown, V., Harris, J. and J. Russell. 2010. *Tackling Wicked Problems: Through the Transdisciplinary Imagination*. London: Earthscan.
Carolan, M. S. 2004. "Ontological politics: Mapping a complex environmental problem." *Environmental Values* 13 (4): 497–522.
Carolan, M. S. 2009. "This Is Not a Biodiversity Hotspot: The Power of Maps and Other Images in the Environmental Sciences." *Society and Natural Resources* 22 (3): 278–86.
Carter, T., Fronzek, S., Inkinen, A., Lahtinen, I., Mela, H., O'Brien, K., Rosentrater, L., Ruuhela, R., Simonsson, L. and E. Terämä. 2014. "Characterising vulnerability of the elderly to climate change in the Nordic region." *Regional Environmental Change*. doi:10.1007/s10113-014-0688-7.
Cope, M. and S. Elwood, eds. 2009. *Qualitative GIS: A Mixed Methods Approach*. London: Sage.
Danermark, B., Ekström, M., Jakobsen, L. and J. Karlsson. 2002. *Explaining Society: Critical Realism in the Social Sciences*. London: Routledge.
de Boer, J., Wardekker, J. A. and J. P. van der Sluijs. 2010. "Frame-based guide to situated decision-making on climate change." *Global Environmental Change-Human and Policy Dimensions* 20 (3): 502–10.
Eddy, B. 2005. "Integral geography: Space, place, and perspective." *World Futures: The Journal of General Evolution* 61 (1–2): 151–63.
Eddy, B. G. 2006. *The Use of Maps and Map Metaphors for Integration in Geography: A Case Study in Mapping Indicators of Sustainability and Wellbeing*. PhD dissertation, Department of Geography and Environmental Studies, Carleton University, Ottawa, Canada.
Eddy, B. 2008. "AQAL topology: An introduction to integral geography and spatiality." *Journal of Integral Theory and Practice* 3 (1): 184–98.
Esbjörn-Hargens, S. 2005. "Integral ecology: The what, who and how of environmental phenomena." *World Futures* 61 (1–2): 5–49.
Esbjörn-Hargens, S. 2009. "An overview of integral theory: an all inclusive framework for the 21st century." Resource Paper 1. Denver, CO: Integral Institute.
Hulme, M. 2009. *Why We Disagree about Climate Change: Understanding Controversy, Inaction and Opportunity*. Cambridge: Cambridge University Press.
Jankowski, P. 2009. "Towards participatory geographic information systems for community-based environmental decision making." *Journal of Environmental Management* 90 (6): 1966–71.
Knigge, L. and M. Cope. 2006. "Grounded visualization: Integrating the analysis of qualitative and quantitative data through grounded theory and visualization." *Environment and Planning A* 38 (1): 2021–37.
Kwan, M. P. 2002. "Feminist visualization: Re-envisioning GIS as a method in feminist geographic research." *Annals of the Association of American Geographers* 92 (4): 645–61.
Kwan, M. P. and G. X. Ding. 2008. "Geo-narrative: Extending Geographic Information Systems for narrative analysis in qualitative and mixed-method research." *Professional Geographer* 60 (4): 443–65.

McCall, M. K. and C. E. Dunn. 2012. "Geo-information tools for participatory spatial planning: Fulfilling the criteria for 'good' governance?" *Geoforum* 43 (1): 81–94.

McCarthy, J. J., Canziani, O. F., Leary, N. A., Dokken, D. J. and K. S. White, eds. 2001. *Climate Change 2001: Impacts, Adaptations, and Vulnerability*. Cambridge: Cambridge University Press.

Metzger, M. J. and D. Schroter. 2006. "Towards a spatially explicit and quantitative vulnerability assessment of environmental change in Europe." *Regional Environmental Change* 6 (4): 201–16.

Norgaard, K. M. 2011. *Living in Denial: Climate Change, Emotions, and Everyday Life*. Boston: MIT Press.

O'Brien, K. 2009. "Responding to climate change: The need for an integral approach." Resource Paper 4. Denver, CO: Integral Institute.

O'Brien, K. L. and J. Wolf. 2010. "A values-based approach to vulnerability and adaptation to climate change." *Wiley Interdisciplinary Reviews – Climate Change* 1 (2): 232–42.

O'Brien, K., Eriksen, S., Nygaard, L. P. and A. Schjolden. 2007. "Why different interpretations of vulnerability matter in climate change discourses." *Climate Policy* 7 (1): 73–88.

O'Brien, K., Leichenko, R., Kelkar, U., Venema, H., Aandahl, G., Tompkins, H., Javed, A., et al. 2004. "Mapping vulnerability to multiple stressors: Climate change and globalization in India." *Global Environmental Change – Human and Policy Dimensions* 14 (4): 303–13.

Preston, B. L., Brooke, C., Measham, T. G., Smith, T. F. and R. Gorddard. 2009. "Igniting change in local government: Lessons learned from a bushfire vulnerability assessment." *Mitigation and Adaptation Strategies for Global Change* 14 (3): 251–83.

Preston, B. L., Yuen, E. J. and R. M. Westaway. 2011. "Putting vulnerability to climate change on the map: a review of approaches, benefits, and risks." *Sustainability Science* 6 (2): 177–202.

PROVIA. 2013. *PROVIA Guidance on Assessing Vulnerability, Impacts and Adaptation to Climate Change*. Nairobi, Kenya: United Nations Environment Programme.

Ramsey, K. 2008. "A call for agonism: GIS and the politics of collaboration." *Environment and Planning A* 40 (10): 2346–63.

Reason, P. and H. Bradbury, eds. 2008. *The SAGE Handbook of Action Research: Participative Inquiry and Practice*. London: Sage.

Rinner, C., Kessler, C. and S. Andrulis. 2008. "The use of Web 2.0 concepts to support deliberation in spatial decision-making." *Computers Environment and Urban Systems* 32 (5): 386–95.

Rosentrater, L. D. 2010. "Representing and using scenarios for responding to climate change." *Wiley Interdisciplinary Reviews – Climate Change* 1 (2): 253–59.

Rosentrater, L. D., Saelensminde, I., Ekstrom, F., Bohm, G., Bostrom, A., Hanss, D. and R. E. O'Connor. 2013. "Efficacy trade-offs in individuals' support for climate change policies." *Environment and Behavior* 45 (8): 935–70.

Ruud, C. 2010. "'Vi har det fint her nede' Klimasårbarhet blant norske eldre – oppfatninger av klimaendringer og implikasjonene for tilpasning" ["We're just fine down here": Climate vulnerability among Norwegian seniors – perceptions of climate change and the implications for adaptation]. MA thesis, Department of Sociology and Human Geography, University of Oslo.

Schuurman, N. 2000. "Trouble in the heartland: GIS and its critics in the 1990s." *Progress in Human Geography* 24 (4): 569–90.

Schuurman, N. 2004. *GIS: A Short Introduction*. Malden, MA: Blackwell.

Tembo, M. D. 2013. *A Dynamic Assessment of Adaptive Capacity to Climate Change: A Case Study of Water Management in Makondo, Uganda*. PhD dissertation, Department of Geography, National University of Ireland, Maynooth.

Turner, N. J., Gregory, R., Brooks, C., Failing, L. and T. Satterfield. 2008. "From invisibility to transparency: Identifying the implications." *Ecology and Society* 13 (2): 7. http://www.ecologyandsociety.org/vol13/iss12/art17/.

UKCIP. 2013. "The UKCIP Adaptation Wizard v 4.0." http://www.ukcip.org.uk/wizard/.

Wilson, M. 2009. "Towards a genealogy of qualitative GIS." In *Qualitative GIS: A Mixed Methods Approach*, edited by Cope, M. and S. Elwood, 156–70. London: Sage.

Yuen, E., Jovicich, S. S. and B. L. Preston. 2013. "Climate change vulnerability assessments as catalysts for social learning: Four case studies in south-eastern Australia." *Mitigation and Adaptation Strategies for Global Change* 18 (5): 567–90.

16

There Must Be More

Communication to Close the Cultural Divide

SUSANNE C. MOSER AND CAROL L. BERZONSKY

Politicization of science and polarization of public opinion is a common phenomenon around climate change. Many countries have seen a hardening of opposing positions on climate change, accompanied by considerable cynicism, distrust, blame, and negative sentiment (e.g., Hulme 2009; Norgaard 2006; Poortinga et al. 2011; Reser et al. 2012; Whitmarsh 2011). Nowhere is the phenomenon arguably as severe and of such devastating global significance as in the United States. This chapter focuses on the political and ideological divides on climate change in the United States to explore what lessons they may hold for communication and engagement across cultural and ideological divides more generally. Overcoming this impasse constitutes an adaptive challenge par excellence: forward movement will only happen if significant actors and segments of the population reach out to each other, engage, and mobilize for dialogue and action.

The premise from which we launch our argument is a seemingly intractable situation whereby politicization of science and ideological polarization are becoming nearly as solidly established as the ever-clearer scientific consensus on human-caused climate change. To those who are convinced that climate change is a serious problem, this dismal state of public discourse is caused by powerful political influences intentionally distorting public understanding and aiming to stall policy progress (e.g., Lahsen 2005; Oreskes and Conway 2010). To them this is deeply frightening in light of the climate crisis they perceive. From the other end of the political spectrum, accepting climate change as a fact looks like the "beginning of the end" of personal freedoms; it is to give in to the left's alleged attempt to control all aspects of our private lives, and so climate change is declared a "hoax" (first stated by U.S. Senator J. Inhofe in 2003; see Inhofe 2011) or at least an unproven theory, maybe a natural phenomenon, and in any case is too uncertain to demand serious attention or action. As Frank (2011) recently noted, conservatives view the state of the world and of national politics in an equally apocalyptic way as the left. Each side, however, sees different problems and different villains at work causing them (Frank 2011).

Repeated public opinion polls and analyses over the last few years have confirmed this polarizing trend in U.S. climate change opinions (Antonio and Brulle 2011; Dunlap and McCright 2008; Leiserowitz et al. 2011b, 2011c; McCright and Dunlap 2011; Mooney 2012). This trend must be seen in the context of a broader trend toward polarization in American politics (e.g., Poole 2008, 2012) and considerable changes in the economic and media landscapes in the United States (Boykoff 2011; Brulle et al. 2012). Serious analysts, unconscious communicators, and intentional polemicists on all sides have felt free and justified to brand those on the other extreme with unfortunate labels and thus to inflame animosities and further increase the polarization (e.g., Clynes 2012; O'Neill and Boykoff 2010; Powell 2011; U.S. Senate Committee on Environment and Public Works 2010; Washington and Cook 2011). For example, former U.S. Vice President Al Gore has been defamed for his stance on climate change; climate scientists are receiving death threats and hate mail; those wishing to see climate action have been called "Communists" or "watermelons" – green on the outside, red on the inside; meanwhile those questioning the reality, human causation, or need for action on climate change have been labeled skeptics, contrarians, deniers, and criminals. Thus, social scientist, Andrew Hoffman (2011, 3) concluded, "the debate appears to be reaching a level of polarization where one might begin to question whether meaningful dialogue and problem solving has become unavailable to participants."

In this chapter, we are interested in exploring openings for meaningful and constructive communication. After summarizing trends in recent opinion polls and the range of explanations for the current state of affairs in the second section, we explore the opportunities and goals of communicating with those who are not heard in public discourse or who don't have enough forums to explore the climate change issue, along with their own views and those of others. The fourth section then turns to the currently opposed sides in the climate debate and asks what possibilities exist for them to come together. The following section explains how dialogue may be the most appropriate forum for deliberate attempts in finding common ground and describes both the forms and ways in which we may move from cultural and ideological divides to interpersonal opening and understanding. Finally, we offer a brief conclusion, emphasizing the common denominator we see in engaging all segments of the population, whether they are or are not currently productively engaged.

While dialogue may not be possible with everyone, and some almost certainly will not be interested in sincere exchange, we believe that further diatribe and debate between the extreme factions will only fortify the political stalemate. Conversely, greater engagement of the American public, especially of those holding less extreme views, through dialogue is possible and can be scaled up to change the social and political climate and enable action.

American Attitudes on Climate Change: Polarization and Silence

Evidence of polarized views on climate change is easily found in the news and on the internet. The portrayal found there would lead one to believe that there is nothing but extreme pro or con opinions on the topic. Opinion polling indeed finds that Americans largely occupy two increasingly separate camps, the Democratic, liberal-leaning, climate-change-accepting wing and the Republican, conservative-leaning, climate-change-skeptical wing, with a quiet, seemingly unimportant contingent in the middle (Dunlap and McCright 2008; McCright and Dunlap 2011).

In actuality, it is more accurate to state that the two polarizing camps in the United States make up the extreme ends of a more diverse opinion spectrum.

Since 2008, researchers at Yale University and colleagues at George Mason University have tracked public opinion and divided the American public into six distinct opinion segments, "Global Warming's Six Americas," based on underlying value commitments and beliefs: the Alarmed, Concerned, Cautious, Disengaged, Doubtful, and Dismissive. These six segments vary by degree along the more egalitarian–communitarian versus hierarchist–individualist value spectra at the heart of cultural theory (e.g., Kahan 2007; Leiserowitz et al. 2009; Thompson et al. 1990). While segment sizes have varied in important ways in the last five years, according to data from September 2012, the Alarmed, constituting 16 percent of the population, fall on one of the extreme ends of the spectrum. People in this category are those most engaged on climate change. They are strongly convinced global warming is happening, that it is human caused, and that it is a serious and urgent threat. This segment of the American public is also most likely to be behaviorally, civically, communicatively, and politically involved in climate change and – either through votes or support for pro-action interest groups – support an aggressive national policy response (Leiserowitz et al. 2012).[1] The next, and typically largest, segment – the Concerned (29 percent) – is also convinced that global warming is a serious problem and supports a strong policy response, but individuals in this group are less personally involved in the issue and are taking fewer actions. The third segment is referred to as the Cautious (25 percent), which includes those who believe global warming is a problem but are less certain that it is happening than the Alarmed or the Concerned, less certain that it is human caused, and do not view it as a personal threat or feel a sense of urgency to respond. The fourth group, the Disengaged (9 percent), does not know or think much about the issue and holds no firm beliefs about climate change. The fifth category

---

[1] It is important to note that being categorized as "Alarmed" does not mean that everyone is publicly vocal about her views and concerns, out in the streets demonstrating, or alarmist in his public rhetoric. The segmentation underlying the Six Americas is based on value commitments and beliefs, not actions. Climate-relevant behaviors (such as energy savings, transportation choices) vary remarkably little across all segments of the Six Americas (Leiserowitz et al. 2009, 2011b). However, the Alarmed, like the Dismissive (discussed later), are more likely than all others to be civically and politically active on the issue. This is one of the reasons why their voices are more readily represented in the mass media.

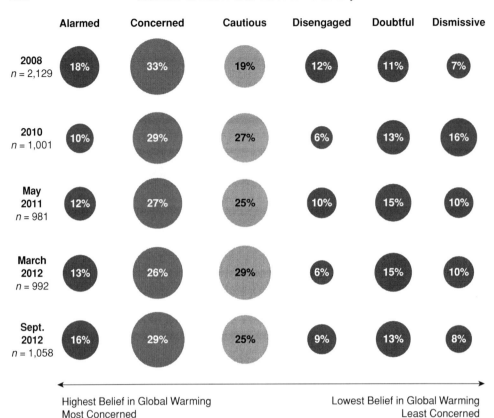

Figure 16.1. Proportion of the U.S. adult population in the Six Americas, 2008–12. *Source*: Composite constructed from findings in Leiserowitz et al. (2008, 2010, 2011c, 2012, 2013).

is made up of the Doubtful (13 percent), who have mixed beliefs in the reality of climate change. To the extent they believe the problem exists at all, they see it as a natural phenomenon and of little concern in the near term. Finally, at the other end of the spectrum, and in 2012 comprising the smallest (but highly visible and outspoken) segment of the population, are the Dismissive (8 percent). People in this category, similar to the Alarmed, are very engaged in the issue but insist that global warming is not happening. And even if it is recognized to exist, it is considered natural and not a threat to either people or the environment, thus it does not warrant a national policy response. Figure 16.1 is a composite of findings from Leiserowitz and colleagues on how the proportions of the "Six Americas" have changed between 2008 and 2012. It illustrates how there have been repeated shifts over time into and out of the middle of the spectrum, with politically significant shifts in the proportion of the most visible segments – the Alarmed and the Dismissive. While statistically the situation at present is an almost complete return to the status in 2008 (prior to the scandal of stolen e-mails from the University of East Anglia, the discovery of several mistakes in the Fourth Assessment report of the IPCC, and the effects of the global economic crisis of 2008), the numbers hide the fact that the stances taken on both ends of the spectrum have hardened and seem politically and culturally more polarized than five years earlier.

Despite important variations over time, the existence and demographic makeup of the Six Americas has been remarkably stable over time (see Figure 16.1). In late 2012, just like any other year that the survey has been conducted, a large majority of Americans (76 percent) does *not* fall into the extreme categories of responses to global warming. However, the 24 percent who are most engaged and most vocal, sometimes called the issue public (Converse 1964; Han 2009; Krosnick 1990), is the portion of the population whose views are heard most frequently in the media, while the less extreme, more malleable views of the majority are not represented and thus essentially absent from public discourse (Balbus 2012).

Social scientists have taken up the issue of polarization, in no small part driven by the fact that those most vocal and engaged on the dismissive end of the spectrum have had an undeniably chilling impact on public policy and action at all levels of government (e.g., Fears 2011; Kaufman and Zernike 2012; Peach 2012). Coming from different disciplines and theoretical backgrounds, analysts have put forward various complementary explanations of the polarization, particularly, of "contrarian" and "denialist" opinions:[2]

- the political organization of powerful industry interests, particularly in the context of neoliberal globalizing goals (Dunlap and McCright 2011; Jacques et al. 2008; Lahsen 2005, 2013; McCright and Dunlap 2003, 2010; Oreskes and Conway 2010)
- the role of the media in creating controversy and skepticism (Boykoff and Boykoff 2004; Butler and Pidgeon 2009; Carvalho 2007; Painter 2011)
- the importance of the social organization of belief systems held by different groups in society (Norgaard 2006, 2011)
- the significance of human psychological development toward greater maturity, explaining the deeper psychological patterns underlying attitudes and beliefs (e.g., Cook-Greuter 2008; Hedlund-de Witt and Hedlund-de Witt 2012; Plotkin 2008)
- the psychological responses to existential threats (Dickinson 2009; Fritsche et al. 2010; Pienaar 2011)
- the cognitive underpinnings of strongly held beliefs, particularly "motivated reasoning") (e.g., Hart and Nisbet 2012; Jost et al. 2003, 2009; Mooney 2012; Roseman 1994; Whitmarsh 2011)

---

[2] With the exception of research driven by an interest in cultural values, notably less attention has been paid in social science circles to why certain individuals or groups hold "alarmist" or "green extremist" views. Remarkably little has been said about how to broaden the issue public and engage those with less strongly held views.

- closely related, the cultural value commitments (and possible evolution in cultural values) that allow or prevent us from accepting certain information and beliefs (Crompton 2011; Hulme 2009; Kahan 2007, 2010; Kahan and Braman 2008; Kahan 2010; Kahan et al. 2007a, 2007b, 2012; Kasser 2009; Lahsen 2008, 2013; McIntosh et al. 2012)

Few concrete ideas have been advanced as to how to continue to communicate once an issue has become so polarized, so ideologically driven, and involving such high stakes literally and psychologically. Moreover, there has been limited attention to the communication needs of the three quarters of the "silent," overlooked population in the middle. Thus we see a need to connect with the ignored middle while fostering the collective capacity to retain a functioning democracy, to collaborate and find solutions together, and to rebuild a civic and humane conversation (Palmer 2011). Our own assessment of the situation leads us to prioritize the reengagement of the less ideologically committed (see next section) and subsequently explore ways to bring the more extreme voices into dialogue.

Communicating with the Missing Middle: Possibilities of Greater Engagement

Nothing opens up the mind like the glimpse of new possibility.
– John O'Donohue (2004, 139)

The largest opportunity for an expanded and fuller conversation among Americans of different persuasions lies with the three quarters in the middle. But where to start? Developing possibilities to reach and engage those not represented in the ideologically driven public debate begins with two fundamental questions all communicators must grapple with: *who* are we communicating with, and *what is the goal* of that communication? Leiserowitz and colleagues have developed a compendium of insights about the Six Americas – their demographics, values, political identifications, voting preferences, levels of pro-environmental behavior, civic engagement, energy and climate policy support, beliefs, concerns, key questions, climate literacy, and emotional responses to climate change. What we know from this body of work is that the Six Americas are quite complex. Assuming, for example, that the Alarmed are always highly informed, knowledgeable, and politically, civically or behaviorally active is true relative to the Cautious or Disengaged but disregards the finding that the Dismissive are sometimes even better informed on factual knowledge, they just draw different conclusions from it; it would equally discount that almost all of the subgroups take at least some pro-environmental actions (but surprisingly few, even among the Concerned and Alarmed) (Leiserowitz et al. 2009, 2011b).

There is one thing, however, that the Four Americas in the middle all have in common: they are not heard from in the media. Their opinions, concerns, and viewpoints are not mirrored and their questions are not addressed in the cross-fire of heated debate. Their level of science literacy in general and climate literacy in particular is rather low, leaving them uncomfortable or unable to participate effectively in a debate that is supposedly over facts (Hill 2010; Leiserowitz et al. 2010b). Many have not bothered to learn more because the issue has long been presented as a scientific, uncertain, and complex topic best left to the experts (Marx et al. 2007). Even those who tried have found it difficult, depressing, or overwhelming to contemplate (Meijnders et al. 2001; Moser 2007; Myers et al. 2012). Still not enough understand just how compelling the scientific consensus has grown over the past twenty years (Ding et al. 2011). Because global warming is caused by humans, it requires policy responses in addition to individual actions, but what those large-scale responses are is unclear to many (Leiserowitz et al. 2011b). Moreover, if it were a truly urgent problem, the argument goes, political leaders would address it; but apparently there are more immediate matters to attend to (Brulle et al. 2012). In short, there are personally compelling reasons *not* to engage.

This then raises the question, what would be the goal of "engaging" the middle Four Americas? And what do we really mean by "engagement"? This is not a simple question to answer. The Six Americas segmentation study – based on cultural value commitments and beliefs – is not primarily about behaviors, but rather about convictions, attitudes and opinions. Thus even the most alarmed and dismissive Americans are rarely out in the streets protesting; few of the Alarmed have given up all fossil fuel consumption, and equally few of the Dismissive are actively feeding the disinformation campaign. Some of the Concerned or Cautious may not want to learn more or let themselves be any more disturbed by the facts, but if they saw people like themselves speak out in favor of climate action and were told what to do, they might take or support the right actions. Some of the Doubtful may well come around to understanding that human-caused climate change is real, but may not vote for a carbon tax or a Democratic candidate promising climate action. In other words, engagement can mean many things, and what one may wish to achieve is specific to the group in focus and the intention behind the communication effort.

Table 16.1 lists key types of engagement which differ in depth and direct impact on climate-relevant actions. Importantly, they are not necessarily hierarchical or sequential (i.e., one kind of engagement is not a precondition for the next), though it is unlikely that one-time engagement will achieve more than superficial and impermanent goals. Repeated engagement is often necessary to achieve more significant shifts. While the types of engagement are listed separately, they often occur simultaneously.

One could develop a "cumulative engagement index" (CEI) from empirical data that reflect these eight fundamental types of engagement. Because some but not all of these types of engagement have been surveyed for the Six Americas over the years, we can only offer a *hypothetical* depiction of such a CEI in Figure 16.2 (partial information available in the Six Americas citations listed earlier). Such a graphic serves as an aid in answering the question – at least generically – what one might want to achieve with communication in terms of "engagement." For example, if the overarching goal is to shift toward a broad, widely visible social norm that accepts

Table 16.1. *Typology of engagement with climate change*

Type of engagement	Description with examples
Cognitive	Focus of engagement is internal, in one's mind • Thinking about climate change • Seeking information and learning/teaching about the issue • Grappling with the complexities of climate change (solutions)
Emotional	Focus of engagement is mostly internal, in one's psyche, but may be shared with others • Allowing emotional responses (e.g., fear, anxiety, concern, grief, anger, guilt, passion, disappointment, despair, hope, empathy) to surface • Consciously or unconsciously coping with the emotional impacts of climate change
Behavioral	Focus of engagement is mostly on day-to-day actions • Making periodic or permanent changes in energy consumption in one's home • Shifting travel and transportation-related behavior • Shifting food and eating habits • Reducing material consumption
Professional	Focus of engagement is on climate-related decisions in one's profession, business, work • Making periodic or permanent changes in energy consumption in one's work place • Developing and implementing strategic plans to guard against negative impacts of climate change (or policy) • Developing and implementing strategic plans to take advantage of business opportunities arising from climate change (or policy) (in mitigation and adaptation)
Social	Focus of engagement is on known others, peers, or a social reference group • Communicating with others about climate change • Enacting solutions together with others, supporting each other • Making one's publicly visible behavior help shape new social norms
Moral/spiritual	Focus of engagement is on cultural norms and the transcendent • Being motivated to take action by one's belief system and underlying values • Developing a sense of responsibility toward nature, others, the future • Finding solace in a moral/spiritual conception of the world • Prayer
Civic	Focus of engagement is primarily on the commons • Speaking out about climate change in public • Attending hearings or public meetings • Writing letters to the editor of a newspaper • Participating in protests
Political	Focus of engagement is on the political process • Getting involved in political organizing and campaigning • Voting for candidates representing one's climate-related position • Voting for local/state climate-related initiatives • Running for office to influence policies and decisions

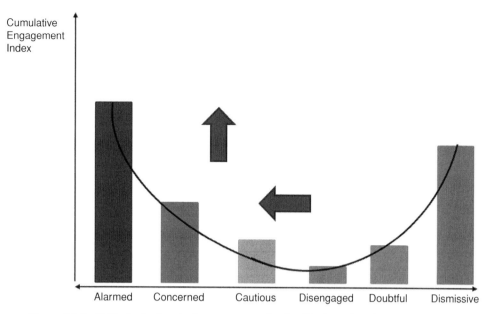

Figure 16.2. Shifts in the level of engagement for the Six Americas based on a hypothetical cumulative engagement index (see text for explanation). *Source*: Concept developed by authors, informed by Leiserowitz et al.'s findings on the Six Americas' engagement around climate change.

and demands more action on climate change, one might want to help elevate the CEI overall to a higher level for the Alarmed, Concerned, Cautious, and Disengaged (arrow pointing upward). The primary focus of effort might be on the cognitive, emotional, social, and behavioral dimensions of engagement. If by contrast (or in addition), the goal is to generate a political momentum for climate action (generically or a particular policy), the focus would be on moving more people into the Alarmed and Concerned columns (arrow pointing toward the left), activating particularly in the political, civic, and moral dimensions of engagement. This is not to make any judgment about the inherent goodness or preference for any particular climate solution (e.g., regulation of CO_2, a market-based cap-and-trade system, or a particular set of energy technologies) but simply to reflect the general preference of the Alarmed and Concerned for action sooner rather than later. We see both of these goals as essential for large-scale climate policy in light of the history of social movements and environmental policy making in the United States: it never started in Washington, but in local communities and state capitols, where progressive ideas were developed first and eventually "trickled up" to Congress when a patchwork of different policies across the nation made business cumbersome (Isham and Waage 2007; Rabe 2004, 2009).

Pragmatically, (re)engaging the middle Four Americas might begin by simply paying attention to them. If the immediate goal is a "good conversation," looking for cognitive, emotional, social, and moral forms of engagement, disengagement can be effectively countered by sincerity and genuine interest in the other: interest in them as individuals, in their lives, their local concerns and worries. Such an approach offers a glimpse not only of "new possibilities" (as John O'Donohue says) and thus opens the door to real connection but also insights into the values and issues that can help make the connection to climate change. It offers possible alternative framings that make the problem more salient and the possible solutions more prominent and relevant. It also opens the floor for a values-based conversation – something we all can participate in (compared to a conversation about atmospheric science). In fact, the tit-for-tat debate over scientific facts is often no more than a thinly disguised debate over the underlying values debaters hold (Sarewitz 2004).

Once people are engaged in a conversation over local concerns and values, no one is "right" or "wrong," but a possibility arises to explore commonalities, differences, and ambiguities. Of course, such conversations can be messy and heated (and are easier when professionally facilitated), but at least they can be held directly. By practicing foundational skills of dialogue such as being fully present, deep listening, respect, self-responsibility, clarity, authenticity, speaking one's own truth, and suspending judgment (Bohm 1996; Brown et al. 2005; Isaacs 1999), a real exchange and learning from and about each other becomes possible. In fact, having such an open and engaging conversation is more likely with the middle Four Americas as their values are closer to each other, less stuck, and more inclusive and compatible than with those who hold extreme positions. A practical example of such a conversation is described in Box 16.1.

Possibilities for a New Start

As we work together to restore hope to the future, we need to include a new and strange ally – our willingness to be disturbed.

– Margaret Wheatley (2002, 34)

Clearly, developing communication channels to the middle Four Americans is not easy. The preceding example, however, provides good evidence that it is possible to bring more people into the climate change conversation. There is no doubt that this opportunity is underutilized. It is also quite possible that the difficulty of doing so fades in comparison with bringing the most extreme sides of the opinion spectrum – for the first time or maybe again – into true dialogue. Here we turn to that more difficult challenge.

"True dialogue" is much more than bringing people together for a shouting match or debate (e.g., Malone 2009). By dialogue we mean "a process for talking about tension-filled topics" (Schirch and Campt 2007, 5) "that aims to build relationships between people as they share experiences, ideas, and information about a common concern" (Schirch and Campt 2007, 6). It is fundamentally different from discussion or debate, which aim at analyzing and breaking apart a big issue into smaller parts. It is about

Box 16.1
Climate Conversations: Opportunities for Meaningful Engagement

A practical example may help illustrate the possibility for meaningful engagement. One of us (S.M.) has been involved in a project titled "Climate Conversations"[3] (held in four U.S. locations in 2012) with individuals who can be described as falling in the Concerned and Cautious categories of the Six Americas. Belonging to these segments was not formally assessed prior to participation, but was judged on the basis of the questions and concerns participants raised during the events (e.g., few if any were truly convinced climate change is happening, but had noticed changes; the human causation was questioned; people generally knew very little about climate change science or of the magnitude of the scientific consensus). Participants were recruited through the Keystone Center (http://www.keystone.org/), typically with the help of local partners.

In a "World Café" style conversation with about forty to fifty participants per event (Hurley and Brown 2009; World Café 2008), attendees were neither already persuaded of climate change's importance and urgency nor did they think of it as a political Trojan horse, hoax, or conspiracy. Attention was paid to achieving gender, racial, professional, and sectoral diversity in the conversations (e.g., by including participants from local governments, the military, garden clubs, farmers, clergy, private firms). Participants came out of curiosity or because of rising concerns over extreme events; some came because a trusted source convinced them to join, while others saw an opportunity to participate in a conversation they might not otherwise get to have. A small number of experts were on hand to answer questions and relate some basic scientific information about historical and projected climatic changes in the region. This helped place people's tacit experience into a broader, credible context, and as such was validating to participants. Their questions fell into a number of categories – and they were deeply appreciative that they could ask them and that answers were given in an open, inviting, and non-judgmental environment:

- *Problem existence*: Is global climate change happening or not (sometimes hinging on different lines of evidence or counter-intuitive weather events that confused people)?
- *Causation/attribution*: Is climate change caused by human action or not (maybe the most contested aspect in discussion)?
- *Climate models*: How sophisticated, reliable, and thus believable are existing global climate models?
- *Degree of danger*: Are the risks from climate change really that significant, dangerous, and urgent – or not?
- *Scientific consensus*: Do climate scientists actually agree that climate change is happening, dangerous, and human caused?
- *Scientific uncertainty*: How much uncertainty is there in projecting future climate changes and related impacts and does it justify action or delay (or inaction)?
- *Cost*: Isn't it too expensive to national or local economies to take mitigation or proactive adaptation actions?

[3] See https://www.keystone.org/policy-initiatives-center-for-science-a-public-policy/environment/climate-conversations.html.

Box 16.1
(continued)

- *Policy alternatives*: What are the possible interventions and are they really acceptable (due to, for example, the degree of government involvement, the locus of control, the manner of implementation, or the risks and benefits involved in alternative approaches)?

It is precisely these questions that are hotly debated by pundits and echoed in the media, but many in the broader public (just like the participants in the Climate Conversations) have not had or not taken the opportunity to deeply examine relevant information and make up their own minds.[4]

Along with learning about these issues from the experts present, conversation participants were glad to be able to voice their concerns and unease about climate extremes and changes observed, and for the connections they could make with others in their region. They asked for credible sources to learn more, wanted to stay connected with each other beyond the event and were eager to discuss possible mitigation and adaptation strategies. As is common for conversations among nonexperts, these questions were not particularly technical but reflected concerns over costs and more generally participants' values such as "doing our part," "responsibility," and "stewardship." Participants also wanted to know how to "talk about climate change without mentioning climate change" – clear indications of the frustration they shared about the loaded term and a strong desire to find ways to enact the right responses regardless (see also Furth and Gantwerk 2013). Maybe most importantly, participants expressed a hunger to have a "real" conversation about the issue (see project Facebook page at the preceding link). The facilitated conversation events, lasting about four hours, were conducted during the workday, and participants lingered long after the official adjournment.

The experience highlights some important insights about the opportunities and challenges of attempting to reach those currently less involved in the climate debate. First, the greatest challenge was to bring people to the event. Significant effort was required to make the issue resonant in locally relevant ways and non-inflammatory language, typically by a locally trusted partner who was not perceived as having a particular agenda. Logistical challenges can make participation difficult as well (e.g., not getting time away from work or family) even though travel support and stipends were offered. Second, at least with those in the population that lean toward accepting climate change's reality and seriousness, there is potential to have a very engaged, meaningful

---

[4] The nature of the media industry is a key factor in the state of affairs. Most mainstream media outlets are for-profit businesses invested in finding news that sells. One implication of that vested interest is that they report on what they believe the public is most interested in (e.g., the immediate and tangible troubles of the economy), probably one of the key reasons why overall coverage on climate change in mainstream media has actually declined in recent years (Boykoff 2014; Brulle et al. 2012). Another implication of the for-profit nature of the media is that they look for the extreme, sensational, and controversial. Apart from weather extremes, controversy is the most interesting aspect of climate change to report on. As long as the two extreme ends of the opinion spectrum are slinging mud at each other, that battle is more newsworthy than the incremental progress of science or of climate change.

> **Box 16.1**
> **(continued)**
>
> conversation. Once involved in dialogue, people pay attention and process information carefully. They get the questions addressed they most care about, and walk away feeling more educated and empowered, particularly through the connections they made with others. Third, the importance of moving quickly from the "problem" to "solutions" cannot be overstated. This became quickly apparent in the Climate Conversations and has been documented in countless studies (e.g., Dietz 2013; Gifford 2011; Lorenzoni et al. 2007; Swim et al. 2011; Weber 2011). It suggests that the Four Americas may well be "tuning out" because it is psychologically challenging to stay in a conversation about an increasingly worrisome problem, when solutions seem ineffectual, out of reach, or infeasible and when leadership is elusive.
>
> This then points to the need for scaling up such conversations, that is, spreading them across the country. While challenging, it is not impossible to reach far more people with such events, as has been demonstrated elsewhere (e.g., Involve 2008; Lehr et al. 2003; Tan and Brown 2005). Many hundreds of individuals were involved in such a deliberative process in Texas, which ultimately led to greater understanding of the opportunities and challenges around wind energy, strong support for wind power, and providing the needed public signal to move state policy forward which, in turn, contributed to Texas becoming a leading developer of wind energy in the United States (Lehr et al. 2003).

exploring issues and meanings together (Dietz 2013). It is not about persuasion (by rational argument or savvy moral claim), but about understanding. Or, as David Bohm (1996, 3) says, "in a dialogue, each person does not attempt to *make common* certain ideas or items of information that are already known to him [or her]. Rather, it may be said that the two people are making something *in common*, i.e., creating something new together." Dialogue is also first and foremost not about problem solving, though often finding commonly accepted solutions to a problem can arise organically from the relationships, trust, and understanding that are developed. Figure 16.3 illustrates a continuum of discourses, where dialogue is the most involved, most demanding, and potentially most fruitful and rewarding for all involved.

What would make it possible for those currently wildly opposed to come together for true dialogue? For those who believe the climate policy stalemate can be reduced to fossil fuel interests stalling progress, or to environmentalists aiming to establish an all-empowered world government, this question may invite only ridicule. It seems quite impossible to even imagine that the two sides would want to come together to be in dialogue. Clearly the politics of climate change cannot be understood without a clear-eyed look at the influence of money in U.S. policy making. Yet outside capitols and beyond policy circles, Americans do indeed have deep differences in values and opinions that currently prevent them from productive exchange. What, then, are the

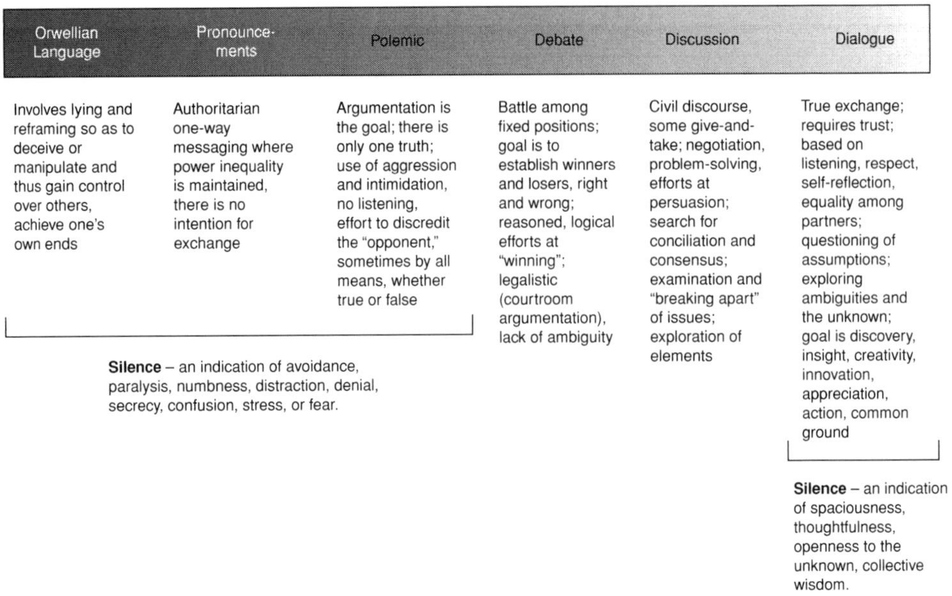

Orwellian Language	Pronouncements	Polemic	Debate	Discussion	Dialogue
Involves lying and reframing so as to deceive or manipulate and thus gain control over others, achieve one's own ends	Authoritarian one-way messaging where power inequality is maintained, there is no intention for exchange	Argumentation is the goal; there is only one truth; use of aggression and intimidation, no listening, effort to discredit the "opponent," sometimes by all means, whether true or false	Battle among fixed positions; goal is to establish winners and losers, right and wrong; reasoned, logical efforts at "winning"; legalistic (courtroom argumentation), lack of ambiguity	Civil discourse, some give-and-take; negotiation, problem-solving, efforts at persuasion; search for conciliation and consensus; examination and "breaking apart" of issues; exploration of elements	True exchange; requires trust; based on listening, respect, self-reflection, equality among partners; questioning of assumptions; exploring ambiguities and the unknown; goal is discovery, insight, creativity, innovation, appreciation, action, common ground

Silence – an indication of avoidance, paralysis, numbness, distraction, denial, secrecy, confusion, stress, or fear.

Silence – an indication of spaciousness, thoughtfulness, openness to the unknown, collective wisdom.

Figure 16.3. Discourse continuum from Orwellian manipulation to true dialogue. *Source*: Adapted from Buie (2010, 100–101).

possibilities for people on the extreme ends of the opinion spectrum to even consider doing the demanding work of dialogue?

We see essentially three reasons that could open possibilities for a new start. By themselves they are not the way to overcome deep cultural divides, but they constitute necessary preconditions for a dialogue to even take place. The first is maybe the most likely: necessity. More and more weather- and climate-related crises constitute such necessities. Whether or not they would change any skeptical or dismissive person's mind and bring a revelation or recognition of something previously denied is not the point; rather, crises may simply necessitate that people communicate and collaborate directly to find pragmatic solutions. Another form of necessity may arise from a legal mandate (e.g., a state law requiring local adaptation planning or mitigation actions). And a third may arise from the stalemate itself: if positions are so deeply dug in that progress on anything becomes impossible, political leaders at any level may be at risk of losing their political mandate, and thus feel compelled to come together with previous opponents.

A second reason is gradual marginalization of extreme views as unacceptable. This may occur, for example, as a result of the turnover of leadership in political and business organizations, the defection of key voices, or the loss of financial support of the increasingly marginalized camp. Deeper structural changes such as improvements in education, changes in the voting system, or recuperation of better reporting standards could also foster a shift toward the public unacceptability of the extreme views (particularly, the unacceptability of outright climate denial), even if such processes are slow and less likely at present. Maybe most likely is that climate change manifests in increasingly notable ways and public opinion shifts toward greater awareness and acceptance, thus making doubt and denial simply the marginalized position to hold. Those marginalized then might feel pressure to join the new social norm. While innovation studies suggest that there are always some that will not adopt the new norm, their influence will be lessened over time.

The final reason why those in hardened positions may come together is less tangible, less driven by external forces and yet can be even more compelling and has precedent in historical conflicts: an awakening or revelatory shift in stance, a higher calling. Rather than a "conversion experience," individuals may feel compelling personal reasons to at least talk to each other, for example, "for our grandchildren" or because of who they wish to be and how they wish to be seen by others.

Whatever the reason, there may come a point when the benefits of stalemate are overpowered by the promises of talking with each other. This will not be easy, given previous self-righteousness, name calling (and sometimes threats), hostile sentiments, and the strongly held opinions and underlying values. In some cases it may be necessary to begin such a dialogue with a neutral and trained facilitator. Regardless, it requires at least a "willingness to be disturbed," as Margaret Wheatley says, to be shaken out of our solidly held convictions and to acknowledge the possibility that we may be biased, blind, and afraid, yet also hold the power to create or destroy the civic fabric of society.

From Cultural Divide to Interpersonal Opening: The Case for Dialogue

We have an obligation to have difficult dialogues in a way we really never had before. That obligation is deep and... acute.

– Cynthia Enloe (2008, 65)

True dialogue – difficult dialogue – among those who hold very different positions from each other such as those on the extreme ends of the Six Americas opinion spectrum is a psychologically demanding, time-consuming investment that requires commitment, perseverance, and personal vulnerability (Figure 16.3). It may take years or even generations to bridge across cultural divides, and among those most set in their beliefs and unwilling to self-examine and reflect, it may not succeed at all. But where there is willingness to interpersonal opening and connection, both personal and sometimes collective benefits can unfold. This has been shown in compelling cases in the past, ones that seemed entirely intractable at the outset, for example, in the case of abortion, same-sex marriage, the Israeli–Palestinian conflict, race relations, electoral and educational reforms, and over retaliatory violence against Muslims after the terrorist attacks of September 11, 2001 (Gastil et al. 2006; Schirch and Campt 2007; see also http://www.publicconversations.org/). The work just referenced offers important insights: for example, the longer the interaction among participants, the more profound and lasting the effect is; dialogue may occur in a number of formats: one-on-one, one-on-many, several-on-several, and several-for-many; and the higher the stakes and the preexisting animosities, the more helpful it is to involve a neutral and skilled facilitator.

Surely, as in the Climate Conversations described, specific questions about climate science and action can be addressed among those holding polarized but not necessarily deeply educated views (e.g., evidence for the reality and human causation of climate change, scientific approaches to studying climate changes and the extent of scientific agreement, the scope and likelihood of future risks, the options to respond to climate change mitigation and adaptation). But the purpose of dialogue aims deeper: it gives people an opportunity to explore their emotional responses to climate change impacts, their feelings and thoughts about response strategies, the pros and cons and trade-offs of taking certain types of actions. It invites people to self-scrutinize their views and values, find deeper motivations, examine positions and power differences. When it works, people may come to see how their needs can be met in new ways, allowing them to accept a policy; or they may see that their livelihoods are assured or come to better understand how different things they hold dear are protected, allowing them to agree to share the cost or responsibility for a given action. In other instances, the dialogue format allows new or modified proposals to surface that involve elements of both sides' concerns and thus become acceptable.

Of far greater significance is the potential for dialogue to shift participants' entrenched positions, and particularly their views of each other (Buie 2010; Palmer 2011). As examples on climate and in other polarized issue areas show, dialogue holds the promise to help us transcend what usually keeps us apart, not by giving us a space to learn about the issue at hand, but to learn about those "others" that hold such seemingly unacceptable views. In short, dialogue gives us an opportunity and asks us to become curious about that other. Although we rarely listen to others now, dialogue asks us to listen intently to the other while withholding judgment of what is being said. If we allow silence at all in our current debates, it is typically an indication of withdrawal, whereas in true dialogue, silence is a space for ambiguity, thoughtfulness, taking something in deeply, and for letting feelings rise and settle again. Dialogue creates a space for us to become less certain in the comfort of our convictions. If we are used to quickly forming opinions and judgments in most day-to-day exchanges, dialogue asks that we consider the other's perspective, maybe even empathetically understand it, even if we do not accept it as true for ourselves. Typically, we maintain our differences and separation with anger, cynicism, disdain, name calling, and language that further feeds inflamed responses, yet in dialogue we are asked to begin from a place of goodness, truth, and respect – allowing a more humanized picture of the other to arise and ourselves to become more vulnerable. It involves recognizing that the thing that inflames us most about the other is most likely our own unconscious, unexplored shadow. Thus, instead of either-or thinking, dialogue makes space for the finer gray shades of our lives and convictions, allowing us to move from defensive posturing to reflective openness and empathy. Ad hominem attacks are set aside for nonviolent speech and action. It is from this openness and curiosity about the other that change in thinking, opinion, and attitudes can occur. When we truly commit and engage in the emotional and cognitive work it takes to be in true dialogue, we evolve, grow, and irrevocably change somehow.

These basic rules of engagement are not unique to climate dialogues (as the examples from different issue areas indicate) but common to all attempts to come into true connection through deep conversation (Bohm 1996; Brulle 2010; Isaacs 1999; Lohmann and Til 2011; Palmer 2011; Patterson et al. 2002; Schirch and Campt 2007; Stone et al. 1999).[5] As Gastil et al. (2006, Section 2, paragraph 9) argue, contrary to what many believe, most people do not want to impose their views on the whole of society; they simply want to align with the "policy, party, or person [who] will best help them make ends meet and keep them reasonably safe." Because most people lack time and experience to become experts on the science and policies related to climate change, they tend to turn to leaders of their cultural group for guidance. In other words, "citizens use cultural affinity as a heuristic, or mental shortcut, for figuring out which politicians and policies are most likely to put food on their tables" (Gastil et al. 2006, Section 3, paragraph 1). Dialogue and deliberation counteract the use of such cultural heuristics. When engaged in "earnest face-to-face deliberation under conditions that convey the good faith and trustworthiness of all participants ... individuals form strong emotional bonds" (Section 3, paragraph 4). "Under these conditions, citizens interested in pragmatic solutions to common problems can achieve a degree of knowledge that relieves them of the need to lean on culture as a heuristic crutch" (Section 3, paragraph 4).

Slowly, space is made for questions, reflection, and learning, which in turn allows for common understanding to emerge and commonalities in visions, goals, values and strategies to be discovered. As Herzig and Chasin (2006, 1) note, in properly guided dialogues, even people who "seem intractably opposed, often change the way they view and relate to each other – even as they maintain the commitments that underlie their views." Eventually, the issue at stake and the divide between groups can be reframed.

Importantly, however, dialogue engages not only cognition but also feelings, spirit, and imagination and as such is always more generative than debate or monologue. An effective use of dialogue thus helps to validate emotional needs (such as one's

---

[5] A number of organizations have begun using dialogue as an engagement format around climate change. See, for example, the National Coalition for Dialogue and Deliberation (http://ncdd.org/); Alberta Climate Dialogue (ABCD) (http://www.albertaclimatedialogue.ca/) uses well-designed citizen deliberations to shift climate change politics; and a National Climate Conversation series of community events was held in 2009 (http://www.climateconversation.org/).

identity) and passions associated with climate change while promoting empathy and understanding for where these come from within oneself and in the other. When we recognize how our own and others' motivations are influencing positions, they can become less fixed and change becomes possible. Finally, dialogue can help stop the mutual demonizing because those who hold different perspectives get to actually know each other. It is easy to vilify another via the safe distance of virtual space; it is much harder to maintain such prejudice and judgment when the other becomes a real person with a name, a face, and a story. All of us have stories of pride, love, loss, and suffering. Dialogue more than other forms of communication can thus foster deep caring and connectedness. We witness each other in our ethical dilemmas, or – as Parker Palmer says – we stand together in the "tragic gap" between how the world is and how it could or should be (Palmer 2011, 26, 189–93). It is from that greater familiarity with each other that defenses can come down, alternative ideas and perspectives can be considered, and spaces for joint problem solving open up.

Conclusion: There Must Be More

In this chapter, we argued that the basic attitude that underlies engaging the unheard majority of Americans is the same as that underlying dialogue across deep cultural differences: curiosity. We should expect that there is more to the "other" than we previously knew or assumed: more than the two extreme positions we usually hear in the media, more than ignorance and disengagement among those we tend not to hear at all, and more humanity than we usually presume among those who think so differently from ourselves. None of us are as simple and homogenous in our stances as our loud opinions might make us believe. Clearly there is a need and possibility of deeper engagement, driven by this curiosity and the desire to discover more about those currently disengaged, unengaged, or wildly engaged on the opposite ends of the opinion spectrum. More, faster, and louder one-way messaging will only add to polarization rather than reduce it. We therefore suggest that engagement strategies for the Six Americas need to be rethought and resources redirected accordingly. Continued polarization of some and disenfranchisement of most others will result in the further erosion of the civic fabric, and that in turn will make addressing the challenges of climate change only harder. Given existing polarization, political disenfranchisement and dissatisfaction, and the very demanding work of true dialogue many may not be willing to engage in, it may not be possible to completely overcome the cultural divides and bring everyone into a productive conversation. But we do not have to perpetuate conventional communication practices, and circumstances may compel us to change them. The promise and track record of dialogue to engage people, initiated by smart, courageous, and trustworthy leaders who are willing to face the adaptive challenge before us, offers a true and promising alternative.

References

Antonio, R. J. and R. Brulle. 2011. "The unbearable lightness of politics: Climate change denial and political polarization." *The Sociological Quarterly* 52 (2): 195–202.

Balbus, A. 2012. *Increasing Public Understanding of Climate Risks and Choices: Learning from Social Science Research and Practice*. Cambridge, MA, and Ann Arbor, MI: Union of Concerned Scientists and Erb Institute and University of Michigan.

Bohm, D. 1996. *On Dialogue*. London: Routledge.

Boykoff, M. T. 2011. *Who Speaks for the Climate: Making Sense of Media Reporting on Climate Change*. New York: Cambridge University Press.

Boykoff, M. T. 2014. *Media Coverage of Climate Change/Global Warming: USA Media Coverage*. http://sciencepolicy.colorado.edu/media_coverage/us/index.html.

Boykoff, M. T. and J. M. Boykoff. 2004. "Balance as bias: Global warming and the U.S. prestige press." *Global Environmental Change* 14 (2): 125–36.

Brown, J., Isaacs, D. and World Café Community. 2005. *The World Café: Shaping Our Futures through Conversations That Matter*. San Francisco: Berrett-Koehler.

Brulle, R. J. 2010. "From environmental campaigns to advancing the public dialog: Environmental communication for civic engagement." *Environmental Communication: A Journal of Nature and Culture* 3 (1): 82–98.

Brulle, R., Carmichael, J. and J. Jenkins. 2012. "Shifting public opinion on climate change: An empirical assessment of factors influencing concern over climate change in the U.S., 2002–2010." *Climatic Change* 114 (2): 169–88.

Buie, S. 2010. *Inviting Dialogue: Renewing the Deep Purposes of Higher Education*. Worcester, MA: Higgins School of Humanities, Clark University.

Butler, C. and N. Pidgeon. 2009. "Media communications and public understanding of change – Reporting scientific consensus on anthropogenic climate change." In *Climate Change and the Media*, edited by Boyce, T. and J. Lewis, 43–58. New York: Peter Lang.

Carvalho, A. 2007. "Ideological cultures and media discourses on scientific knowledge: Re-reading news on climate change." *Public Understanding of Science* 16 (2): 223–43.

Clynes, T. 2012. "The battle over climate science: Climate scientists routinely face death threats, hate mail, nuisance lawsuits and political attacks. How much worse can it get?" *Popular Science*. http://www.popsci.com/science/article/2012-06/battle-over-climate-change.

Converse, P. E. 1964. "The nature of belief systems in mass publics." In *Ideology and Discontent*, edited by D. E. Apter, 206–61. New York: Free Press of Glencoe.

Cook-Greuter, S. R. 2008. "Mature ego development: A gateway to ego transcendence?" *Journal of Adult Development* 7 (4): 227–40. http://www.cook-greuter.com/GatewaytoTransc.2000%202008%20updated.pdf.

Crompton, T. 2011. "Values matter." *Nature Climate Change* 1 (6): 276–77.

Dickinson, J. L. 2009. "The people paradox: Self-esteem striving, immortality ideologies, and human response to climate change." *Ecology and Society* 14 (1): 34. http://www.ecologyandsociety.org/vol14/iss1/art34/.

Dietz, T. 2013. "Bringing values and deliberation to science communication." *Proceedings of the National Academy of Sciences of the United States of America* 110 (Suppl. 3): 14081–87.

Ding, D., Maibach, E. W., Zhao, X., Roser-Renouf, C. and A. Leiserowitz. 2011. "Support for climate policy and societal action are linked to perceptions about scientific agreement." *Nature Climate Change* 1 (9): 462–66.

Dunlap, R. E. and A. M. McCright. 2008. "A widening gap: Republican and Democratic views on climate change." *Environment* 50 (5): 26–35.

Dunlap, R. E. and A. M. McCright. 2011. "Climate change denial, Sources: actors and strategies." In *Routledge Handbook of Climate Change and Society*, edited by C. Lever-Tracy, 240–59. London: Routledge.

Enloe, C. 2008. "Reflections." In *Inviting Dialogue: Renewing the Deep Purpose of Higher Education*, edited by S. Buie, 65. Worcester, MA: Higgins School of Humanities, Clark University.

Fears, D. 2011. "Virginia residents oppose preparations for climate-related sea-level rise." *Washington Post*, December 17.

Frank, T. 2011. *Grace Lee Bogs on Detroit and "The Next American Revolution: Sustainable Activism for the Twenty-First Century."* Transcript of an interview of Grace Lee Boggs and Thomas Frank with Amy Goodman and Juan Gonzalez on April 14. New York: Democracy Now!

Fritsche, I., Jonas, E., Kayser, D. N. and N. Koranyi. 2010. "Existential threat and compliance with pro-environmental norms." *Journal of Environmental Psychology* 30 (1): 67–79.

Furth, I. and H. Gantwerk. 2013. *Citizen Dialogues on Sea Level Rise: Start with Impacts/End with Action*. San Diego, CA: Viewpoint Learning.

Gastil, J., Kahan, D. and D. Braman. 2006. "Ending polarization: The good news about the culture wars." *Boston Review* 30 (2). http://bostonreview.net/BR31.2/gastilkahan braman.php.

Gifford, R. 2011. "The dragons of inaction: Psychological barriers that limit climate change mitigation and adaptation." *American Psychologist* 66 (4): 290–302.

Han, H. 2009. *Moved to Action: Motivation, Participation and Inequality in American Politics*. Stanford, CA: Stanford University Press.

Hart, P. S. and E. C. Nisbet. 2012. "Boomerang effects in science communication: How motivated reasoning and identity cues amplify opinion polarization about climate mitigation policies." *Communication Research* 39 (6): 701–23.

Hedlund-de Witt, A. and N. Hedlund-de Witt. 2012. "Reflexive communicative action for climate solutions: Towards an integral ecology of worldviews." Paper presented at the Culture, Politics, and Climate Change Conference, Boulder, CO.

Herzig, M. and L. Chasin. 2006. *Fostering Dialogue across Divides: A Nuts and Bolts Guide from the Public Conversations Project*. Watertown, MA: Public Conversations Project.

Hill, D. 2010. "Science and technology: Public attitudes and understanding." In *Science and Engineering Indicators 2010*, ed. National Science Board, 7.1–7.49. Arlington, VA: NSF.

Hoffman, A. J. 2011. "Sociology: The growing climate divide." *Nature Climate Change* 1: 195–96.

Hulme, M. 2009. *Why We Disagree about Climate Change: Understanding Controversy, Inaction, and Opportunity*. Cambridge: Cambridge University Press.

Hurley, T. J. and J. Brown. 2009. "Conversational leadership: Thinking together for a change." *The Systems Thinker* 20 (9): 2–7.

Inhofe, J. M. 2011. "Climate Change Update, Senate Floor Statement by U.S. Sen. James M. Inhofe (R-Oklahoma)." January 4. http://inhofe.senate.gov/pressreleases/climateupdate.htm.

Involve. 2008. *Deliberative Public Engagement: Nine Principles*. London: National Consumer Council.

Isaacs, W. 1999. *Dialogue: The Art of Thinking Together*. New York: Currency Books.

Isham, J. and S. Waage, eds. 2007. *Ignition: What You Can Do to Fight Global Warming and Spark a Movement*. Washington, DC: Island Press.

Jacques, P. J., Dunlap, R. E. and M. Freeman. 2008. "The organisation of denial: Conservative think tanks and environmental scepticism." *Environmental Politics* 17 (3): 349–85.

Jost, J. T., Glaser, J., Kruglanski, A. W. and F. J. Sulloway. 2003. "Political conservatism as motivated social cognition." *Psychological Bulletin* 129 (3): 339–75.

Jost, J. T., Kay, A. C. and H. Thorisdottir, eds. 2009. *Social and Psychological Bases of Ideology and System Justification*. New York: Oxford University Press.

Kahan, D. M. 2007. "Cultural cognition as a conception of the cultural theory of risk." Cultural Cognition Project Working Paper 73. New Haven, CT: Yale University Law School.

Kahan, D. M. 2008. "Cultural cognition as a conception of the cultural theory of risk." Cultural Cognition Working Paper. New Haven, CT: Yale University Law School.

Kahan, D. M. 2010. "Fixing the communications failure." *Nature* 463 (7279): 296–97.

Kahan, D. M. and D. Braman. 2008. "The self-defensive cognition of self-defense." *American Criminal Law Review* 45 (1): 1–65.

Kahan, D. M., Braman, D., Gastil, J., Slovic, P. and C. K. Mertz. 2007a. "Culture and identity-protective cognition: Explaining the white male effect in risk perception." *Journal of Empirical Legal Studies* 4 (3): 465–505.

Kahan, D. M., Braman, D., Slovic, P., Gastil, J. and G. L. Cohen. 2007b. "The Second National Risk and Culture Study: Making sense of – and making progress in – the American culture war of fact." http://ssrn.com/paper=1017189.

Kahan, D. M., Peters, E., Wittlin, M., Slovic, P., Ouellette, L. L., Braman, D. and G. Mandel 2012. "The polarizing impact of science literacy and numeracy on perceived climate change risks." *Nature Climate Change* 2 (1): 732–35.

Kasser, T. 2009. "Shifting values in response to climate change." In *2009 State of the World: Into a Warming World*, edited by Engelman, R., Renner, M. and J. Sawin, 122–25. New York: W. W. Norton.

Kaufman, L. and K. Zernike. 2012. "Activists fight green projects, seeing U.N. plot." *New York Times*, February 3.

Krosnick, J. A. 1990. "Government policy and citizen passion: A study of issue publics in contemporary America." *Political Behavior* 12 (1): 59–92.

Lahsen, M. 2005. "Technocracy, democracy, and U.S. climate politics: The need for demarcations." *Science, Technology and Human Values* 30 (1): 137–69.

Lahsen, M. 2008. "Experiences of modernity in the greenhouse: A cultural analysis of a physicist 'trio' supporting the backlash against global warming." *Global Environmental Change* 18 (1): 204–19.

Lahsen, M. 2013. "Anatomy of dissent: A cultural analysis of climate skepticism." *American Behavioral Scientist* 57 (6): 732–53.

Lehr, R. L., Guild, W., Thomas, D. L. and B. G. Swezey. 2003. *Listening to Customers: How Deliberative Polling Helped Build 1000 MW of New Renewable Energy Projects in Texas*. Golden, CO: National Renewable Energy Laboratory.

Leiserowitz, A., Maibach, E. and C. Roser-Renouf. 2008. *Global Warming's "Six Americas": An Audience Segmentation*. New Haven, CT, and Fairfax, VA: Yale Project of Climate Change, Yale School of Forestry and Environmental Studies and Center for Climate Change Communication, George Mason University.

Leiserowitz, A., Maibach, E. and C. Roser-Renouf. 2009. *Saving Energy at Home and on the Road: A Survey of Americans' Energy Saving Behaviors, Intentions, Motivations, and Barriers*. New Haven, CT, and Fairfax, VA: Yale Project of Climate Change, Yale School of Forestry and Environmental Studies, and Center for Climate Change Communication, George Mason University.

Leiserowitz, A., Maibach, E. and C. Roser-Renouf. 2010a. *Global Warming's Six Americas, January 2010*. New Haven, CT, and Fairfax, VA: Yale University, Yale Project on Climate Change and George Mason University, Center for Climate Change Communication.

Leiserowitz, A., Maibach, E., Roser-Renouf, C., Feinberg, G. and P. Howe. 2013. *Global Warming's Six Americas, September 2012*. New Haven, CT, and Fairfax, VA: Yale University, Yale Project on Climate Change Communication and George Mason University.

Leiserowitz, A., Maibach, E., Roser-Renouf, C. and J. D. Hmielowski. 2011a. *Politics and Global Warming: Democrats, Republicans, Independents, and the Tea Party*. New Haven, CT, and Fairfax, VA: Yale University, Yale Project on Climate Change Communication and George Mason University.

Leiserowitz, A., Maibach, E., Roser-Renouf, C. and N. Smith. 2011b. *Americans' Actions to Conserve Energy, Reduce Waste, and Limit Global Warming: May 2011*. New Haven, CT, and Fairfax, VA: Yale University, Yale Project on Climate Change Communication and George Mason University.

Leiserowitz, A., Maibach, E., Roser-Renouf, C. and N. Smith. 2011c. *Global Warming's Six Americas, May 2011*. New Haven, CT, and Fairfax, VA: Yale University, Yale Project on Climate Change Communication and George Mason University.

Leiserowitz, A., Maibach, E., Roser-Renouf, C. and J. Hmielowski. 2012. *Global Warming's Six Americas in March 2012 and November 2011*. New Haven, CT, and Fairfax, VA: Yale University, Yale Project on Climate Change Communication and George Mason University.

Leiserowitz, A., Smith, N. and J. R. Marlon. 2010b. *Americans' Knowledge of Climate Change*. New Haven, CT: Yale University, Yale Project on Climate Change Communication.

Lohmann, R. A. and J. V. Til, eds. 2011. *Resolving Community Conflicts and Problems: Public Deliberation and Sustained Dialogue*. New York: Columbia University Press.

Lorenzoni, I., Nicholson-Cole, S. and L. Whitmarsh. 2007. "Barriers perceived to engaging with climate change among the UK public and their policy implications." *Global Environmental Change* 17 (3–4): 445–59.

Malone, E. L. 2009. *Debating Climate Change: Pathways through Argument to Agreement*. London: Earthscan.

Marx, S. M., Weber, E. U., Orlove, B. S., Leiserowitz, A., Krantz, D. H., Roncoli, C. and J. Phillips. 2007. "Communication and mental processes: Experiential and analytic processing of uncertain climate information." *Global Environmental Change* 17 (1): 47–58.

McCright, A. M. and R. E. Dunlap. 2003. "Defeating Kyoto: The conservative movement's impact on U.S. climate change policy." *Social Problems* 50 (3): 348–73.

McCright, A. M. and R. E. Dunlap. 2010. "Anti-reflexivity: The American conservative movement's success in undermining climate science and policy." *Theory, Culture and Society* 27 (2–3): 1–34.

McCright, A. M. and R. E. Dunlap. 2011. "The politicization of climate change and polarization in the American public's views of global warming, 2001–2010." *The Sociological Quarterly* 52 (2): 155–94.

McIntosh, S., Phipps, C., Debold, C. and M. E. Zimmerman. 2012. *Campaign Plan for Climate Change Amelioration*. Boulder, CO: Institute for Cultural Evolution.

Meijnders, A. L., Midden, C. J. H. and H. A. M. Wilke. 2001. "Role of negative emotions in communication about CO_2 risks." *Risk Analysis* 21 (5): 955–66.

Mooney, C. 2012. *The Republican Brain: The Science of Why They Deny Science – and Reality*. Hoboken, NJ: John Wiley.

Moser, S. C. 2007. "More bad news: The risk of neglecting emotional responses to climate change information." In *Creating a Climate for Change: Communicating Climate Change and Facilitating Social Change*, edited by Moser, S. C. and L. Dilling, 64–80. Cambridge: Cambridge University Press.

Myers, T. A., Nisbet, M., Maibach, E. and A. Leiserowitz. 2012. "A public health frame arouses hopeful emotions about climate change." *Climatic Change* 113 (3–4): 105–12.

Norgaard, K. M. 2006. "'We don't really want to know': Environmental justice and socially organized denial of global warming in Norway." *Organization and Environment* 19 (3): 347–70.

Norgaard, K. M. 2011. *Living in Denial: Climate Change, Emotions, and Everyday Life*. Cambridge, MA: MIT Press.

O'Donohue, J. 2004. *Beauty: The Invisible Embrace. Rediscovering the True Sources of Compassion, Serenity, and Hope*. New York: Harper Perennial.

O'Neill, S. J. and M. Boykoff. 2010. "Climate denier, skeptic, or contrarian?" *Proceedings of the National Academy of Sciences of the United States of America* 107 (39): E151.

Oreskes, N. and E. M. Conway. 2010. *Merchants of Doubt: How a Handful of Scientists Obscured the Truth on Issues from Tobacco Smoke to Global Warming*. New York: Bloomsbury Press.

Painter, J. 2011. *Poles Apart: The International Reporting of Climate Change Scepticism*. Oxford: Oxford University, RSI.

Palmer, P. J. 2011. *Healing the Heart of Democracy: The Courage to Create a Politics Worthy of the Human Spirit*. San Francisco: Jossey-Bass.

Patterson, K., Grenny, J., McMillan, R. and A. Switzler. 2002. *Crucial Conversations: Tools for Talking When Stakes Are High*. New York: McGraw-Hill.

Peach, S. 2012. "Sea level rise, one more frontier for climate dialogue controversy." *Yale Climate Media Forum*. http://www.yaleclimatemediaforum.org/2012/02/sea-level-rise-one-more-frontier-for-climate-dialogue-controversy/.

Pienaar, M. 2011. "An eco-existential understanding of time and psychological defenses: Threats to the environment and implications for psychotherapy." *Ecopsychology* 3 (1): 25–39.

Plotkin, B. 2008. *Nature and the Human Soul: Cultivating Wholeness and Community in a Fragmented World*. Novato, CA: New World Library.

Poole, K. T. 2008. *The Roots of the Polarization of Modern U. S. Politics*. San Diego: University of California.

Poole, K. 2012. "Picture of a polarized Congress." *UGAresearch*. ftp://voteview.com/wf1/ViewpointPolarization.pdf.

Poortinga, W., Spence, A., Whitmarsh, L., Capstick, S. and N. F. Pidgeon. 2011. "Uncertain climate: An investigation into public scepticism about anthropogenic climate change." *Global Environmental Change* 21 (3): 1015–24.

Powell, J. L. 2011. *The Inquisition of Climate Science*. New York: Columbia University Press.

Rabe, B. G. 2004. *Statehouse and Greenhouse: The Emerging Politics of American Climate Change Policy*. Washington, DC: Brookings.

Rabe, B. G. 2009. "Second-generation climate policies in the States: Proliferation, diffusion, and regionalization." In *Changing Climates in North American Politics: Institutions, Policymaking, and Multi-Level Governance*, edited by Selin, H. and S. Vandeveer, 67–85. Cambridge, MA: MIT Press.

Reser, J. P., Bradley, G. L., Glendon, A. I., Ellul, M. C. and R. Callaghan. 2012. *Public Risk Perceptions, Understandings and Responses to Climate Change and Natural Disasters in Australia and Great Britain: Final Report*. Gold Coast, Australia: National Climate Change Adaptation Research Facility.

Roseman, I. J. 1994. "The psychology of strongly held beliefs: Theories of ideological structure and individual attachment." In *Beliefs, Reasoning and Decision-Making: Psycho-Logic in Honor of Bob Abelson*, edited by Schrank, R. C. and E. Langer, 175–208. Hillsdale, NJ: Lawrence Erlbaum Associates.

Sarewitz, D. 2004. "How science makes environmental controversies worse." *Environmental Science and Policy* 7 (5): 385–403.

Schirch, L. and D. Campt. 2007. *The Little Book of Dialogue for Difficult Subjects: A Practical Hands-on Guide*. Intersource, PA: Good Books.

Stone, D., Patton, B. and S. Heen. 1999. *Difficult Conversations: How to Discuss What Matters Most*. New York: Penguin Books.

Swim, J. K., Stern, P. C., Doherty, T. J., Clayton, S., Reser, J. P., Weber, E. U., Gifford, R. and G. S. Howard. 2011. "Psychology's contributions to understanding and addressing global climate change." *American Psychologist* 66 (4): 241–50.

Tan, S. and J. Brown. 2005. "The World Cafe in Singapore: Creating a learning culture through dialogue." *Journal of Applied Behavioral Science* 41 (1): 83–90.

Thompson, M., Ellis, R. and A. B. Wildavsky. 1990. *Cultural Theory*. Boulder, CO: Westview Press.

U.S. Senate Committee on Environment and Public Works. 2010. *"Consensus" Exposed: The CRU Controversy*. Minority staff report. Washington, DC: U.S. Senate.

Washington, H. and J. Cook. 2011. *Climate Change Denial: Heads in the Sand*. London: Earthscan.

Weber, E. U. 2011. "Climate change hits home." *Nature Climate Change* 1 (4): 25–26.

Wheatley, M. J. 2002. *Turning to One Another: Simple Conversations to Restore Hope to the Future*. Berkeley, CA: Berrett-Koehler.

Whitmarsh, L. 2011. "Scepticism and uncertainty about climate change: Dimensions, determinants and change over time." *Global Environmental Change* 21 (2): 690–700.

World Café, The. 2008. *Café to Go: A Quick Reference Guide for Putting Conversations to Work*. San Francisco: The World Café. http://www.theworldcafe.com/pdfs/cafetogo.pdf.

17

Social Transformation

The Real Adaptive Challenge

KAREN O'BRIEN AND ELIN SELBOE

We started this book by posing the question, *How do we adapt to changes for which we ourselves are responsible?* The subsequent chapters provided diverse examples of how and why adaptation to climate change is a social, cultural, political, and human process, intricately linked to wider processes of change. Embedded in these processes are issues related to beliefs, values, and worldviews, as well as to power, interests, identities, and loyalties. Understanding how such "adaptive elements" influence interpretations and approaches to climate change adaptation is critical to understanding the potential for social transformations to sustainability.

One of the core messages from this book is that adaptation to climate change is unlikely to have long-term effects if it is treated as only a technical problem. Engaging with climate change as an adaptive challenge involves acknowledging the beliefs, values, and worldviews that influence how different individuals and groups approach change. These factors also influence perceptions of justice, equity, and sustainability and shape the systems, structures, and norms that guide relationships among people, with nature, and with future generations. Adaptation, we contend, has to be redefined to include not only adapting to observed and near-term impacts of climate change but also adapting to the idea that humans are capable of transforming systems at a global scale. While this notion has been captured in the concept of the Anthropocene (see Dalby 2013; Steffen et al. 2011), the personal and political aspects of adaptation and their implications for social transformations have not been adequately explored.

We argue here for a *broader* interpretation of adaptation, recognizing that many of the current approaches to dealing with climate change are insufficient for meeting the complex social-ecological challenges of the twenty-first century and beyond (see Pelling 2011). Of particular concern is that adaptation to the impacts of climate change seldom challenges the drivers of risk and vulnerability, including the structures and systems that contribute to increasing greenhouse gas emissions and growing social vulnerability (Eriksen et al. 2015; Pelling 2011). Indeed, adaptation may not always have long-term, positive effects, particularly for vulnerable groups (Eriksen et al. 2011).

Pelling et al. (2012) and Manuel-Navarrete et al. (2012) refer to such adaptations as reformist approaches that do not challenge fundamental systems, including their norms, rules, functioning, or power relations. In contrast to more radical responses, reformist approaches maintain established economic and social relations, including the dominance of the capitalist mode of production (Pelling et al. 2012). There are important political dimensions to meeting climate change as an adaptive challenge, especially when moving beyond reformist approaches.

We also argue here for a *deeper* interpretation of adaptation, recognizing that individual and shared beliefs, values, worldviews, and knowledge systems influence how people or institutions approach change itself. The personal dimensions of change, which represent the subjective or "interior" worlds of individuals and groups, mediate relationships, including social interactions and political responses to climate change. Neither adaptation nor transformation is a neutral concept: they both mean different things to different people and groups, and can be used for different interests and purposes. Addressing climate change as a technical problem often fails to recognize the significance of adaptive elements, including habits, loyalties, emotions, identities, and power, which influence decisions about which risks are acceptable, what types of adaptations to prioritize, and who should bear the costs.

Confronting the adaptive challenge of climate change involves acknowledging that the problem is both systemically conditioned and socially constructed, which calls for new ways of engaging with the interlinked political and personal dimensions of climate change. This is quite different from technical and managerial approaches that tend to normalize the drivers and impacts of climate change or make them appear inevitable, then promote adaptation through better technology, more knowledge, or improved know-how, skills, and management practices. Most systemic changes require more than structural or operational redesign: they require transformations in mind-sets (Kegan and Lahey 2009). Facing climate change as an adaptive challenge opens up greater possibilities for affecting deliberate changes in social, political, and ecological systems, not only by adapting to the impacts of climate change but through social transformations to sustainability (O'Brien 2012; Pelling 2011).

How do we then engage with adaptation to climate change? What has to be done differently? These are not easy questions to answer, as adaptive challenges by definition cannot be resolved by following a clear, linear path (Heifetz et al. 2009). In this final chapter, we consider the types and qualities of responses that may be needed to adapt successfully to climate change. Acknowledging the importance of the processes, initiatives, and approaches to adaptation described in the preceding chapters, we take a step back now and consider the challenge from a different perspective. Adaptation, we argue, also involves repoliticizing climate change by "unveiling" and challenging the "given," particularly those structures and systems that have been normalized or taken for granted, thus maintaining business as usual and the status quo. Adaptation calls for a political critique that uncovers the elements driving climate change risk and vulnerability, and it includes recognizing and legitimizing alternatives and aligning these with human emancipation processes to achieve a real and democratic politics of inclusion and plurality. Such processes depend on reflexivity, collaboration, and an ability to engage in true dialogue, learning, and leadership. Skillful engagement with climate change that integrates the political and personal dimensions of the problem can turn climate change adaptation into an opportunity for realizing societal transformations.

Challenging the "Given": The Politics of Social Transformation

Climate change has deep-rooted social, cultural, political, and human dimensions. While climate change is often presented as the consequence of greenhouse gas emissions and the use of fossil fuel energy, many authors have drawn links between climate change and globalization processes, social practices, and values, identities, and emotions (Leichenko and O'Brien 2008; Norgaard 2011; Shove 2010; Urry 2011). The challenge of climate change reflects the deep structures of contemporary society, characterized by a disassociation between the risks we face and our daily lives, and by our alienation from nature and from ourselves (Beck 1992; Manuel-Navarrete et al. 2012; Pelling et al. 2012; Swyngedouw 2013). The scale and temporal dimensions of climate change in particular make it invisible in everyday life, "yet simultaneously, and increasingly formative of it and configured by it" (Pelling et al. 2012, 1). This distancing is related to socially produced strategies for collectively "living in denial," disassociating climate change from the larger political economy, everyday life, or social action (Norgaard 2011). Alienation, separation, and denial make it more difficult to connect with existing social movements, mobilization, and politics (Pelling et al. 2012).

In contemporary society, it is argued that that these processes take place in postpolitical or postdemocratic conditions, under which predominant representations of society tend to be consensual and technocratic, and most often power, conflict, and exclusion are rendered invisible in climate change politics (Kenis and Mathijs 2014; Swyngedouw 2010a, 2013). Thus, despite the fact that there is extensive debate and disagreement related to climate change adaptation and mitigation, and an apparent politics of climate change where these issues are negotiated and put high on the policy agenda (Giddens 2009), there is in reality no radical dissent, critique, or fundamental conflict that truly challenges the status quo (Swyngedouw 2010a). What is needed, then, is more critical engagement and a repoliticization that brings in power and injustices, unequal distribution of resources, and networks of control and influence that not only push for reform of systems and action but demand real transformation of both power holders and social systems. This requires acknowledging how and why the systemic features of current arrangements are not "given" but rather socially produced and perpetuated.

A first step for social transformation involves questioning and challenging systems and structures to uncover the underlying power relations and the particular ideas and interests that underpin and uphold them. Engagement with social transformation means repoliticizing climate change by directly raising questions of "who decides," "whose values count," and, perhaps most important, "why?" This means drawing attention to the values, interests, and mind-sets underlying the systems and structures that uphold the status quo and contribute to or reinforce climate change risks and vulnerabilities. As discussed in Chapter 1, dominant actors and groups impose meaning in such a way that it is taken for granted; consensus is established around a certain view of the world that makes current social, economic, political, or environmental systems and arrangements seem natural (Bourdieu 1977, 1991, 1998). Misrecognizing their arbitrariness as natural provides legitimacy for domination and reproduces existing power relations and social systems (Stokke and Selboe 2009).

It is often in times of crises when the arbitrariness of these previously undiscussed and commonsensical notions and arrangements can be exposed, brought into discussion, and confronted with alternative discourses (Bourdieu 1991, 1998). Indeed, both in academia and in "real life," there have been attempts to question and challenge the legitimized order of the current capitalist and liberal-democratic systems that constitute and normalize practices that lead to climate change. This can uncover how mitigation and adaptation decisions and actions or inactions that are presented as inevitable and undisputable are not in fact natural or "given" (Manuel-Navarrete 2010; Swyngedouw 2010a, 2010b, 2013; Klein 2014). For example, Pelling et al. (2012) see the crises of capitalism and climate change as results of unsustainable and unjust development trajectories. They argue that economic globalization and the carbon economy have come at high costs for human welfare, not only through climate change but also through growing inequality, the global economic downturn, and public disengagement and marginalization from political processes.

Undressing the status quo and challenging "the given" inevitably leads to a repoliticization of climate change, moving beyond postpolitical or postdemocratic approaches (Kenis and Mathijs 2014; Swyngedouw 2010a, 2010b, 2011, 2013). It is through such anti-hegemonic struggle against domination through consent that spaces for critique and dissent are created and where what was earlier either invisible or impossible becomes both visible and possible. Introducing opportunities for "real" or authentic politics can thus create new room for alternative visions, voices, and action (Swyngedouw 2010b). This includes formulating, discussing, enabling, and enacting possibilities that are considered to be unrealistic within mainstream discourses.

Yet it is here where political movements often get stuck. Indeed, when critique and politicization become ends in themselves, antagonistic discourses and a focus on the present can inhibit mobilization around new visions and real transformation (Kenis and Mathijs 2014; Kenis and Lievens 2014). Analyzing the grassroots movement Climate Justice Action (CJA), Kenis and Mathijs (2014) found that although CJA fought hard to challenge the consensual, postpolitical logic governing the debate on climate change and was able to open spaces for previously unheard voices, they nonetheless failed to engage with alternatives. The lack of attention to alternatives, along with a discourse that reinforced we/them distinctions, limited the movement's outreach and subsequent capacity to repoliticize the issue of climate change (Kenis and Mathijs 2014).

Realizing Alternatives: Practicing the Politics of Difference

Launching alternatives to the hegemonic systems, structures, and practices that are being challenged and criticized is part of social transformations. Critical discourses and antihegemonic struggles recognize a need to be inclusive, avoiding the perpetuation of distinctions between we/them or us/other that can alienate people and reduce the potential for mobilization. This is particularly relevant in the case of climate change, where the participation of a diversity of actors is considered essential, as climate change responses involve transformations in all strands of social life. The question is, how should alternatives be formulated and realized, and how can an inclusive politics of difference be practiced, particularly in contexts where the rights and voices of many have been traditionally silenced?

Adaptive responses to climate change, we argue, need to include a plurality of visions and actors through democratic processes. This entails promoting participation and inclusion while acknowledging a diversity of values, beliefs, worldviews, interests, and aspirations, taking the differences seriously and recognizing how they relate to issues of justice, power, and exclusion. Although such differences may be widely recognized, they must also be closely interrogated to surface the explicit and implicit beliefs and assumptions that maintain rather than evolve current systems. For example, although diverse worldviews and values are acknowledged in climate change debates, seldom are the underlying beliefs and assumptions exposed, discussed, and challenged, particularly those that limit the possibilities for visualizing and realizing alternatives. This is not to argue for the harmonious or consensual democracy of postpolitics or postdemocracy, nor for the perpetuation of a polarized politics of difference. It is rather to argue for a politics where participation, inclusion, and plurality are stressed, yet where it is also recognized that all actors and groups hold certain beliefs, values, and interests that can either limit or expand alternatives.

Navigating the differences and conflicts that arise from political processes of change can be challenging. Importantly, not all perspectives are equally inclusive of the rights of others, as they all represent situated knowledge and partial views. When calling for a plurality of actors and visions in the elaboration and discussion of radical alternatives for social transformation, there are likely to be conflicts or clashes between competing ideas about alternative future trajectories and development paths – this is the nature of social relations, power, and politics. Also, when progressing to making priorities and decisions, it is inevitable that these will profit some actors, values, interests, and visions more than others (Swyngedouw 2013). Thus, a deeper and more enduring democracy is required to continuously make relationships between power and exclusion visible so that they can enter the terrain of contestation (Mouffe 2002a, 33–34, cited by Kenis and Mathijs 2014, 148).

Questioning the given and developing alternative futures is inherently about the vital transformation of social systems and power structures to become more inclusive and fair (Pelling 2011). However, skillful engagement with systems involves both individual and collective agency and should therefore also focus on the internal transformations through which people liberate themselves (Kapoor 2007; Manuel-Navarrete 2010, 784; Manuel-Navarrete and Buzinde 2010). Manuel-Navarrete (2010, 784) links this internal transformation to the acknowledgment of how structures and institutions "condition us to treat each other and ourselves as objects rather than as free self-creating subjects," leading to alienation from the processes through which we create and experience the world. The diversity and creativity of human agency for human emancipation cannot be underestimated, and a "co-transformation" of self and society is seen as necessary for transformation to be radical and lasting (Kapoor 2007). Like Manuel-Navarrete, Kapoor (2007, 475) recognizes the need for struggles to transform oppressive social structures, but argues that this must be coupled by a deeper inner change:

While focusing on the power of the macro forces, we seem to have forgotten the power of the individual visionary and the micro-level on-the-ground initiatives and efforts, spread all over the world, involving the everyday activities of real human being, which are creatively, positively and often very quietly transforming the lives and future of millions, if not billions of human beings today.

This relates to Freire's (1970, 167) dialogical theory of action, which is based on empowering "subjects who meet to *name* the world in order to transform it." This transforms individuals into conscious social agents, creating the world by promoting alternative and sustainable futures that nurture rather than destroy relationships. Adaptation from this perspective can be considered a process of evolving relationships toward human flourishing and emancipation to encompass a larger circle of care.

The relationship between structure and agency has long been discussed in the social sciences (Bourdieu 1977, 1990, 1998; Giddens 1984), and the relationship and interactions between the two have been the focus of meta-theories such as integral theory and critical realism (Bhaskar et al. 2010; Wilber 2001). What these theories show is that the personal is political, just as the political is personal. As Kapoor (2007, 478) notes, "transformed individuals become key actors in bringing about social change. Transformed societies or social groups enable and nurture more conscious, sensitive and empowered individuals." For action to be liberating, it must be human, empathetic, loving, communicative, and humble (Freire 1970; Kapoor 2007). For action to make a large-scale difference, however, it must challenge hegemonic systems, structures, and social norms. Facing climate change as an adaptive challenge thus calls for both political and personal transformations (O'Brien and Sygna 2013). From a broader and deeper perspective, adaptation to climate change is about social transformations that come about through a liberating and inclusive politics of difference that opens up for alternative views and responses.

The Challenge of Change

Technical problems can usually be identified and addressed through knowledge, expertise, innovation, political will, and resources. Adaptive challenges, in contrast, are often less clear, less linear, and more uncomfortable to deal with, for they often create anxiety, conflict, and a sense of disequilibrium (Heifetz et al. 2009). According to Kegan and Lahey (2009), humans have developed complex ways of dealing with anxiety, leading to what they refer to as an immune system that protects against change. These include not only individual immune systems but collective systems that unknowingly protect groups from making changes that may be desired or required based on new information, knowledge, or understandings (Kegan and Lahey 2009). Norgaard (2011), for example, discusses how cultural norms set the standard for what individuals "ought" to feel in a given context or situation. In her study of climate change responses in a small Norwegian town, she notes that emotions such as guilt and helplessness in relation to climate change are not only unpleasant to experience personally but also inappropriate to reveal publicly or follow up through action.

There is no doubt that technical problems can also create disequilibria, but in many cases this can be alleviated rapidly through technical solutions. For example, a leaky roof that causes water damage can be repaired, or when there is too much runoff, water pipes and drainage systems can be cleaned out or redimensioned (if resources are available), eventually reducing the problem. Adaptive challenges, in contrast, have no easy solutions and typically involve an extended period within a productive yet tolerable zone of discomfort before they can be resolved. Individuals and groups often experience limits to the amount of anxiety, disequilibrium, and feelings of loss that they can tolerate, depending on their emotional capacities, social norms, and systems of meaning making. If the discomfort or disequilibria become too great, it is easy to fall into what Heifetz et al. (2009) call "work avoidance," characterized by distraction, displacement of responsibility, indifference, or denial. Importantly, if the limit of tolerance is not crossed, it is in the productive zone of discomfort that the "adaptive work" of realizing alternative solutions and social transformation can be carried out.

Resistance to change often stems from a fear of losing something important; underlying these fears are deeply held beliefs and expectations about the way that the world is or should be, including ideas about right or wrong, good or bad, costs or benefits, and winners and losers. Most often, this resistance reflects a desire to maintain what is stable, predictable, and familiar. If difficult trade-offs and potential losses are at stake – a key characteristic of climate change responses – interpretations of problems are likely to be conservative, favoring deeply held beliefs, values, and interests and dominant knowledge systems. There is, in general, a tendency for people to gravitate toward interpretations that are "technical rather than adaptive, benign instead of conflictual, and individual rather than systemic" (Heifetz et al. 2009, 116). Not surprisingly, policy makers tend to prefer technical solutions that use existing expertise to address easily identified problems (Rowson 2011).

Yet, ironically, resistance to the changes associated with substantial climate change mitigation and adaptation efforts is likely to contribute to "severe, widespread, and irreversible" impacts by the end of the twenty-first century (IPCC 2014). While immune systems help people or groups to avoid anxiety by not contesting the "given," the given here is contributing to environmental changes that are increasing the sense of uncertainty and disequilibria worldwide. Adaptive responses to climate change would acknowledge anxiety thresholds and legitimize the fear of loss by engaging with these emotions directly. They would also pay attention to competing commitments and the underlying assumptions and beliefs that support or hinder change (Kegan and Lahey 2009). Indeed, identifying the "givens," including the beliefs associated with them, can be an important first step in pursuing an inclusive politics of difference that skillfully engages with rather than diminishes the anxiety and conflict that may result from these differences.

Skillful Engagement for Social Transformation

Political and personal transformations that challenge hegemonic systems *and* actively pursue alternatives will require new ways of approaching change. Skillful engagement with adaptive challenges involves enhanced capacities for perceiving, understanding and engaging with complexity (Jordan 2011). One key practice is to contest the "natural, normal, and given" by questioning assumptions, seeking new information, or considering the object from new perspectives. This calls for reflexivity, or an awareness of one's own perspective, including its strengths and limitations, and an understanding of how this perspective relates to interests, identities, social norms, or habits of thought. Being aware of one's own situated knowledge can prompt one to consider more options and to modify stories or narratives about the role of humans in relation to change processes. Reflexive engagement leads to greater awareness and capacity to engage politically with hegemonic discourses and systems and to conscious changes that advance human security through social transformations (O'Brien et al. 2010; O'Brien 2013). According to Heifetz et al. (2009), the people who lead adaptive change most successfully tend to be reflexive, focusing both on themselves and the problem, recognizing how they interact with, affect, and are influenced by larger systems.

Skillful engagement with an adaptive challenge includes acknowledging that there are different ways of making meaning and that systems that appear dysfunctional to some may be considered normal and safe by others. Research on action logics and leadership capacities recognizes the significance of mind-sets in confronting adaptive challenges and that a change in the view of a system often leads to changes in systems (Boiral et al. 2009). Action logics influence problem comprehension as well as the identification of solutions and the preferred strategies to solve them (Torbert et al. 2004). Yet, for the reasons discussed in Chapter 1, it is not easy or ethical to forcefully change other people's mind-sets, and collaborating successfully with those who do not share the same subjective reality is a key skill in meeting climate change as an adaptive challenge.

The capacity to work collectively to achieve a larger goal can be described as "collaborative power," defined by Slaughter (2011) as "the power of many to do together what no one can do alone." She notes that it can take many forms: (1) mobilization through a call to action; (2) connection by linking people to one another and toward a common purpose; and (3) adaptation, which includes a willingness to shift views to enter into meaningful dialogue.

Given that collective action on climate change is essential, the relationship between individual and collective change must be (re)considered. As Jordan (2011, 48) points out, "capacities to manage complex societal issues do not necessarily reside in individuals. Skillfully designed methods and organizations and/or networks of people working together in constructive ways may allow us to create capacities that reside in collectives rather than in individuals with exceptional talents." Research on social networks, for example, shows that "interconnections between people give rise to phenomena that are not present in individuals or reducible to their solitary desires and actions" (Christakis and Fowler 2009, 303), and research on quantum social theory suggests that collective self-consciousness can be a driver of progressive change (Wendt 2006).

There is, however, little doubt that conflicts will arise and have to be managed when challenging and launching alternatives to existing norms, structures, and consensual realities. Dissensus and conflicts must be made visible rather than hidden under the guise of a global consensus on climate change as a technical problem that can be addressed through business as usual and politics as usual (Swyngedouw 2010a, 2010b, 2013). This may involve difficult discussions about which losses are negotiable versus nonnegotiable, and why (Heifetz et al. 2009). Yet whereas conflicts and struggles among different interests and groups are expected, they may arise even among those with common goals and interests. For example, collaborations within project groups can go sour if motivations and modes of engagement are fundamentally different, particularly if the adaptive aspects are ignored or treated as technical issues.

Dialogues and deliberations can be skillful means for engaging with conflicts and should include a plurality of actors with different visions and beliefs, in particular the often "silent majority" and those currently marginalized, sanctioned, and/or silenced. In making an argument for dialogue, Bohm (1996) contends that collective thought is more powerful than individual thought, especially if people can think together in a coherent way. Like others, he was convinced that shared meaning is the glue that holds people and societies together and that dialogue could be used to achieve a coherent movement of thought and communication. Bohm (1996) holds that in a dialogue, participants let go of their assumptions: "You don't believe them, nor do you disbelieve them; you don't judge them as good or bad. You simply see what they mean – not only your own, but the other people's as well." From this position, there is room for new and original thoughts to emerge, leading to new ways of seeing both problems and solutions.

Yet this collective thinking is rare, and political discussions about climate change responses often take the form of fragmented debates, where few are interested in hearing other perspectives. With so much at stake, it is no surprise that discussions and negotiations about climate change become emotional and often lead to gridlock. Paradoxically, dialogue depends on both acknowledging and allowing for plurality, difference and conflict, and involves the coming together of different viewpoints, values and visions to be able to discuss the multiple paths that may lead toward transformation. This is neither easy nor straightforward and suggests that in addition to dialogue, skillful engagement with an adaptive challenge requires both transformative learning and leadership.

Transformative learning refers to "the process by which we transform our taken-for-granted frames of reference (meaning perspectives, habits of mind, mind-sets) to make them more inclusive, discriminating, open, emotionally capable of change, and reflective so that they may generate beliefs and opinions that will prove more true or justified to guide action" (Mezirow 2000, 7–8). This includes the capacity to become critically aware of one's own assumptions and expectations and their context, as well as those of others, when making interpretations. The challenges of managing diversity have brought attention to "triple-loop" (reflexive) learning and identifying individual and collective "blind spots" in current thinking (Flood and Romm 1996; Scharmer 2009; Torbert 2004).

Facing an adaptive challenge may also require new types of leadership. Transformative leadership is characterized by the capacity to reflect, reinterpret, reframe, and intervene in such a way that not only creates results but inspires others to move forward (Heifetz et al. 2009; Olsson et al. 2006). Heifetz (2009) describes how leaders can help groups in society make progress on difficult issues by creating a supportive environment for this work. In this regard, leadership for adaptation involves creating environments for developing alternative solutions, including learning sites, policy laboratories, and shadow spaces and networks (Nguyen et al. 2011; Olsson et al. 2006; Pelling et al. 2008). Although individual leaders play an important role in transformative change, the magnitude necessary for achieving global sustainability and justice calls for collective change. There is, however, a transition under way, from a paradigm where leadership resides in a person or role to one in which leadership is a collective process that is spread through networks of people (Slaughter 2011; Petrie 2011).

A New Approach to Adaptation

The main message from this book is that climate change is more than a technical problem that can be addressed through more expertise, better management, and more resources. Neither mitigation nor adaptation to climate change is a neutral process that can be separated from understandings of the underlying drivers of risk and vulnerability. Adaptive challenges involve questioning rather than accommodating the very systems and structures that perpetuate risk and vulnerability, including the environmental changes to which societies and sectors are currently adapting. Questioning these systems draws attention to issues of power and to the individual and shared beliefs, values, and worldviews that often lie behind them. It also involves the promotion of innovative democratic processes that unleash the possibilities for social transformations that are latent in every society.

Drawing from the research presented in this book, we argue that it is time to redefine adaptation to consider it as an adaptive response to change – one that recognizes that humans are capable of contributing not only to global climate change but also to global transformations to sustainability. The adaptive challenge of climate change is essentially about relationships, where the challenge is to live in the present with other people, other species, and future generations in mind. A broader and deeper approach to adaptation acknowledges vital elements such as power, identity, values, mind-sets, habits, and beliefs. Because climate change problems and responses are systemically conditioned and socially constructed, the solution is connected to a collective, collaborative power to create equitable and sustainable alternatives. In the face of a crisis that is often felt as inevitable and unavoidable, it becomes clear that social transformations represent the real adaptive challenge for humanity.

Addressing climate change as an adaptive challenge involves some new and potentially uncomfortable conversations about change and calls for more courageous and radical approaches that challenge the drivers, structures, and systems contributing to climate change and other environmental and social problems. Promoting social transformations requires an inclusive politics of difference – a transformation that is both political and personal, skillfully engaging *with* a diversity of beliefs, values, knowledge systems, and worldviews rather than against them. In describing successful adaptation to climate change, Moser (2013, 301) includes "the ability to hold on to or create a positive vision of the future and being engaged in shaping it, rather than standing helpless and unheard on the side lines, watching an imposed future unfold." This is likely to involve a radical transformation in mind-sets, lived social experiences, knowledge systems, and power relations directed toward human flourishing and emancipation. This transformation is, in fact, the adaptive challenge of climate change.

Acknowledgments

We would like to thank the authors of this book as well as all the researchers of the "Potentials for and Limitations to Adaptation to Climate Change in Norway" (PLAN) project. The project was funded by the Research Council of Norway through the NORKLIMA program and led by Karen O'Brien at the Department of Human Geography and Sociology at the University of Oslo. We would also like to thank Johanna Wolf, Robin Leichenko, and Linda Sygna for valuable comments on earlier drafts of this chapter.

References

Beck, U. 1992. *Risk Society*. London: Sage.
Bhaskar, R., Frank, C., Høyer, K. G., Næss, P. and J. Parker. 2010. *Interdisciplinarity and Climate Change: Transforming Knowledge and Practice for Our Global Future*. Abingdon, UK: Routledge.
Bohm, D. 1996. *On Dialogue*. Edited by L. Nichol. London: Routledge.
Boiral, O., Cayer, M. and C. M. Baron. 2009. "The action logics of environmental leadership: A developmental perspective." *Journal of Business Ethics* 85 (4): 479–99.
Bourdieu, P. 1977. *Outline of a Theory of Practice*. Cambridge: Cambridge University Press.
Bourdieu, P. 1990. *The Logic of Practice*. Stanford, CA: Stanford University Press.
Bourdieu, P. 1991. *Language and Symbolic Power*. Cambridge: Polity Press.
Bourdieu, P. 1998. *Practical Reason: On the Theory of Action*. Cambridge: Polity Press.
Christakis, J. and J. Fowler. 2009. *Connected: The Amazing Power of Social Networks*. New York: HarperPress.
Dalby, S. 2013. "Biopolitics and climate security in the Anthropocene." *Geoforum* 49: 184–92.
Eriksen, S., Aldunce, P., Bahinipati, C., Martins, R., Molefe, J., Nhemachena, C., O'Brien, K., Olorunfemi, F. Park, J., and K. Ulsrud. 2011. "When not every response to climate change is a good one: Identifying principles for sustainable adaptation." *Climate and Development* 3 (1): 7–20.
Eriksen, S., Inderberg, T. H., O'Brien, K. and L. Sygna. 2015. "Introduction: Development as usual is not enough." In *Climate Change Adaptation and Development: Transforming Paradigms and Practices*, edited by Inderberg, T. H., Eriksen, S., O'Brien, K. and L. Sygna, 1–18. Abingdon, UK: Routledge.
Flood, R. L. and N. R. A. Romm. 1996. *Diversity Management: Triple Loop Learning*. Chichester, UK: Wiley.
Freire, P. 1970. *The Pedagogy of the Oppressed*. New York: Pantheon.
Giddens, A. 1984. *The Constitution of Society: Outline of a Theory of Structuration*. Cambridge: Polity Press.
Giddens, A. 2009. *The Politics of Climate Change*. Cambridge: Polity Press.
Heifetz, R., Grashow, A. and M. Linsky. 2009. *The Practice of Adaptive Leadership: Tools and Tactics for Changing Your Organization and the World*. Boston: Harvard Business Press.

IPCC. 2014. *Climate Change 2014: Mitigation of Climate Change. Contribution of Working Group III to the Fifth Assessment Report of the Intergovernmental Panel on Climate Change.* Edited by Edenhofer, O., Pichs-Madruga, R., Sokona, Y., Farahani, E., Kadner, S., Seyboth, K., Adler, et al. Cambridge: Cambridge University Press.

Jordan, T. 2011. "Skillful engagement with wicked issues: A framework for analysing the meaning-making structures of societal change agents." *Integral Review* 7 (2): 47–91.

Kapoor, R. 2007. "Transforming self and society: Plural paths to human emancipation." *Futures* 39 (5): 475–86.

Kegan, R. and L. Lahey. 2009. *Immunity to Change.* Boston: Harvard Business Press.

Kenis, A. and M. Lievens. 2014. "Searching for 'the political' in environmental politics." *Environmental Politics* 24 (3): 531–48.

Kenis, A. and E. Mathijs. 2014. "Climate change and post-politics: Repoliticizing the present by imagining the future?" *Geoforum* 52: 148–56.

Klein, N. 2014. *This Changes Everything: Capitalism vs the Climate.* London: Allan Lane.

Leichenko, R. M. and K. L. O'Brien. 2008. *Environmental Change and Globalization: Double Exposures.* Oxford: Oxford University Press.

Manuel-Navarrete, D. 2010. "Power, realism, and the ideal of human emancipation in a climate of change." *WIREs Climate Change* 1 (6): 781–85.

Manuel-Navarrete, D. and C. N. Buzinde. 2010. "Socio-ecological agency: From "human exceptionalism" to coping with "exceptional" global environmental change." In *The International Handbook of Environmental Sociology*, edited by Redclift, M. and G. Woodgate, 136–49. Cheltenham, UK: Edward Elgar.

Manuel-Navarrete, D., Pelling, M. and M. Redclift. 2012. "Conclusions: Alientation, reclamation and radical vision." In *Climate Change and the Crisis of Capitalism: A Chance to Reclaim Self, Society and Nature*, edited by Pelling, M., Manuel-Navarrete, D. and M. Redclift, 189–98. Abingdon, UK: Routledge.

Mezirow, J. 2000. *Learning as Transformation: Critical Perspectives on a Theory in Progress.* San Francisco: Jossey-Bass.

Moser, S. C. 2013. "Navigating the political and emotional terrain of adaptation: Community engagement when climate change comes home." In *Successful Adaptation to Climate Change: Linking Science and Policy in a Rapidly Changing World*, edited by Moser, S. C. and M. T. Boykoff, 289–305. Abingdon, UK: Routledge.

Nguyen, N. C., Bosch, O. J. H. and K. E. Maani. 2011. "Creating 'learning laboratories' for sustainable development in biospheres: A systems thinking approach." *Systems Research and Behavioral Science* 28 (1): 51–62.

Norgaard, K. M. 2011. *Living in Denial: Climate Change, Emotions and Everyday Life.* Cambridge, MA: MIT Press.

O'Brien, K. 2012. "Global Environmental Change II: From Adaptation to Deliberate Transformation." *Progress in Human Geography* 36 (5): 667–76.

O'Brien, K. 2013. "The courage to change: Adaptation from the inside-out." In *Successful Adaptation: Linking Science and Practice in Managing Climate Change Impacts*, edited by Moser, S. and M. Boykoff, 306–19. London: Routledge.

O'Brien, K. L., St Clair, A. L. and B. Kristoffersen. 2010. "The framing of climate change: Why it matters." In *Climate Change, Ethics, and Human Security*, edited by O'Brien, K., St. Clair, A. L. and B. Kristoffersen, 3–22. Cambridge: Cambridge University Press.

O'Brien, K. and L. Sygna. 2013. "Responding to climate change: The three spheres of transformation." In *Proceedings of Transformation in a Changing Climate*, 19–21. Oslo: The University of Oslo.

Olsson, P., Gunderson, L. H., Carpenter, S. R., Ryan, P., Label, L., Folke, C. and C. S. Holling. 2006. "Shooting the rapids: Navigating transitions to adaptive governance of social-ecological systems." *Ecology and Society* 11 (1): 18.

Pelling, M. 2011. *Adaptation to Climate Change: From Resilience to Transformation*. Abingdon, UK: Routledge.

Pelling, M., High, C., Dearing, J. and D. Smith. 2008. "Shadow spaces for social learning: A relational understanding of adaptive capacity to climate change within organisations." *Environment and Planning A* 40 (4): 867–84.

Pelling, M., Manuel-Navarrete, D. and M. Redclift. 2012. "Climate change and the crisis of capitalism." In *Climate Change and the Crisis of Capitalism: A Chance to Reclaim Self, Society and Nature*, edited by Pelling, M., Manuel-Navarrete, D. and M. Redclift, 1–17. Abingdon, UK: Routledge.

Petrie, N. 2011. *Future Trends in Leadership Development*. Center for Creative Leadership. http://www.ccl.org/Leadership/pdf/research/futureTrends.pdf.

Rowson, J. 2011. "Transforming behavior change: Beyond nudge and neuromania." RSA Report, November. http://www.thersa.org/__data/assets/pdf_file/0006/553542/RSA-Transforming-Behaviour-Change.pdf.

Scharmer, C. O. 2009. *Theory U: Leading from the Future as It Emerges*. San Francisco: Berrett-Koehler.

Shove, E. 2010. "Beyond the ABC: Climate change policy and theories of social change." *Environment and Planning A* 42 (6): 1273–85.

Slaughter, A.-M. 2011. "A new theory for the foreign policy frontier: Collaborative power." *The Atlantic*, November 30. http://www.theatlantic.com/international/archive/2011/11/a-new-theory-for-the-foreign-policy-frontier-collaborative-power/249260/.

Steffen, W., Persson, Å., Deutsch, L., Zalasiewicz, J., Williams, M., Richardson, K., Crumley, C., et al. 2011. "The Anthropocene: From global change to planetary stewardship." *Ambio* 40 (7): 739–61.

Stokke, K. and E. Selboe. 2009. "Symbolic representation as political practice." In *Rethinking Popular Representation*, edited by Törnquist, O., Webster, N. and K. Stokke, 59–78. New York: Palgrave Macmillan.

Swyngedouw, E. 2010a. "Apocalypse forever? Post-political populism and the spectre of climate change." *Theory, Culture and Society* 27 (2–3): 213–32.

Swyngedouw, E. 2010b. "The impossible sustainability and the post-political condition." In *Making Strategies in Spatial Planning: Knowledge and Values*, edited by Cerreta, M., Concilio, G. and V. Monno, 185–205. Dordrecht: Springer.

Swyngedouw, E. 2011. "Interrogating post-democratization: Reclaiming egalitarian political spaces." *Political Geography* 30 (7): 370–80.

Swyngedouw, E. 2013. "The non-political politics of climate change." *Acme* 12 (1): 1–8.

Torbert, W., et al. 2004. *Action Inquiry: The Secret of Timely and Transforming Leadership*. San Francisco: Berrett-Koehler.

Urry, J. 2011. *Climate Change and Society*. London: Polity Press.

Wendt, A. 2006. "Social theory as Cartesian science: An auto-critique from a quantum perspective." In *Constructivism and International Relations: Alexander Wendt and His Critics*, edited by Guzzini, S. and A. Leander, 181–219. London: Routledge.

Wilber, K. 2001. *A Brief History of Everything*. 2nd ed. Dublin: Gateway.

Index

adaptation
  actors, 18, 230–233, 235, 240–246, 262, 264, 266, 273, 275, 278, 280–281, 283
  as a technical problem, 2, 3, 6–7, 13, 19, 234, 311–312, 317, 319, 321
  as an adaptive challenge. *See* adaptive challenge
  needs, 18, 258, 261, 264, 271–272, 273, 275, 280–283
  opportunistic, 15–16, 141–145, 147–148, 152, 154–156
  planning, 16, 46, 98–101, 109, 144, 190, 243, 271–272, 275, 300
  policy, 14–15, 41, 42–43, 45–46, 47, 52–54, 81, 85–87, 89–93, 108, 144, 167, 171, 245, 271
  redefinition, 2, 311, 321
  stakeholders, 271, 278, 281
  urban, 13, 14, 63–66, 68, 70–78
adaptation processes, 16, 17, 36, 64, 81–83, 85, 93, 119, 233, 237, 243, 245–246, 252, 254, 265, 272, 281–282
  biological, 4
  evolutionary, 3–6
  human, 3, 13, 311
  political, 2, 119–121, 315–316
  social, 119, 168
adaptation strategies, 14, 15, 41, 42–44, 47–48, 52–58, 72, 81, 88, 100, 103, 108, 109, 112, 114, 124, 132, 142, 261, 298
adaptive
  effects, 3–6
  elements, 2–3, 9, 13, 17, 311–312
  responses, 8, 185, 189, 206, 208, 315
  work, 10, 19, 317
adaptive capacity, 5, 13–14, 16–17, 64, 66, 81, 86–88, 92–93, 100–101, 113, 126, 131, 133, 176, 194, 209, 213–216, 221, 223–226, 233, 235–236, 238, 240–242, 252–253, 272, 274–275, 279, 281–282
adaptive challenge, 1–3, 6–19, 32, 99, 118, 132, 160, 168, 234, 287, 304, 311–312, 317–322
adaptive collaborative risk management (ACRM). *See* risk management
agency, 12, 26, 93, 132, 155, 208, 225, 316
  human, 12, 316

agriculture, 85, 87, 98, 101–104, 113, 120, 123–128, 131, 133, 161, 195, 197–198, 200, 202–203, 205, 206, 240–241, 243, 255
Ålesund, 68–69, 71–72, 76
alternatives, 298, 313, 315–319, 321
  sustainable, 11, 19, 321
Anthropocene, 25–26, 33, 144, 149, 154, 311
Arctic, 12, 140–149, 153, 154, 156, 171, 196
aspirations, 7, 15, 118–120, 124–129, 132–134, 315
attitudes, 2, 8, 9, 30, 35, 126, 172, 175–176, 216, 262, 289–291, 293, 303
  and values, 35, 172–176
  change, 262, 303
  how Q-methodology reveals attitudes. *See* Q methodology
  instrumental perspective, 216
  on climate change, 289–293
Australia, 14–15, 81, 85–93

behavior, 4, 9, 17, 65, 160–163, 168, 172, 174, 194–196, 200, 208, 213
  organizational, 213, 216, 224–226
  pro-environmental, 168
behavioral change, 2, 4, 9, 11, 130, 143, 161
Beitostølen, 124, 126, 129–132
beliefs, 2, 6–10, 18–19, 83, 118, 171, 172, 176, 190, 194, 196, 271, 274, 278, 279, 283, 289–293, 301, 315, 317, 318
Bodø, 68–69, 72, 75

Canada, 16, 17, 98, 152, 171–172, 173, 176
  climate change responses, 230–231, 237, 239–240
capitalism, 11, 314
climate change adaptation leadership (CCAL). *See* leadership
climate scenarios, 2, 26–27, 33, 47, 50, 57, 63, 75, 140, 199, 206, 236, 241, 256–258, 261, 263, 275
coastal zone(s), 15, 87, 98, 101, 106–110, 113
collaborative power, 319, 321
collective action, 12, 29–30, 67, 319
collective thinking, 320

325

communication, 11, 13, 18, 30, 47–48, 57, 66–67, 76, 87, 98, 101, 184, 186, 206–207, 227, 231, 253, 262, 265, 272, 275–276, 287–288, 292–294, 296, 300, 304, 320
  across cultural and ideological divides, 18, 287–288, 292, 293, 300–301, 304
  frames, 275
  GIS, 276
  technology, 186
complexity, 6–7, 9–10, 17, 82, 89–90, 118, 168, 230, 234, 236, 238, 254, 258, 260, 318
conflict, 7, 12, 15, 19, 32, 35, 103, 119–120, 125, 129–134, 151, 154, 173, 190, 203, 259, 265, 272, 277, 280, 301, 313, 315–320
  interests and aspirations, 118–121
  land use, 203, 259
  of interests, 119–120, 125, 129, 131, 134, 151, 154
consciousness, 11, 52, 53, 319
consensus, 11, 53–54, 66, 141–143, 179, 182, 287, 293, 297, 314, 319
crisis, 8, 43, 47, 54, 112, 123–124, 220–221, 224, 287, 291, 321
  climate, 8, 287
  economic, 54, 123, 291
  financial, 124
critical realism, 276, 316
critique, 11, 12, 149, 312, 313–315
cultural theory, 194–196, 205–206, 208–209, 289
  of risk, 195, 205
curiosity, 18, 174, 297, 303, 304

debate, 1, 2, 24–32, 36, 48, 53, 71, 92, 141, 149–150, 153, 276, 288, 292–293, 296, 298, 302–303, 313–315, 320
  global justice, 28, 30, 36
  justice, 29–32
  true dialogue, 296, 300, 303
  USA, climate, 288, 293, 296, 298
decision making, 16, 44, 82–83, 134, 160, 163, 166, 196, 236, 244–246, 271
  frameworks, 271
deliberation, 17, 18, 176, 303, 319
democracy, 11, 31–32, 292, 315, 316
depoliticization, 10, 15, 133
development, 2, 5, 7, 11, 14–16, 31, 52–58, 63–65, 68–73, 74–77, 85–86, 89, 92, 118–121, 123, 124–134, 142, 144–156, 162, 165, 173, 175, 199, 204, 206, 209, 217, 233, 235, 237–241, 242–244, 246, 258, 261, 291, 314, 315
  discourse, 15, 132, 133
  pathways, 129–134
  sustainable. See sustainable development
dialogue, 3, 10, 18–19, 76, 163, 233, 253, 278, 287–288, 292, 296, 299–304, 313, 319–320
discourse, 12, 15–17, 24, 26, 46, 54, 130, 132–133, 141, 143, 145, 149, 150–151, 153, 154–155, 160, 172, 176, 178–190, 206–209, 218, 220–221, 234, 276, 287–288, 291, 299–300, 314–315, 318
  continuum, 299–300
disequilibria, 317, 318
drivers, 14, 19, 43–45, 55, 56, 58, 88, 118, 149, 311–312, 321

Earth system, 13, 25–26, 36
economic growth, 125, 128, 129, 130, 131, 132, 133, 144, 148, 151, 217
elderly, 112, 161, 281–283
electricity, 17, 130, 213–223, 225
electricity grids, 17, 214, 225
embeddedness, 244
emotions, 7, 9, 279, 283, 312–313, 317–318
energy, 6, 7, 45, 47, 70, 83, 98–99, 103, 105–106, 128–130, 132–133, 141–146, 147, 151, 153–154, 217–220, 313
energy security, 141, 150–151, 154
engagement, 7, 10, 18, 85, 131, 171, 209, 231, 233, 235, 241, 243, 253, 262, 265, 266, 278–280, 287, 288, 292–294, 295, 297, 303–304, 313–315, 316, 318–320
entrepreneurship, 45, 51, 125, 129, 133, 205, 235
environmental security, 141, 144–147, 149–151, 153–155
equity, 3, 5, 8–9, 12, 15, 42, 90, 98–103, 108–114, 118, 140–141, 147, 155, 165, 252–253, 264, 311
  and justice, 3, 15, 42, 98–101, 113
ethics, 13, 16, 24, 26, 27, 34–36, 155
exclusion, 19, 175, 313, 315, 316

fair, 12, 16, 28–30, 36, 73, 100, 113, 160, 162, 165, 166, 258, 316
fairness, 8–9, 28, 113–114, 160, 162–163, 165
farmers, 16, 101–105, 113, 120, 123–127, 131–132, 161, 167, 194, 196, 198, 202–206, 208, 297
farming. See agriculture
Finnmark, 196, 197
focus group, 100, 179–180, 253, 257, 260–262, 277
food production, 103, 127, 131
forestry, 18, 87, 175, 253, 255–266
frames, 11, 44, 82, 93, 205, 275, 282, 320
Fredrikstad, 68–69, 74, 76
future generations, 24, 27, 30, 31, 35, 282–283, 311, 321

geographic information systems (GIS), 18, 272, 275–284
  integral GIS, 280–284
  qualitative GIS, 277–279
geo-narrative, 279
geopolitics, 141–142, 144–148, 154
global agreements, 144, 151, 153, 155, 156
Global Warming's Six Americas, 289–297, 301, 304
  Alarmed, the, 289–291, 292–295
  Cautious, the, 289, 292
  Concerned, the, 289, 292–293, 297
  Disengaged, the, 289
  Dismissive, the, 289
  Doubtful, the, 289–291, 293
globalization, 25, 141, 149, 154, 313, 314
governance, 6, 42, 64, 68, 72, 81, 83, 85, 90, 93, 141, 156, 160, 164, 168, 198, 213, 226, 232, 235, 252, 254, 265, 275
  polycentric, 83
greenhouse gas emissions, 2–3, 5, 7, 46, 69, 141–142, 144, 147, 150, 153, 243, 245, 311, 313
grounded visualization, 280

habits, 6, 9, 234, 294, 312, 318, 320, 321
Hammerfest, 68–69, 71–72, 74–75, 197, 199, 200, 204, 209
hegemony, 2, 5, 10, 15, 118, 129, 131, 133–134, 149, 315, 317–318
heuristic, 303
human emancipation, 313, 316, 321–322
human flourishing, 12, 35, 316, 321

identity, 16, 30, 32–33, 35, 50, 52, 119–120, 125–126, 128, 130, 133, 160–168, 175, 182, 188, 197, 201–202, 208, 216, 234, 271, 273, 275, 304, 321
  and place attachment, 160, 163, 166, 168
  cultural, 32–33, 123, 126–128, 130, 133, 197
  group, 16
  local, 126, 133
  personal, 16
  place, 119, 163–165, 166
  processes, 162
  social, 120, 161
ideological divide, 18, 287, 288
immune system, 317–318
inclusion, 18, 76, 99, 162, 199, 204, 313, 315
inclusive politics, 315, 317, 318, 321
inequality, 55, 103, 151, 300, 314
institutional, 1, 2, 6, 17, 66, 87, 90, 214–217, 219, 221–226, 235, 241, 253
  logics, 216–217, 223–226
instrumental, 35, 125, 154, 199, 214, 216, 221, 225–226, 253, 276
Integral Theory, 273–276, 278–279, 316
  GIS, 278
  quadrants, 273–275
interests, 1, 2, 3, 5, 7, 9–11, 16, 18–19, 119–120, 125, 126, 129, 131–134, 141, 144–148, 151–155, 174, 197, 205–206, 208, 213, 231, 244–246, 257, 282–283, 291, 299, 311, 312, 314–319

justice, 3, 7, 8, 12–13, 19, 24, 27–33, 36, 42, 85, 99–101, 103, 110, 113, 118, 147, 154, 160, 162–163, 174, 311, 315, 320
  global, 28, 30, 36, 149
  intergenerational, 29–31, 36

knowledge
  local, 52, 124, 127, 131–132, 206, 236, 246, 253
  situated, 9, 315, 318
  systems, 312, 318, 321
  tacit, 84, 272, 279, 297
  traditional, 173, 187
Kronoberg, 256–261
Kyoto protocol, 29, 46, 141, 143, 151

Labrador, 16, 171–175, 179–180, 182–190
leadership, 15, 19, 51, 84, 92, 93, 233–238, 242–246, 299–300, 313, 319, 320–321
  Climate Change Adaption Leadership (CCAL), 236, 237, 240, 241, 242, 246
  Complexity Leadership Theory (CLT), 234
  constraints, 233, 234
  contextual intelligence, 234, 236
  policy, 235–236
  super-agents, 236, 246

  sustainability, 236
  vision and meaning-making, 233, 235
learning, 4, 7, 9, 13, 17–19, 72, 75–78, 82–83, 219, 224, 231, 233–235, 252–255, 260, 262, 271, 280, 294, 296, 298, 303, 313, 320
  collective, 14, 64–68, 70, 74–75, 76–78, 266
  individual, 14, 18, 65, 77, 265
  organizational, 14, 65, 78
  process, 17, 64–65, 253–254, 264
  social, 231, 240–241, 243–245, 254–255, 265, 278, 281
  transformative, 254, 320
Lebesby, 197, 199–200, 202–204, 209
Lofoten, 197, 200
London, 14, 41–58
loss, 1, 19, 27, 33, 35, 47, 68, 102, 106–107, 126, 130, 179, 183, 186, 188–190, 202, 221, 245, 300, 304, 317–319

maladaptation, 100, 166, 234
mindsets, 2, 6, 9, 17, 262–263, 265, 275, 312, 314, 319–321
mitigation, 5, 7, 15, 24, 31, 33, 36, 42, 45–46, 48, 51, 53, 55, 57, 83, 85, 89, 92, 127, 141–146, 149, 151–152, 154, 156, 196, 235, 238, 242, 255, 294, 297–298, 300, 302, 313–314, 318, 321
mountain agriculture. *See* agriculture
mountain farming, 126–129
multiplicity, 1, 5, 90, 129, 134, 154, 168, 198, 209, 234, 246, 254, 272–274, 275, 281, 320
municipal planning, 16, 45, 67–68, 70–76, 78, 119, 129, 133, 195, 197–198, 200, 203–204, 208

negotiations, 15, 64, 119–120, 125, 129, 131, 134, 141, 143, 151, 155, 219, 320
networks, 13–14, 16, 17, 25, 45, 48, 57, 64–69, 71–73, 75–77, 84, 103, 120, 124–125, 127, 131–133, 161–162, 213–214, 218, 230, 233, 241–245, 253, 255, 264, 313, 319–321
New Public Management (NPM), 218–219
New York City, 14, 41–50, 56–58, 106–107, 109–112
New York State, 15, 47, 98–99, 101–104, 106, 108–109, 113–114
Niagara Region, 17, 230–231, 237, 239–246
norms, 17, 66, 74, 84, 92, 126, 161, 174, 213–216, 220, 253–254, 265, 294, 311–312, 317–319
Northern Norway, 16, 147–148, 153, 194–198, 203–205, 208
Norway, 15–17, 63, 69, 121–124, 126, 129–131, 140–143, 145, 166, 195–198, 203–205, 208, 215, 217, 219–224, 281

oil and gas, 7, 15–16, 141–143, 147, 148–150, 152–153, 197
Ontario, 17, 230–231, 239, 241
opportunistic adaptation, 15, 140–145, 147–148, 152, 154–156
organization, 2, 6, 8, 17, 44, 48, 50, 64–67, 70–72, 75–76, 83, 93, 124, 132, 206, 213–214, 216–217, 226, 233–236, 241–245, 282, 291, 300, 303, 319
organizational change, 14, 17, 81, 83–84, 91, 214, 216, 221–222, 225–226
Øystre Slidre, 119–124, 126, 128–134

Index

participation, 4, 15, 17–19, 76, 99–100, 113, 162, 177, 232, 245, 253–255, 297–298, 315
participatory approaches, 3, 18, 230–231, 237, 246, 265–266, 272
participatory processes, 15, 18, 230, 233, 235, 236, 253, 262, 264, 265
petroleum, 15, 142, 146–147, 150–156
philosophy, 30, 31, 263
place
   attachment, 16, 160, 163–168
   sense of, 127, 164, 167, 275
plurality, 8, 10, 15, 19, 134, 313, 315, 319, 320
polar areas, 141, 153
polarization, 287–289, 291, 304
policy, 1–2, 12, 14–15, 31, 41–48, 50–58, 63, 67, 81–83, 85–93, 99, 109, 130, 133, 141–144, 146–148, 154–156, 167–168, 207, 226, 231, 236, 238, 245–246, 253–254, 259, 261, 273, 276, 289, 290–295, 298–299, 313, 320
   public, 41, 43–45, 56, 58, 276, 291
   transitions, 14, 41–45, 55–56. *See also* transition
   urban, 14, 56, 58, 63
   window, 43, 47, 50, 53, 92
political action, 12, 156
political elite, 11, 92, 141, 146, 148, 150, 155
politicization, 12, 133, 287, 313–314
politics, 8, 12–13, 18–19, 31–32, 90, 120, 123, 134, 141–142, 145–146, 155–156, 160, 288, 299, 313–319, 321
postdemocratic, 313, 314
postnormal science, 276
postpolitical, 313–315
power, 2, 5, 10–12, 15–19, 66–68, 100, 119, 131–132, 134, 145–147, 155, 163, 164, 165, 173–174, 181, 184, 185, 233, 236–238, 244, 245, 254, 299, 300, 302, 311–314, 315–316, 319–321
   spaces for action, 236, 242
power relations, 11–12, 15, 19, 119, 134, 147, 233, 312, 314, 321
priorities, 3, 6–7, 12, 15, 17, 51, 54, 56–57, 87, 90, 108, 120, 132, 134, 141, 150, 155, 183, 186, 188, 197, 199, 204, 208, 223, 231, 271, 272, 274–275, 283, 312, 315
public opinion, 44, 287–289, 301

Q methodology, 172, 176–178
   to reveal attittudes, 176
quantum social theory, 319

reflexivity, 10, 19, 254, 273, 278–279, 313, 318, 320
repoliticization, 12, 312–315
resilience, 5, 14, 41, 44–45, 48–49, 54–55, 58, 82, 106, 110, 127, 162, 188, 205, 207–209, 215, 216, 223, 232, 252, 282
resistance, 9, 19, 84, 131, 164, 216, 317–318
resource dependent regions, 106
Rigolet, 172, 173–186, 188–189
risk management, 48, 53, 54, 86, 93, 231, 236, 237, 238, 240, 243, 245
   adaptive collaborative risk management (ACRM), 231–233, 236, 237, 240, 245

Sarpsborg, 68–69, 76
scientific knowledge and information, 50, 56–58, 194–195, 205–207, 208–209, 242, 244, 257–258, 260, 262, 297
skiing, 121, 128, 130
snow, 121, 126, 128, 130–131, 166, 172, 179, 182, 184, 219, 240, 260, 262
social contract, 160
social coordination, 14, 66–67, 70, 77
   hierarchy, 70–72
   markets, 73–75
   networks, 75–76
social learning. *See* learning, social
social networks, 16, 120, 124, 127, 133, 161, 245, 253, 264, 319
social-ecological inventories, 17, 230, 232, 238, 242
St. Lewis, 174–176, 179–180, 184, 188
stakeholder, 7–8, 17–18, 48, 57, 70, 100, 199, 231, 233, 246, 253–255, 257–258, 261–266, 277, 282–283
stakeholders silence, 289–292
structures and systems, 2, 9, 311, 314, 321
subjective aspects, 7
subjective filters, 272
subjective judgments, 272
sustainability, 14, 18, 41–42, 45, 58, 119–120, 124, 126, 132, 134, 160, 233, 234, 236, 238–241, 243–245, 253–254, 311, 312, 320–321
   urban, 45, 58
sustainable adaptation and mitigation (SAM), 235, 240, 245
sustainable development, 31, 53, 54–55, 134
Sweden, 215, 219–226, 255, 257, 259
systems
   global, 3, 6
   hegemonic, 315, 317, 318
   social, 65, 313, 314, 316
systems, structures and change, 1–4, 6–7, 9, 11–12, 311–318

tipping points, 14, 41, 44
tourism, 120, 121–124, 126, 128–134, 140, 243
transdisciplinary framework, 273
transformation, 1–2, 11, 12, 118, 121, 132–134, 167, 214, 234–235, 311–317, 320, 321
   deliberate, 118, 124, 134
   internal, 12, 316
   personal, 317, 318
   political, 321
   social, 3, 12, 18, 45, 311–322
   structural, 132
transition, 14, 168, 224, 230, 239, 243, 321
   urban policy, 14, 41–45, 55–57
   urban policy transition frame, 14, 41

understanding, 2, 25, 84, 120, 124–125, 145, 174, 194, 206, 216, 226, 246, 255, 257, 262, 264, 274–275, 279, 282, 287, 293, 299, 303–304
   mutual, 231
   vulnerability, 275
urban planning, 14, 63–65, 75, 77–78, 86
urban policy, 14, 41, 56, 58, 63, 70
urban policy transition frame. *See* transition

value
    change, 24, 35, 130
    commitment, 18, 289, 292, 293
    conflict, 15, 119, 132, 133, 173, 190
values, xi, 2–3, 6–10, 13, 15–19, 24, 32, 34–36, 66, 83–84, 90, 118–120, 124–127, 131–134, 166, 168, 171–174, 176–177, 181, 183–190, 194, 196–197, 202, 205–209, 213–216, 230, 232, 254, 264, 265, 271–275, 277, 283, 292, 294–296, 298, 299, 301, 302, 303, 311–314, 315, 318, 320–321
    organised in systems, 173
    -what are, 172–173
Vestvågøy, 197, 199, 200, 202–204, 209
vulnerability, 2, 5, 9, 12, 15, 16–18, 19, 46–47, 55, 71, 73, 82, 86, 88–91, 93, 98–100, 105–114, 118–121, 131–132, 146, 162, 171, 172, 176, 190, 194–210, 215, 217–221, 223–224, 238, 243, 256, 264, 271–275, 278–284, 301, 311–313, 321
    "objective", 274, 278, 280–282
    "subjective", 274, 280, 282
    assessment, 86, 88–89, 280–283
    contextual, 18, 107, 275, 278, 282–283
    determinants, 55, 201
    frames, 275, 282
    indicators, 278–279, 281
    mapping, 272, 278, 284
    narratives, 194, 201, 204, 207–209
    of the elderly, 161–162, 281–283, 284
    outcome, 274
    representation, 275, 276–277
Västerbotten, 257–261

ways of life, 16, 33, 119, 195, 196, 206–207
worldviews, 2, 7, 8–10, 13, 16, 19, 194, 196, 207, 275, 276, 311–312, 315, 321–322

Printed in the United States
By Bookmasters